DECOMPOSITION IN
TERRESTRIAL ECOSYSTEMS

STUDIES IN ECOLOGY

GENERAL EDITORS

D.J. ANDERSON BSc, PhD
Department of Botany
University of New South Wales
Sydney

P. GREIG-SMITH MA, ScD
School of Plant Biology
University College of North Wales
Bangor

and

FRANK A. PITELKA PhD
Department of Zoology
University of California, Berkeley

STUDIES IN ECOLOGY VOLUME 5

DECOMPOSITION IN TERRESTRIAL ECOSYSTEMS

M. J. SWIFT, O. W. HEAL &
J. M. ANDERSON

BLACKWELL SCIENTIFIC PUBLICATIONS

OXFORD LONDON EDINBURGH MELBOURNE

First published 1979

British Library Cataloguing in Publication Data

Swift, M J
    Decomposition in terrestrial ecosystems.—
    (Studies in ecology; vol. 5).
    1. Biodegradation
    I. Title   II. Heal, O W   III. Anderson,
    Jonathan Michael
    574.5′264   QH541

ISBN 0-632-00378-2

Distributed in the United States of America by
the University of California Press
Berkeley, California

Set, printed and bound in Great Britain.
Text set by William Clowes (Great Yarmouth) Ltd, Norfolk.
Printed by Galliard (Printers) Ltd, Great Yarmouth, Norfolk
and bound by Mansell Bookbinders Ltd, Witham, Essex.

Dedicated to the memory of

IAN HEALEY

He planned this book and his
influence can be seen throughout

CONTENTS

Preface   ix

Acknowledgements   xi

1   Decomposition Processes in Terrestrial Ecosystems

1.1   Ecosystem definition and classification   1
1.1.1   The significance of decomposition processes within ecosystems   5

1.2   Decomposition processes and ecosystem structure   8
1.2.1   Organic turnover within ecosystems   8
1.2.2   Soils and soil organic matter   15

1.3   Decomposition processes and nutrient cycling   24
1.3.1   The plant subsystem   27
1.3.2   The decomposition subsystem   34
1.3.3   The herbivore subsystem   39
1.3.4   Nutrient turnover in ecosystems   41

1.4   Decomposition processes and ecosystem development   44

2   The Decomposition Subsystem

2.1   The structure of the decomposition subsystem   49

2.2   Models of the decomposition subsystem   58

3   The Decomposer Organisms

3.1   The diversity of decomposer organisms   66

3.2   The functional ecology of decomposer organisms   66
3.2.1   Body size   71
3.2.2   Trophic function   75
3.2.3   Saprotrophic micro-organisms   77
3.2.4   Saprotrophic invertebrates   84
3.2.5   Saprotrophic vertebrates   95
3.2.6   Necrotrophs   96
3.2.7   Biotrophs   101
3.2.8   Conclusions: the structure of the decomposer community   101

3.3   Distribution patterns of decomposer organisms   106

3.4	The quantitative ecology of decomposer organisms	111
3.4.1	Decomposer communities in contrasting biomes	112

4   **The Influence of Resource Quality on Decomposition Processes**

4.1	The effect of resource quality on decomposition rate	118
4.2	The time-course of decomposition processes	122
4.2.1	Changes in resource composition during decomposition	123
4.3	Chemical attributes of resource quality and their influence on decomposition	131
4.3.1	Carbon and energy sources	132
4.3.2	Nutrient sources	139
4.3.3	Modifiers	148
4.4	Physical attributes of resource quality and their influence on decomposition	157
4.4.1	Surface properties	158
4.4.2	Toughness	158
4.4.3	Particle size	159
4.5	The influence of resource quality on the structure of decomposer communities	163

5   **Decomposition Processes at the Molecular Level**

5.1	Community metabolism of the decomposition subsystem	167
5.2	The chemistry of decomposer resources	169
5.2.1	Primary resources	169
5.2.2	Secondary resources and faeces	178
5.3	Depolymerisation reactions	181
5.3.1	Polysaccharase enzymes	182
5.3.2	Depolymerisation of lignin and cutin	186
5.3.3	Protease enzymes	187
5.3.4	The ecological significance of depolymerisation reactions	189
5.4	Immobilisation and mineralisation	193
5.4.1	Immobilisation reactions and energy transfer	196
5.4.2	Mineralisation reactions and elemental cycles	200
5.5	Humus chemistry	206
5.5.1	The composition and structure of humus	207
5.5.2	Formation of humus	210
5.5.3	Decomposition of humus	213

5.6   Decomposition at the molecular level   215

6   The Influence of the Physico-Chemical Environment on Decomposition Processes

6.1   The environments of decomposer organisms   220

6.2   Moisture   221
6.2.1 The influence of moisture on decomposer organisms   226
6.2.2 The influence of moisture on decomposition rate   230

6.3   Aeration   231
6.3.1 Oxygen   232
6.3.2 Carbon dioxide   235
6.3.3 Combined effects of O_2 and CO_2 237

6.4   pH   237
6.4.1 The effects of pH on the activity of decomposer organisms   244

6.5   Temperature   247

6.6   Leaching and other direct effects of climate   254

6.7   Analysis of the effects of the physico-chemical environment on decomposition processes   256

7   The Decomposition Subsystem; Synthesis and Summary

7.1   The fungus-ant symbiosis   268

7.2   The vertebrate rumen   271

7.3   Coastal tundra   276

7.4   Cold temperate peat bog   280

7.5   Temperate cultivated grassland   284

7.6   Temperate deciduous forest   291

7.7   Summary; global patterns of decomposition in terrestrial ecosystems   302

Appendix: Methods of study for decomposition ecology   318

References for Appendix   334

References   342

Index   363

PREFACE

You ask what is the use of classification, arrangement, systematisation? I answer you: order and simplification are the first steps towards the mastery of a subject—the actual enemy is the unknown.

Thomas Mann
The Magic Mountain (1924)

The study of decomposition is not a unified scientific discipline—it draws upon the subject matter of ecology, soil science, agriculture, forestry, microbiology, physiology, biochemistry and zoology. In writing this book we have attempted to draw information from all these disciplines to provide a coherent view of decomposition processes and their role in ecosystems. In doing so we have adopted the view that ecology is the discipline which most appropriately provides the link between the different fields of study. Indeed the main gain of modern research with decomposition has been the application of ecological principles to its investigation.

The book is not intended as a comprehensive review of research into decomposition processes. On the contrary, we have been deliberately selective in our choice of examples. Our main emphasis throughout has been to try and establish some basic concepts and principles which help the reader to form a coherent view of decomposition processes. Ecology tends to suffer from a plethora of theory. This does not seem to us to be the case in decomposition studies. Although man has manipulated decomposition by agricultural practice throughout his history this is still done by an essentially pragmatic approach. In comparison with the sophisticated understanding of genetics, physiology, biochemistry and ecology of higher plants, our understanding of the processes of nutrient regeneration in the soil is crude indeed. This is due, not to a lack of observation or experiment, but to a lack of a coherent framework of understanding. Our main ambition is that the account we have given may provide a simple base for such a framework and will stimulate the development of a sound quantitative concept of the ways in which decomposition processes take place.

In the first chapter of the book we provide some examples of decomposition viewed at the ecosystem level of organisation. From there we derive some simple hypotheses of the functional roles of decomposition processes in ecosystems. In the second chapter the hypotheses are formulated in a more abstract manner by the development of a 'modular' system which we propose as a simple description of the basic structure of the decomposition subsystem, common to all terrestrial ecosystems and forming an essential dynamic part of all such systems. The following four chapters are reductionist in character, in contrast to the largely holistic view of Chapter 1. We examine the view of the decomposition subsystem obtained when described in terms of the decomposer organisms and their activities (Chapter 3), in terms of its chemistry (Chapter 5), or the regulatory effects of the environment (Chapter 6), or the nature of the organic materials being decomposed (Chapter 4).

The examples given in each of these analytical chapters are used to amplify or challenge the basic hypotheses of the first two chapters and thence to develop a much more detailed view of the manner in which decomposition processes operate. In the final chapter the holistic view is restated by a series of examples—case studies—which attempt to fuse the general principle with the specific information available from detailed research. In the second part of this chapter we also attempt some very general statements about the major patterns of variation in decomposition processes, on a global scale in the form of a set of charts. We hope that this simple form of summary statement may, in particular, prove useful in the teaching of the ecology of decomposition.

The interpretation of decomposition research requires an understanding of the methods employed in its study. As these again range across a wide number of disciplines this presents a formidable problem. We have tried to solve this problem by providing an Appendix which gives the reader an introduction to the literature on appropriate methods. The reader may find it of value to consult the appropriate parts of the Appendix prior to reading some sections of the main text—particularly parts of Chapters 1 and 3.

M.J. Swift,
Department of Botany and Microbiology,
Queen Mary College, London
now at:
Biology Department,
University of Papua New Guinea,
Port Moresby,
Papua New Guinea

O.W. Heal,
Institute of Terrestrial Ecology,
Merlewood Research Station,
Grange-over-Sands, Cumbria

J.M. Anderson,
Department of Biological Sciences,
The University, Exeter

ACKNOWLEDGEMENTS

Completion of this book has depended on the advice, help and support of many people—but above all on Clare, Elsie and Hilary whose patience has been incredible and whose encouragement was vital.

Myra Barr, Jean Smith and Jo Robertson achieved miracles in rendering mountains of manuscript into readable form and we offer them our deepest thanks.

We should also like to acknowledge our indebtedness to many colleagues all over the world whose work, ideas and advice have stimulated our efforts. We only hope that we have not misrepresented them too badly. We give our special thanks to those who have given us permission to reproduce their illustrations or data; they are acknowledged individually in the legends. The following publishers have also granted permission to reproduce material copyrighted by them and this we gladly acknowledge and extend them our thanks; Macmillan Publishing Co. Inc., New York (Figs. 1.1, 1.8, 1.11, 6.6 and 6.15); Springer-Verlag, Heidelberg (Figs. 1.2, 3.1, 7.5 and 7.13); Dr. W. Junk, The Hague (Figs. 1.4 and 1.5); Cambridge University Press (Figs. 1.6, 3.5(part), 6.14 and 7.1); Blackwell Scientific Publications (Figs. 1.13, 3.13, 3.15, 4.4, 4.10, 4.11, 4.16, 6.8 and 6.11(part)); John Wiley and Sons, Inc., New York (Figs. 1.15, 5.2(part), 5.9 and 6.3); Pergamon Press Ltd., Oxford (Figs. 1.16, 5.13 and 6.18); Ecological Monographs and Duke University Press (Figs. 1.7 and 7.12); Oxford University Press (Figs. 1.19, 4.8 and 6.7); UNESCO (Figs. 2.5 and 7.10); Swedish Natural Science Research Council, Stockholm (Figs. 2.8, 2.10 and 7.4); Edward Arnold Publishers Ltd., London (Figs. 3.16 and 4.12); Academic Press Inc. (London), Ltd. (Figs. 3.7 and 6.11(part)); Acaralogia, Munich (Fig. 3.14); Longman Group Ltd., Harlow (Figs. 3.17 and 6.5); Sté. royale de Botanique de Belgique, Brussels (Fig. 4.1); Forest Research Institute, Helsinki (Figs. 4.2 and 4.5); Oikos, Copenhagen (Fig. 4.3); Pierron Editeur, Sarregueimines (Fig. 4.7); Scientific American Inc., New York (Figs. 5.1(part) and 5.3); Biotechnical Laboratory, Helsinki (Fig. 5.2, part); United States Department of Agriculture, Madison (Fig. 5.6); Applied Science Publishers, Barking (Fig. 5.7); Martinus Nijhoff, The Hague (Fig. 5.8); University Park Press, Baltimore (Figs. 5.11 and 6.12); Williams and Wilkins Co., Baltimore (Fig. 5.12, part); Marcel Dekker, New York (Fig. 5.12, part); Elsevier Publishing Co., Amsterdam (Fig. 5.14); Chapman and Hall Ltd., London (Figs. 6.2, 6.4 and 6.9); Pedobiologia, Berlin (Fig. 6.13);

IBP Tundra Biome Steering Committee, Stockholm (Figs. 6.16, 6.17 and 6.20); University of Illinois, Carbondale (Fig. 6.19); Meadowsfield Press, Edinburgh (Figs. 7.2 and 7.3); American Association for Advancement of Science (Fig. 7.11).

1

DECOMPOSITION PROCESSES
IN TERRESTRIAL ECOSYSTEMS

1.1 Ecosystem definition and classification

The ecosystem can be recognised as a unit of biological organisation, as can the cell, the tissue, the organism, or the population, because of its functional integrity. The ecosystem is characterised by the integrated and largely self-maintained functioning of a diverse community of organisms within a range of physical environments. The limits of ecosystems are sometimes a cause for argument but a useful definition for the purposes of this book is that of an area which is self-contained in terms of its primary production and nutrient cycling—that is to say, the extent of transfer of production and nutrient elements across the ecosystem boundary is small in comparison with the internal fluxes. Transfers of nutrient from land to water by drainage may be a significant aspect of the functioning of the terrestrial ecosystem and the appropriate ecosystem boundary is for this reason often regarded as the catchment area of a drainage stream. Conversely it is not correct to refer to the soil as an ecosystem as the input and output of nutrients from the plants dominates the nutrient cycling through soil.

The vast range of different ecosystems can be categorised into a number of *ecosystem-types*, distinguished in terms of the character of their vegetation. Such broad categories are sometimes termed *biomes*. It should be emphasised that vegetation is only one aspect and the animal, microbial, chemical and physical features may be just as characteristic. Whittaker (1975) distinguished twenty-six such types for the terrestrial zone; of these we have chosen six to analyse and compare in greater detail. The distribution of the ecosystem-types all over the world surface is broadly correlated with differing climatic zones as is illustrated by Fig. 1.1. The climatic details for our six selected types are shown in Fig. 1.2; these are worth studying in some detail for it will become apparent in the course of the ensuing discussion that the climatic pattern is the fundamental determinant of ecosystem structure and dynamics including the nature of the decomposition processes.

Tundra are polar to boreal ecosystems lacking trees and dominated by dwarf-shrubs, grasses, sedges, mosses and lichens. They may be grazed by reindeer (in Europe) or caribou and musk ox (in America) and lemmings.

1

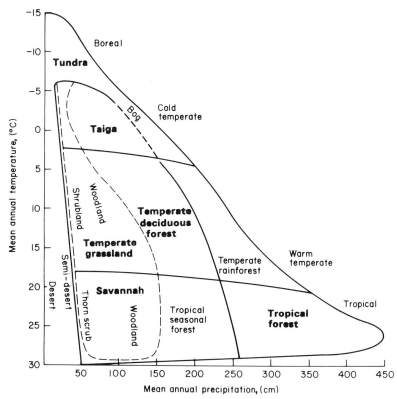

FIG. 1.1. Distribution of major ecosystem-types in relation to annual temperature and rainfall regimes. Ecosystem-types described in the text are given in
bold type; climatic details for these areas are given in Fig. 1.2 and soil types in
Figs. 1.8 and 1.9. The boundaries are approximate only and may be modified
markedly by local conditions such as altitude, soil type, closeness to coasts and
by fire. The dashed line encloses an area in which the extent of importance of
trees and shrubs associated with the grasslands may vary markedly—particularly
in the savannah regions (modified from Whittaker 1975).

Sub-zero temperatures dominate the climate; the surface thaw period may last
from three to six months and is the only time of biological activity (Fig. 1.2B).
Below the surface layers the frost is permanent. Tundra varies markedly in
its precipitation but water availability is determined by temperature. *Taiga*
or *Boreal Forest* is characterised by the coniferous forest-lands of Eurasia
and North America. In the north this type represents the cold extreme for
tree growth and in the warmer south a mixed forest type may develop.
Water is in excess all the year round but the winters are long and cold and
the summers cool (Fig. 1.2A). In the temperate climatic zones we may recognise
the ecosystem types, whose distribution is largely determined, by moisture

availability (compare C and D in Fig. 1.2). *Temperate Deciduous Forest* is found in Western Europe, the Eastern United States, China and Japan and to a limited extent in the southerly parts of the South American, African and Australasian continents. Broad-leaved trees such as oaks, beeches and maples are dominant. Winters are often cold but warm wet summers produce a predominantly humid climate with a long growing season and only rare periods of drought. The *Temperate Grasslands* differ markedly in this respect. The summers are dry and the winter rainfall is associated with lower temperatures. These features combine to produce a summer soil water deficit which excludes the potential for tree growth. The prairies of North America and the steppes of Asia characterise this biome but again there are more limited extents in the southern hemisphere, notably the pampas of South America and high veldt in Southern Africa. These ecosystems support many herbivores such as most of the domesticated ruminants and the bison of North America. In contrast to the boreal regions the vegetation distribution of the tropical zones is largely determined by the seasonal pattern of rainfall; temperature varies little throughout the year (Fig. 1.2, E and F). In the *Savannah* areas of Africa, South America and Australia a long dry season at the cooler period of the year produces a water deficit for at least five months of the year. The resulting vegetation is a tropical grassland with scattered trees. The density of the tree cover varies on a broadly latitudinal basis, i.e. largely in relation to annual rainfall. In modern times much savannah is burnt annually and large areas may be sub-climaxes where forests would normally develop. The African savannahs support the richest fauna of herbivores in the world. The *Tropical Forest* occupies the equatorial zones in a great band stretching from the Amazon, through West Africa, Zaire and Madagascar, to the Far East (Malaysia, Burma and Indonesia) and North East Australia. At the latitudinal margins of its distribution the climate is seasonal with a short dry season limiting production. The greatest development is seen in areas where there is abundant rain throughout the year coupled with a temperature regime that shows greater diurnal than seasonal fluctuation. The trees may exist in several storeys with the largest emergents in excess of 40 m in height. There is a luxuriant flora of epiphytes, creepers and lianas but undergrowth is largely absent in climax forest. Plant and animal diversity reaches its maximum in tropical forest. The latter is dominated by invertebrates and the vertebrates are mainly arboreal (old- and new-world monkeys, sloths, rodents, lizards etc.). More detailed descriptions of these ecosystem types can be found in other reviews (Walter 1973; Eyre 1968, 1971) where references to primary descriptions can also be found.

It should be clear that within each of these types considerable variation in ecosystem structure and production may be found. This is caused by local variations in climate and soil type often associated with topographical variation. The broadly latitudinal zonation of the types is broken, for instance

Chapter 1

(A)

Archangel (10m) 0·4°466

[61-24]

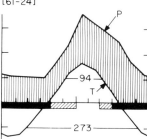

(B)

Thule (37m) −11·1° 64

[3]

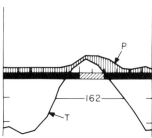

(C)

Kabul (1799m) 11·5° 309

[13-23]

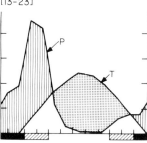

(D)

Washington D.C. (22m) 13·8° 1053

[30]

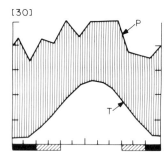

(E)

Colombo (7m) 26·6° 2370

[13-30]

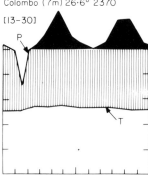

(F)

Salisbury (1472 m) 18·5° 840

[20]

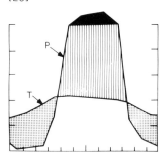

by the presence of mountain ranges. Associated with these may be found an altitudinal zonation which in many ways mimics, at a single latitude, the latitudinal zones of vegetation types. We must also remember that communities of organisms are not defined by sharp boundaries and that continuous gradients of change in ecosystem composition and structure may occur over long or short distances.

We must now turn from attempting to distinguish the boundaries of ecosystems to considering their internal structure.

1.1.1   The significance of decomposition processes within ecosystems

The functioning of all ecosystems can be recognised as occurring within three distinct *subsystems*—the *plant subsystem, the herbivore subsystem* and the *decomposition subsystem* (Fig. 1.3). The integrity of the ecosystem is maintained by the transfers of matter and energy between these three components. The annual gain of energy and matter by the plant subsystem (the *net primary production, NPP*) may be distributed in three ways.

(1)   In some ecosystems part of it is stored in perennial tissues and contributes to net growth, or biomass increase. In grass-dominated ecosystems this increment is usually negligible but in immature (subclimax) forests it may constitute 20–60% of the NPP.

(2)   A minor component of the NPP is consumed by herbivorous animals feeding on leaves, stems or roots. Even in intensively grazed grasslands this fraction rarely exceeds 25% of total NPP although more than half of the above ground production may be removed; in forest systems it probably never exceeds 10%.

Fig. 1.2. Climatic diagrams for six major zones: (A) Boreal zone (Taiga); (B) Arctic zone (Tundra); (C) Arid temperate zone (Temperate grassland); (D) Typical temperate zone (Temperate deciduous forest); (E) Equatorial zone (Tropical rain forest); (F) Tropical zone (Savannah). The legend over each graph gives the station of observation, its height above sea level (m), the mean annual temperature (°C) and the mean annual precipitation; the number of years on which the data is based is given in brackets below the station name. The ordinate gives 10°C or 20 mm rainfall for each interval; the abscissa gives the months of the year from January to December. Curve T is the mean monthly temperature, P the mean monthly precipitation—the area shaded black is rainfall above 100 mm per month and the scale in this range is reduced by a factor of 0·1. The area with stippling is a season of relative drought, that with vertical hatching a humid season. On the abscissa the black shading shows months with a mean daily minimum below 0°C, the diagonal shading shows months with an absolute minimum below 0°C, i.e. with occasional frosts; the number in the centre is the mean number of frost-free days (after Walter 1973).

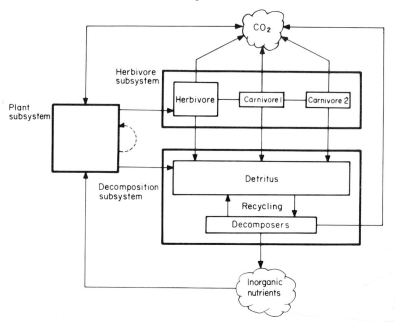

FIG. 1.3. A general model of ecosystem structure. The three subsystems are shown together with their main components. The major pathways of transfer of matter within the ecosystem are shown by the arrows. Organic matter pools are shown as rectangles, inorganic as 'clouds'. Note in particular the links between the herbivore and decomposition subsystems, the recycling of matter *within* the decomposition subsystem and the nett storage of matter that may occur within the plant subsystem (broken arrow).

(3)   In all mature ecosystems the bulk of the NPP is shed as *plant litter* or is secreted as soluble organic matter. This component thus enters the decomposition subsystem as dead organic matter or *detritus*. The faeces and carcasses of the herbivores and predators of the herbivore subsystem also contribute to this detrital input to decomposition (Fig. 1.3).

The detritus is broken down by the combined action of the *decomposer community* which is composed predominantly of micro-organisms (bacteria and fungi) and invertebrate animals. These organisms feed on the detritus and utilise the energy, C and other nutrients for their own growth (*secondary production*). Eventually the decomposers die and their carcasses enter the detritus compartment and are acted upon by other decomposers. This *recycling* of materials within the decomposition subsystem is an important feature of decomposition processes not matched within the herbivore subsystem.

The analysis of the decomposer community, the mechanisms of its functioning and their regulation by environmental factors are the subject matter of Chapters 2 to 6. In this chapter we shall consider only the significance of the decomposition subsystem within ecosystems without attempting to detail its internal structure.

Within ecosystems the decomposition subsystem performs two major functions—the mineralisation of essential elements and the formation of soil organic matter.

To maintain production plants must have continuing access to essential nutrient elements in *available* form. Because roots are not generally permeable to organic molecules the essential elements must be present in the soil solution in one of a small number of inorganic forms. For instance nitrogen is taken up most readily by plants in the form of nitrate ions; ammonium ions are less readily absorbed by plant roots and amino acids hardly at all. As most of the N in soil is combined in organic molecules only a small proportion of the total is available to the plant at any one time. To a varying extent the same is true for other essential elements such as P, S, K, Ca and Mg. The maintenance of primary production is dependent on the replenishment of this pool of available nutrients to balance the nutrient taken up by the plants. Nutrients are released in mineral form by the weathering of parent rocks but this input is generally insignificant in relation to the nutrient demand of the vegetation. The major part of the nutrient replenishment is accomplished by the mineralisation of the elements by the action of decomposer organisms. *Mineralisation* is the conversion of an element from organic to inorganic form. This includes for instance the formation of CO_2 as a result of the respiration of carbohydrates and the formation of ammonia from protein degradation. The central role of the decomposition subsystem in the cycling of essential elements is also illustrated in Fig. 1.3.

The decomposition of any piece of plant detritus is only completed over a time scale extending for hundreds or thousands of years. The residues of decomposition within this period contribute to the formation of *soil organic matter*. In the short term these residues are constituted from particulate matter formed by the action of decomposer organisms—partially digested plant matter, animal carcasses, faeces, microbial cells. This component of the soil organic matter may be termed the *cellular fraction*. A second component of the soil organic matter, *humus*, is less readily identifiable and persists over a longer term. Humus is a mixture of complex polymeric molecules synthesised during decomposition processes. It has an amorphous character and forms colloidal particles which often associate physically with the inorganic soil minerals.

The ways in which these two processes, mineralisation and soil organic matter formation, are accomplished by the decomposition subsystem determine to a large extent the structural and functional features of ecosystems. The balance between primary production and the rate of decomposition determines the amount of organic matter accumulated within ecosystems. The rate of movement of essential nutrient elements through the decomposition subsystem is an important regulator of primary production. We shall now consider these two contributions in more detail. Organic matter turnover and distribution can be regarded as a major determinant of *ecosystem structure* whereas nutrient cycling represents a major aspect of internal *ecosystem dynamics*.

1.2   Decomposition processes and ecosystem structure

1.2.1   Organic turnover within ecosystems

In Table 1.1 a number of parameters are listed which help to define the internal characteristics of ecosystems within each of our biomes and provide some comparisons between them of relevance to the study of decomposition. Ranges of values are given for the two main quantitative parameters used in production ecology—the NPP and the biomass. Both of these include above-ground and below-ground estimates. It should be noted that the below-ground estimates of root biomass and production are notoriously difficult and have been made a great deal less frequently than the comparable above-ground determinations. The chief sources of error in production ecology probably accrue from this factor and the proportions of biomass given in rows five to seven must therefore be regarded as 'best estimates'.

A number of general points of comparison between ecosystems may be derived from these data. NPP shows a rough correspondence to latitude. This reflects the general regulatory importance of climate on production as well as vegetational composition. A number of models have satisfactorily predicted NPP on the basis of climatic indices which combine temperature and moisture characteristics (Lieth 1976). Among the simplest of these is the sum of actual evapotranspiration—the amount of water lost from the soil by evaporation from the surface and from plants (Rosenzweig 1968). At a corresponding latitude forests tend in general to be more productive than grasslands. This is probably expressive of the prime importance of moisture availability. Reference back to Figs. 1.1 and 1.2 reminds us that grasslands typically occupy the drier regions of a particular latitude. This contrast is dramatically expressed in the biomass figure. Forests carry much greater biomass than grasslands even when production is roughly comparable. This is because of the twenty to forty per cent of the production which is diverted to perennial storage in woody tissue during development. The importance of water stress in the drier climates occupied by grasslands and savannah is further emphasised by the proportionately large root biomass they carry.

These general structural features of the plant subsystem have some importance for decomposition processes. Firstly, the difference in the nature of the biomass is reflected in the composition and siting of the litter input. Thus in tundra, grassland and savannah there is relatively less lignified

Table 1.1. Production and decomposition in six ecosystem-types. For definition of the decomposition parameters k and $3/k$ see p. 13. (From data given by Rodin & Basilevic (1967, 1968) and Whittaker (1975).)

	Tundra	Boreal forest	Temperate deciduous forest	Temperate grassland	Savannah	Tropical forest
1. NPP (Mean) t ha^{-1} yr^{-1}	1·5	7·5	11·5	7·5	9·5	30
2. NPP (Range) t ha^{-1} yr^{-1}	0·1–4	4–20	6–25	2–15	2–20	10–35
3. Biomass (Mean) t ha^{-1}	10	200	350	18	45	500
4. Biomass (Range) t ha^{-1}	1–30	60–400	60–600	2–50	2–150	60–800
5. Photosynthetic %	13	7	1	17	12	8
6. Wood %	12	71	74	0	60	74
7. Root %	75	22	25	83	28	18
8. Litter input t ha^{-1} yr^{-1}	1·5	7·5	11·5	7·5	9·5	30
9. Litter standing crop t ha^{-1}	44	35	15	5	3	5
10. k_L yr^{-1}	0·03	0·21	0·77	1·5	3·2	6·0
11. $3/k_L$ yr^{-1}	100	14	4	2	1	0·5

material which tends to be more slowly decomposed, and the major site of input is below ground in contrast to the litter layer of forest floors. Within these mature (i.e. non-incrementing) ecosystems the organic matter fixed in primary production is partitioned between the herbivore and decomposition subsystems. Heal & MacLean (1975) devised a simple model to illustrate the passage of energy through the two subsystems. The model (Fig. 1.4.)

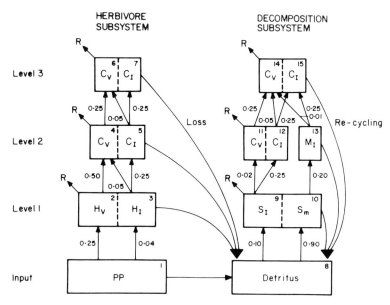

FIG. 1.4. A simple trophic model comparing the organisms of the herbivore and decomposition subsystems (after Heal & MacLean 1975). The symbols are those of the original paper: S = saprovore decomposers; M = microbivore; H = herbivore; C = carnivore; v = vertebrate; i = invertebrate; m = micro-organism; R = loss to respiration; PP = primary production. The transfers linking the compartments are the fractions transferred at any one linear run of the model.

has a three tier trophic structure for both subsystems. Within each level distinction was made between vertebrate and invertebrate consumers and, in the case of the decomposers, micro-organisms. The efficiency of transfers of nett production between trophic levels was based on fairly substantial available documentation of production and assimilation efficiencies. The apportionment of energy between trophic levels (i.e. the relative consumption efficiencies) is not as well documented particularly for the decomposers and these values were arbitrarily determined. Good data are available for NPP and the proportion consumed by herbivores in grassland and forest eco-systems and these were used to predict secondary production and respiration

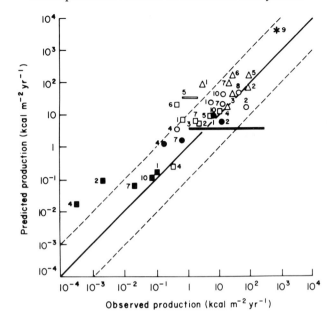

FIG. 1.5. Relationship between predicted and observed heterotroph production. The known NPP from ten sites were entered as inputs to the Heal and MacLean Model (Fig. 1.4) to predict the secondary production for various groups of heterotrophs: ($\bigcirc$) invertebrate herbivores; ($\bullet$) vertebrate herbivores; ($\blacksquare$) vertebrate carnivores; ($\square$) invertebrate carnivores; ($\triangle$) invertebrate saprovores and microbivores; ($*$) microbial saprovores. The predicted values are shown plotted against those actually measured at the sites which range from Tundra (3 and 4), Cold Temperate Moorland (2 and 7), Temperate Grassland (1) to Temperate Deciduous Forests (5, 6, 8, 9 and 10) (see Heal & MacLean 1975, for further details).

in ten sites. The model predictions showed a good agreement with site estimates of secondary production as Fig. 1.5 shows. Validation data for herbivores, carnivores and invertebrate decomposers was extensive but only one estimate of microbial production was available so conclusions with regard to the decomposition subsystem are tentative. Whether this simple structure would be equally valid for other terrestrial ecosystems is as yet untested but there is no reason to suppose otherwise.

The model postulates some interesting contrasts between the two subsystems which are illustrated by the data of Table 1.2. This is based on a grassland with a high herbivore consumption totalling 25% of NPP. The proportion of energy dissipated during respiration by the decomposition subsystem is nonetheless 86·5%. The added dominance of energy flow through the decomposers is due to two features of the model structure; firstly the transfer of energy as carcasses and faeces from the herbivore subsystem to

Chapter 1

Table 1.2. Calculated ingestion, production, respiration and egestion by heterotrophs (k cal m^{-2} yr^{-1}) per 100 k cal m^{-2} net annual primary production in a grassland ecosystem. The efficiencies of consumption, assimilation and production shown in Fig. 1.4 were used in the calculation. Symbols as in Fig. 1.4 (modified from Heal & MacLean 1975).

	Ingestion	Production	Respiration	Egestion
Herbivore subsystem				
Herbivores Vertebrate (H$_v$)	25·000	0·250	12·250	12·500
Invertebrate (H$_i$)	4·000	0·640	0·960	2·400
Carnivores Vertebrate (C$_v$)	0·160	0·003	0·123	0·031
Invertebrate (C$_i$)	0·170	0·040	0·095	0·034
Decomposition subsystem				
Decomposers Invertebrate (S$_i$)	15·153	1·212	1·818	12·122
Microbial (S$_m$)	136·377	54·551	81·826	—
Microbivores Invertebrate (M$_i$)	10·910	1·309	1·964	7·637
Carnivores Vertebrate	0·041	0·001	0·032	0·008
Invertebrate	0·648	0·155	0·363	0·130
Total	192	58	99	35
% passing through				
Herbivore subsystem	15·2	1·6	13·5	42·9
Decomposition subsystem	84·8	98·4	86·5	57·1

the decomposition subsystem; secondly the recycling structure of the latter system. Whereas the herbivores and their associated predators are only able to work over the organic matter once, the secondary production of the decomposition subsystem re-enters the detritus pool and provides further resources for the decomposers. This accounts for the total ingestion being double the NPP and the 99% dissipation of energy from respiration.

We may now consider the dynamics of the detritus entering the decomposition subsystem. Jenny *et al.* (1949) and Olson (1963) have suggested that a useful expression of steady state in an ecosystem is that the ratio of dead organic matter production to total decomposition approximates to unity over a given period of time. Under these circumstances the ratio of the annual input of dead organic matter (I) to the mean annual standing crop of dead organic matter (Xss) gives an index of the rate of decomposition characteristic of the ecosystem i.e. $k = I/Xss$ where k is the annual fractional weight loss. The same authors have also shown that $3/k$ and $5/k$ give estimates of the time taken for 95% or 99% respectively of the standing crop to decompose. Estimates of k thus theoretically represent a useful ecosystem constant. There are however a number of limitations to its usefulness. Firstly, the total detrital input has rarely been estimated, largely because of the problems associated with the measure of below-ground input (i.e. root death and root exudation). An estimate can be made by taking the ratio of the total NPP (assuming no nett increases in biomass and negligible loss to herbivores) to the total dead organic matter. Fig. 1.6 shows a plot of these two characteristics for a wide range of sites corresponding to several of our ecosystem types.

An alternative method of calculation is to confine attention to the above ground input of plant detritus direct to the decomposition subsystem i.e. plant litter fall (L). This can be measured with considerable precision (see Appendix) and data are available for many sites (e.g. see Table 1.1). The standing crop corresponding to the litter fall is that of the above ground plant litter (X_L) i.e. that which originates from litter-fall and the appropriate ratio is $k_L = L/X_L$. Some authors have used a ratio of L to Xss (e.g. Olson 1963) but this is not strictly correct as Xss includes material of below ground origin. The litter turnover ratio k_L clearly differs from the total turnover ratio k for it describes the decomposition of only a component part of the total organic matter in the system and relates only to the earlier stages of decomposition. No account is taken of stages beyond the point at which detritus is no longer recognisable as plant litter. We have already suggested, and considerable evidence will be presented in later sections, that the further processes of decomposition—of particulate matter and humus—are equally of significance. One further word of caution is necessary. Turnover rates calculated in this way depend on the concept of a steady state of the litter standing crop but this cannot be readily assumed,

the soil chemist and by the micromorphologist using soil sectioning techniques. The two approaches are complementary, but not always easy to
reconcile and there is some confusion in terminology. The term humus has
been defined here in a restrictive sense referring to a component of soil
organic matter which is chemically and morphologically distinct from other
fractions. Micro-morphologists may use the term humus in a broader sense
to include all organic matter in soil. Babel (1975) in a recent review describes
the distinguishable components of 'humus' as 'particles. . . easily recognisable
as plant residues,. . .droppings of soil animals,. . .fungal hyphae,. . .often
accompanied by a yellow-to-brown fine substance. . .(which). . .consists of
mineral particles bound together by coloured isotropic materials'. The 'fine
substance' has no recognisable microscopic structure and probably largely
corresponds to the chemical definition of humus, although a direct correspondence is difficult to draw. We shall continue to distinguish between the
cellular fraction of soil organic matter—which we defined earlier as containing
recognisable plant, faecal, animal or microbial remains—from the amorphous
humus. We believe it is conceptually important to do so because these two
products of decomposition are formed in different ways, and are subsequently
decomposed by different pathways (see Chapter 2).

The soil organic matter (SOM) may be fractionated by physical or chemical
methods. Ford *et al.* (1969) used ultrasonic dispersion in heavy liquids of
specific gravity about 2, to float off a so-called 'light fraction' which accounted
for about 20–30% of the soil carbon. This fraction was composed of undecomposed or partially decomposed plant fragments, faecal material and
microbial cells plus a certain amount of partially humified material and is
thus equivalent to the cellular fraction as we have defined it. The humus
sensu stricto sinks in the high gravity solution, because it is largely absorbed
to soil mineral particles. A more usual fractionation is to extract humus from
SOM with dilute aqueous alkali, although an insoluble fraction (known as
humin) is left. This alkali extract provides the basis for the classical chemical
fractionation of humic materials which is described in Section 5.5.1.

Humus forms colloidal particles which are negatively charged probably
due to both hydroxyl and carboxyl groups linked to the largely aromatic
core. As with clay particles that are similarly charged, this means that the
humus particle may attract a swarm of cations (Fig. 6.7). These cations can
exchange for others in the soil solution; the *cation exchange capacity* (CEC)
of a soil is a very important characteristic which may determine a number of
properties, including the pH and the nutrient balance of the soil solution
(see Section 6.4 for a detailed discussion) and is largely determined by the
quantity and type of colloidal particles present. The CEC of humus colloids
is higher than that of any clay and this is one of the ways in which the humus
content of soil can markedly affect its properties. Clays, which are also
negatively charged, and humus colloids may form complexes so that the

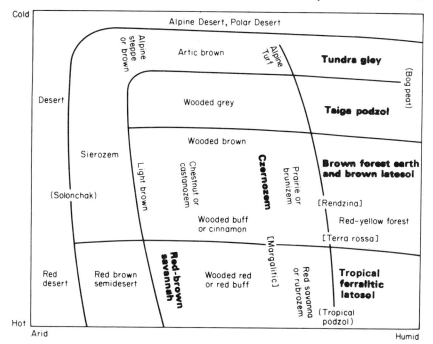

FIG. 1.8. Distribution of soil types in relation to climate. Zonal soils are shown in open script, intrazonal soils in brackets—those formed on limestone in square brackets, those formed in areas of impeded drainage in round brackets. The soils typically found associated with the major ecosystem-types (Figs. 1.1 and 1.2) are given in bold letters. Soils formed under forest are written horizontally; those formed under grassland vertically (redrawn from Whittaker 1975).

highest CECs are found in soils with both clay and organic matter content high.

Accumulation of SOM in different ecosystems. Under similar climatic conditions soil development is sufficiently consistent to form a basis for a classification of soil types. These are the 'zonal soils' or 'great soil groups' as initially defined by Russian soil scientists in the nineteenth and early twentieth centuries. The distribution of these soil types in relation to climate is shown in Fig. 1.8. The correlations with vegetational development are obvious when compared with Fig. 1.1. As with the vegetation, many subdivisions may be recognised within each zonal type and clear boundaries are often difficult to draw. Variations may occur due to the effect of parent rock or topography, including the production of soils of similar type within widely differing climatic zones. These are then termed *intra-zonal* soils and include water-logged soils such as peats and soils derived from limestone such as rendzinas.

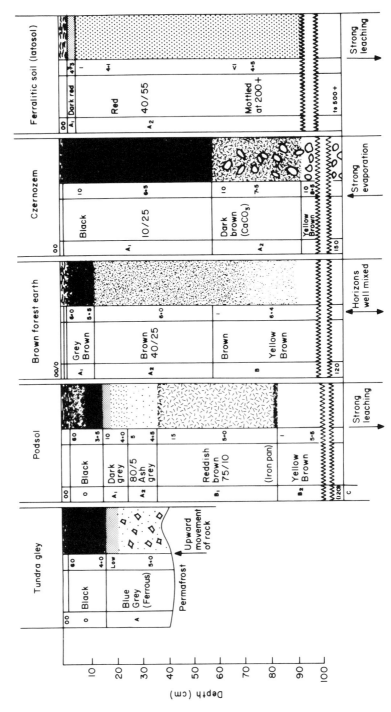

Of other soil classifications the most widely used is the United States System of Soil Taxonomy which has many advantages over the zonal system. The soil names in the latter are so familiar however that it is more convenient to retain its use.

Zonal soils are described mainly on the basis of their characteristics as seen in profile. Soils always have a vertical structure which develops due to the polarity of mineral input from the parent rock at the base of the profile, of organic litter at the surface, and by the movement of water either leaching downward or flushing upward during evaporation. As a result, in most soils a number of distinct *horizons* can be recognised. A standardised letter code has been adopted for these horizons as explained in the legend to Fig. 1.9 which illustrates some characteristics of six zonal soils which broadly correspond to the climatic zones of the six ecosystem types discussed in the previous sections. It should be emphasised that the definition of horizon boundaries is again a difficult procedure and horizons given the same code in different soils may be by no means identical in terms of their composition and properties.

The standing crop of organic matter in a soil is determined by the equilibrium between the input of the products of plant litter decomposition and the output, due to its own decomposition. Estimates for the standing crop are difficult to obtain for a wide range of soil types, partly because of the difficulty in assessing whether the term SOM or 'humus' in one reference is the same as for another and partly because data are usually

---

FIG. 1.9. Selected zonal soils corresponding to the major biome types. The profile symbols indicate the following (e.g. with reference to the Podzol, the most elaborate soil). O = *Organic horizons* formed from litter originating above the soil; Oo, original forms of litter readily recognisable; O, forms of litter less easily recognisable, marked humus formation (= fermentation layer) usually dark in colour. A = *Eluvial horizons*, mineral horizons which are characterised by leaching out of colloids and ions; A_1 is usually darkly stained with humus colloids; A_2 is the area of maximum eluviation and therefore lighter in colour, lower in organic matter, clays, iron and aluminium oxides. B = *Illuvial horizons*, mineral horizons in which there is nett deposition of colloids and mineral oxides from the leachate or occasionally from below; B_1 (if present) is transitional A to B but usually darker than A_2 indicating some deposition; B_2 is the main area of deposition, usually darker in colour with higher organic content plus some iron coloration occasionally forming a hard iron pan under conditions of heavy leaching—podsols, latosols. C = horizon outside biological activity and overlying the parent rock. For each soil the horizon designation and approximate depths are shown in the first column. In the second are descriptions of the colour, the ratio of sand to clay (in percentage of weight) and any special inclusions. In the third column the percentage soil organic matter, as loss on ignition (top), and pH (bottom) is given for each horizon. (From data in Money 1972 and Fitzpatrick 1971.)

presented as soil SOM content rather than as weight per unit area. The relative size of the standing crops in the different soil types can however be indirectly assessed by comparison of the depths of the organic horizons (A_1 in particular) and their organic content (usually determined by loss on ignition). Based on those considerations the importance of climate is readily determined. Organic matter accumulation declines from the tundra to the tropics and shows generally lower levels in the drier parts of similar latitudes. This demonstrates a broad negative correlation with the decomposition rate (Table 1.1). As we have already pointed out, the extent of accumulation is greater than the differences in NPP (i.e. detrital input) between ecosystem types would predict.

Factors other than climate influence the development of SOM quite markedly. The most detailed analysis of this has been the work of Jenny and his colleagues at the University of California, Berkeley. Although ultimately interested in agricultural fertility Jenny's approach has been to consider pedogenesis within an ecosystem framework. He has described the factors influencing soil development in the form of a general equation (Jenny 1961);

$$S = f(cl, o, r, p, t, \ldots)$$

that is the soil state (S) which may be for instance the organic carbon or nitrogen content, is a function of climate (cl), a biotic factor (o) which generally refers to the vegetation-type of the area, topography (r), parent material (p) and time (t). In one example (Jenny *et al.* 1968) he found that whilst the accumulation of organic C in a series of Californian soils was primarily determined by climatic factors and the type of prevailing vegetation, the amount and availability of N was more strongly influenced by the nature of the parent materials, particularly the clay composition.

Variations in topography, soil type or vegetational cover may result in marked differences in the extent of SOM accumulation within climatic zones. The generalisation that tropical soils are all characterised by low organic contents is challenged by an example of the third of these factors. In the Kerangas tropical heath forests of S.E. Asia highly organic horizons up to 50 cm deep may accumulate in regions of optimally high rainfall and temperature. This appears to be associated with a marked degree of resistance of the litter to decay.

Further examples of the influence of non-climatic factors can be seen when a more detailed consideration is given to the nature as well as the amount of SOM formed in different soil types. The brief descriptions given in Fig. 1.9 indicate that the organic horizons differ qualitatively as well as quantitatively; in colour and particle size for instance. Micro-morphological and biological studies have led to the distinction of categories of organic layers (often termed 'humus types') particularly with regard to temperate woodland soils. As with many other aspects of SOM study there is much confusion in nomenclature between various authorities but the classification rests largely on the recognition of gradations between two extreme types—termed *mull* and *mor*. A consideration of these forms of SOM will serve to illustrate some of the features relevant to decomposition processes without the necessity of detailing the wide range of possible intermediate, developmental or deviant forms. References to most of the detailed works of classifications are given by Howard (1969) who also discusses the confusions in terminology.

The major distinctions between mull and mor are given in Table 1.3. From these it is easy to draw the conclusion that conditions that favour mull formation also promote rapid decomposition of plant litter and those that promote mor formation are less favourable in this respect. Mor is characterised by an accumulation of the cellular fraction of SOM. In mull the main accumulation is not of cellular material but of humus. Climate is clearly a major determinant of these factors and explains the broad correlation between the podsolisation and mor organic layer characteristic of Boreal Forest and the mull-type development of many Temperate Deciduous Forests. The intervention of other factors may produce highly local variations

Table 1.3. Characteristics of mull and mor soil types. Unless otherwise mentioned descriptions refer to the A_1 horizon. Based on data from Howard (1969), Russell (1973) and Fitzpatrick (1971).

	Mor	Mull
Vegetation	Conifers Heathland	Deciduous trees Grasses
Moisture relations	Strong leaching—high precipitation or free drainage due to topography or high sand content	Leaching and flushing—gentle topography and/or warmer conditions
Horizons	O—very well defined A_1—sharp, deep and largely organic	O—rarely recognisable Other horizons show good mixing of organic and mineral
Organic matter	High cellular component. A_1 has 20–80% loss on ignition	Low cellular component. A rarely exceeds 10% loss on ignition
Nutrients	C:N > 20 pH 3·5–5·0 CEC[1] 80–120 (me %) BS[2] 20–40%	C:N < 15 pH 5·0–7·0 CEC[1] 20–40 (me %) BS[2] 40–100%
Organisms	Fungal mycelium abundant Bacterial counts low Earthworms absent Mesofauna populations high	Mycelium less obvious Bacterial counts high Earthworms present Lower mesofauna populations

[1] CEC = cation exchange capacity

[2] BS = base saturation

in mull and mor formation within each of these ecosystem types. Nihlgard (1971) has shown how the type of plant litter may dramatically affect the type of SOM formed. Spruce forest planted over areas previously occupied by beech in Sweden resulted in the development of mor organic layers replacing the previous mull within sixty years. In contrast Minderman (1960) found that the development of the different 'humus' forms in the Hackfort Forest of the Netherlands was associated, not with the two tree species, oak and birch, but rather with differences in the soil water regimes which in their turn were influenced by soil texture and topography.

It is clear from this brief consideration of mull and mor formation in forest areas that soils in which large accumulations of SOM are found are typically those with a high cellular component. The possible differences between ecosystems in terms of humus accumulation will be discussed in Chapter 5.

1.3   Decomposition processes and nutrient cycling

A primary distinction that may be made in describing the biogeochemical cycle of any element is that between the *mineral reservoir* and the *biological exchange pool.*

As we have already pointed out the appropriate unit for the study of nutrient cycling is the ecosystem. Ecosystems are not entirely self-contained with regard to nutrients. There is input to all ecosystems from nutrient reservoirs in the lithosphere and atmosphere and from other ecosystems via precipitation, wind blow, ground and soil water flow, and biological migration; ecosystems also lose matter through the same agencies. To recapitulate our earlier definition however the amount of nutrient transfer within the ecosystem—that is within and between organism and soil pools—is much greater than the exchange with other ecosystems.

The origin of most nutrient elements lies in the lithosphere i.e. the primary mineral forms of solid rock. In this form they are generally inaccessible to living organisms and this may thus be distinguished as part of a mineral reservoir separate from the biological exchange pool or biosphere. Some elements, notably C and to a lesser extent N, have gaseous reservoirs in the atmosphere. Exchanges between the atmosphere and the biosphere are much more direct than those between the lithosphere and the biosphere. When referring to the ecosystem nutrient pool we are generally excluding atmospheric and lithospheric components as the boundaries of these are impossible to define. The uptake of C from the atmosphere by plants is clearly a major flux however and so this exclusion creates an anomaly for C in terms of the previous assertion of the predominance of within-ecosystem fluxes. Under some circumstances the same may be true of N.

Our main purpose in this chapter is to describe some of the major nutrient pools within ecosystems, the fluxes between them and the functional role of

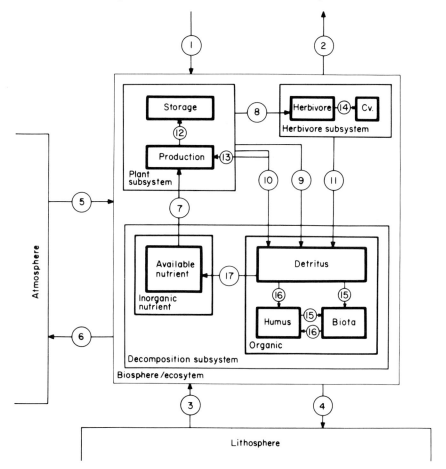

FIG. 1.10. Nutrient pools and fluxes within ecosystems. The model has a hierarchical structure of pools represented by the diminishing sizes of the boxes. The ecosystem is connected to other ecosystems by input (1) and output (2) fluxes which may be of material of either mineral or biological origin. Exchanges with the lithosphere and atmosphere are also pictured as extra-ecosystem transfers as explained in the text. These consist of the formation of secondary minerals (3) and their loss (4); the fixation of C and N by the plants or micro-organisms (5) and volatilisation of elements (e.g. as CO_2, CH_4, H_2S or N_2) (6). Within the ecosystem the three main subsystems are connected by fluxes representing uptake of nutrients by plants (7), losses from them by herbivore consumption (8), leaching (9) and litter production (10). The transfer to the decomposition subsystem from the herbivore subsystems is also shown (11). Within the plant subsystem the main fluxes shown are of storage in perennial tissues (12) and withdrawal from senescent leaves (13). Predation transfers (14) are shown in the herbivore subsystem. Immobilisation transfers from detritus to decomposers (15) or humus (16) are shown within the decomposition sub-system. Mineralisation (17) replenishes the inorganic pool and the pool of available nutrient in particular. Further details in the text.

Chapter 1

decomposition processes in determining the nature and rate of these fluxes. As the chemistry of the different elements essential to living systems differs so do the pathways by which they move in nutrient cycles. So also does their significance in the various pools and fluxes within ecosystems. Detailed description of all these variations would require a volume of its own. The mechanisms of some of these transformations at the molecular level will be described in Chapter 5, and in Chapter 7 we consider the quantitative aspects of the cycling of different elements within selected ecosystems.

For discussions of the role of decomposition processes in nutrient cycling it is convenient to group the nutrient pools within ecosystems into the three main compartments—the plant, herbivore and the decomposition subsystems. The main fluxes between and within these compartments are shown in Fig. 1.10.

Table 1.4. The essential nutrient elements (after Brady 1974).

	The macronutrients		
Element	Mineral forms	Organic forms	Available forms
N	(N_2 gas)	Proteins and derivatives	NH_4^+
		Heterocyclic bases	NO_3^-
P	Apatite (primary)	Phytin	HPO_4^{2-}
	Ca, Al, Fe phosphates (secondary)	Nucleic acid	$H_2PO_4^-$
S	Pyrite	Protein	SO_4^{2-}
	Gypsum		
K	Feldspar, Mica (primary)	—	K^+
	Clays (secondary)		
Ca	Feldspar, Calcite, Horneblende, Dolomite	—	Ca^{2+}
Mg	Mica, Dolomite Horneblende, Serpentine	—	Mg^{2+}

The micronutrients	
Element	Commonest forms in soil
Fe	Oxide, Sulphide, Silicate
Mn	Oxide, Silicate, Carbonate
Zn	Sulphide, Oxide, Silicate
Cu	Sulphide, Hydroxycarbonate
Bo	Borosilicate, Borate
Mo	Sulphide, Molybdate
Cl	Chloride
Co	Silicate

Other elements (such as sodium, fluorine, iodine, silicon, strontium and barium) although essential to many organisms are not universally so.

1.3.1   The plant subsystem

From the earliest days of scientific agriculture the demand of plants for a range of *essential elements* has been recognised. Of these C and O are obtained readily from air and water. The remainder are obtained from the soil and are generally only available to plants in soluble inorganic form. A list of these elements is given in Table 1.4. In terms of the quantities with which they are required for the optimum productivity of plants they may be subdivided into the six *macronutrients* and about eight *micronutrients*. Each of these nutrients may occur in soil in a variety of organic molecular or inorganic ionic or mineral forms (Table 1.4). The transition between these categories will be considered in a following section but it is appropriate to consider here the inorganic nutrient pools of soil and their availability to plants.

Nutrient availability

Nutrient elements are absorbed by plant roots from the *soil solution*. Generally speaking the nutrients in the soil solution only comprise a minor fraction of the total inorganic pool for a particular element; the major component of each element is either insoluble or adsorbed to solid surfaces. In either case it is inaccessible to the plant. The factors determining the size and the rate of input to the pool of available nutrients are thus important. For instance P is only available to plants as the monobasic ($H_2PO_4^-$) or, less commonly,

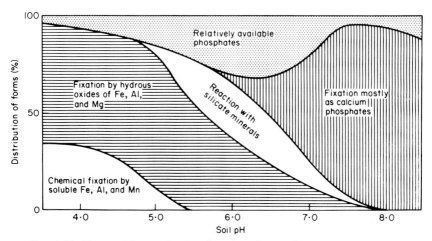

FIG. 1.11. Forms of inorganic phosphate in soil at various soil pH values. At any given time in a given soil the proportions of different components may vary because of the dynamic nature of soil chemistry (e.g. see Fig. 1.12). In particular the degree of immobilisation will affect the size of the available pool. This diagram does however give an impression of the relative sparsity of available P in most soils (after Brady 1974).

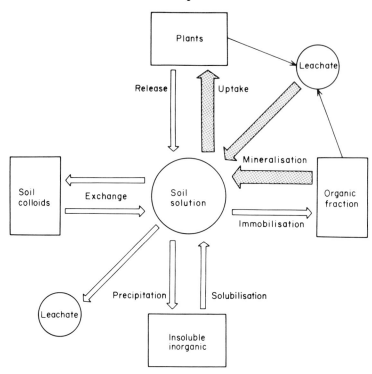

FIG. 1.12. The flux of nutrients through the soil solution. The size of the available pool of nutrients in the soil solution at any one time will depend on the equilibrium between these various inputs and outputs. The importance of different fluxes differs from time to time (for an example see Table 1.5). The shaded fluxes are those that probably predominate in most ecosystems, although at different seasons.

dibasic (HPO_4^{2-}) anion. These soluble inorganic forms of phosphate may comprise only a small percentage of the total inorganic phosphate in a soil. Most of it is in insoluble forms which vary accordingly to the soil pH (Fig. 1.11). It has been calculated however that crop plants may absorb up to 100 times the P present in the soil solution in the course of a growing season. The key factor in nutrient availability is not the size of the nutrient pool in the soil solution but the rate at which it can be replenished.

The soil solution is highly dynamic and subject to a variety of different inputs and outputs (Fig. 1.12). Under differing conditions various of the inputs and outputs may predominate. Thus during a season of high productivity, uptake by plants may markedly diminish the size of the pool but this will be replenished by a corresponding period dominated by decomposition and mineralisation (Table 1.5). The replenishment process may also differ for different elements. Thus the cationic elements, Ca, Mg and K, may be

Table 1.5. The effect of plant growth on the concentrations (ppm) of available ions in the soil solution. Barley growing in agricultural soil, pH 8·2. The uptake of nutrient by the crop during the growing season reduces the pool of available nutrients but by the following Spring this has been replenished (after Russell 1973, from Burd & Martin 1924).

	NO_3	PO_4	SO_4	K	Ca	Mg
Before planting, year 1	173	3·3	671	41	222	97
At harvest, year 1	16	1·2	598	22	192	64
Before planting, year 2	263	2·9	785	35	276	94

largely replenished, in the short term at least, by exchange phenomena controlled by the CEC of the soil colloids which adsorb the cations whereas the anionic forms of N, P and S are more dependent on biological (i.e. decomposer) modes of replacement. Finally the equilibrium of all the inputs and outputs to the soil solution is affected by general soil conditions particularly soil moisture and pH and the amount of soil colloidal material such as clays. The influence of these features is discussed in Chapter 6 but for more detailed accounts of nutrient availability the reader should consult specialist texts such as those by Russell (1973) or Black (1968).

Nutrient uptake

The uptake of nutrient elements in ionic form is an active process which has been much studied but which is still poorly understood. The removal of nutrients from the soil solution by roots is dependent on a wide range of soil factors which affect the uptake process as well as on the availability of the nutrients. These factors include soil pH and oxygen concentration both of which may be affected by the nature and rate of decomposition processes. Nutrient uptake is also directly affected by the activity of micro-organisms which contribute to decomposer functions, particularly the inhabitants of the rhizosphere. At low nutrient availabilities these organisms may compete with the root for nutrient elements. In most circumstances however they probably increase nutrient availability to the plant both by decomposition of organic molecules and by aiding the mineralisation of insoluble inorganic sources. For instance micro-organisms from the rhizosphere have been shown to liberate soluble phosphate from the mineral apatite by means of acid secretions (Nicholas 1965).

Even more significant in nutrient cycling than the rhizosphere inhabitants are the mycorrhizal fungi. It has long been known that the infection of certain temperate forest trees by Basidiomycete fungi can lead to the establishment of a mutualistic association conferring increased efficiency in nutrient uptake to the plant. A much more widespread distribution of mycorrhizal associations involving non-Basidiomycete fungi has been more recently

Table 1.6. Concentrations of nutrient in mycorrhizal infected plants compared with uninfected plants. *Pinus strobus* seedlings grown in a prairie soil (mg dry weight, mean values, n = 4; after Hatch 1937).

	Dry weight	Root/shoot ratio	N		P		K	
			wt	% dw	wt	% dw	wt	% dw
Infected	404·6	0·78	5·00	1·24	0·789	0·196	3·02	0·744
Uninfected	302·7	1·04	2·69	0·85	0·236	0·074	1·38	0·425

documented and shown to have similar significance for nutrient uptake dynamics. Indeed it has been predicted that practically all natural populations of plants, except those on very nutrient rich soils, will be shown to be infected by one or other of these forms of mycorrhiza. Whilst the form of association varies, the nature of the relationship appears to be fundamentally similar. The fungus acts as a 'bridge' between the plant and soil solution and nutrients largely enter the plant by translocation through the fungal hyphae (Harley 1972; Mosse, Tinker & Sanders 1976). The consequence of infection is increased productivity and a greater concentration of nutrients by the plant (Table 1.6). The uptake by the fungus and the transfer from the fungus to the host both involve active transport mechanisms which are probably fuelled by plant photosynthate. The quantitative importance of mycorrhiza in nutrient flow can be assessed from a calculation made by Harley (1972) that they constitute about 4% of the root biomass in temperate forests. From Table 1.1 it follows that this constitutes about 4 t ha^{-1} dry matter. Assuming about 2% N and 0·5% P for fungal nutrient contents this suggests a standing crop of nutrient in mycorrhizal fungus of 80 and 20 kg ha^{-1} for N and P respectively. In the case of N this equals and for P is four times the estimated annual uptake for this type of forest given in Table 1.8.

The involvement of heterotrophs in nutrient uptake processes serves to emphasise the close relationship between decomposition and nutrient availability. It is unclear however whether mycorrhizal fungi are ever directly involved in decomposition. Most mycorrhizal fungi are reported to lack any of the enzymatic capacity for degrading plant litters which is characteristic of decomposer fungi (Harley 1969). The 'trade-off' in the mutualistic association is the provision of sugars from the host photosynthate. Thus the mycorrhiza must depend on the operation of the normal decomposer food web for the actual mineralisation of the plant material. The increase in efficiency in uptake derives from the greatly enhanced surface area for absorption provided by the fungal mycelium, and its much more intimate contact with the sites of mineralisation. The generality of this view is brought into question by the hypothesis of Went & Stark (1968). They suggested that on the nutrient-poor forest soils of the South American rainforest the mycorrhizal fungi are directly involved in decomposition and mineralisation. This establishes a very rapid passage of nutrient direct from the organic form in the litter to the organic translocate in the fungus and thence to the root without ever entering a pool of inorganic soil nutrient. They termed this 'direct mineral cycling'. Recently Herrera *et al.* (1978) have produced evidence to show the passage of labelled phosphate from leaf litter through mycorrhizal fungus to host root in tropical forest seedlings. There is however no evidence to support the notion that enzymes from the mycorrhizal fungi participate directly in the decomposition process. Mycorrhizal uptake probably only succeeds the activity of a number of other decomposer fungi (Herrera, personal communication). We may picture this rainforest situation as differing quantitatively rather than qualitatively from that of the temperate forest. That is to say, the 'tightness' of the litter-decomposer-mycorrhiza-root association is greater. It is difficult to do much more than speculate in this area as so little is known of the processes involved at the interface between the mycorrhizal fungus and the litter and soil from which it derives its nutrient. This is an area deserving an intensive research effort for its results could have significance for improving crop yields.

It is worth re-emphasising that, in temperate ecosystems at least, the nutrient capture activity of the mycorrhizal fungi is carried out in return for a C subsidy from the NPP of the host plant. The rhizosphere flora is probably largely maintained in the same manner. If nutrient recovery is part of decomposition processes then this part of C flow should be included in input to the decomposer system. The outcome of this uptake activity, whether by direct activity of the roots, or mediated by mycorrhizal fungi, is the concentration of nutrients into the plant roots. We have now to consider briefly what happens to those nutrients before they re-enter the decomposition subsystem.

Nutrient allocation and release

The nutrients taken up by the roots are distributed within the actively growing plant by translocation. The distribution of the nutrients is determined by the relative activity of the plant tissues the most productive tissues accumulating the greater concentrations of nutrients (Table 1.7). Thus photosynthetic tissues always have the highest concentrations followed by the young roots. Perennial tissue has the lowest and the association between cell proliferation and nutrient concentration is shown by the much higher content of nutrients found in bark (including the cambium), compared with the adjacent sapwood. This is the first evidence of 'nutrient conservation' by the plant of which we shall see further features later. The most marked conservation is seen with regard to N and P as can be seen from a comparison of the leaf and wood concentrations in Table 1.7. This results in marked changes in the relative proportions of different elements in different types of

Chapter 1

Table 1.7. Concentrations of macro-nutrients (% dw) in different tissues of a forest tree, *Quercus robur* (after Duvigneau & Denaeyer de Smet 1970).

		N	P	K	Ca	Mg
Leaf		2·4	0·15	1·0	1·0	0·14
Fraction of total		0·19	0·12	0·16	0·04	0·03
Twig + bud (1 yr)		1·3	0·097	0·36	1·4	0·15
Twig + bud (3 yr)		1·0	0·078	0·26	1·5	0·18
Branch (10–15 cm)	Heart	0·13	0·003	0·17	0·07	0·015
	Sapwood	0·16	0·019	0·17	0·09	0·028
	Bark	0·57	0·030	0·20	2·9	0·37
Root (1–3 cm)	Wood	0·45	0·024	0·21	0·10	0·032
	Bark	0·52	0·027	0·39	2·3	0·070

tissue. Thus for instance the ratio of $N:K:Mg:P$ in leaf is about $16:7:1:1$ but in sapwood is about $8:9:2:1$ indicating the markedly greater relative decline in N and P concentrations.

In Table 1.8 plant nutrient data are presented from a detailed study of temperate deciduous forest ecosystems by Duvigneau & Denaeyer de Smet (1970). The amount of each nutrient element taken up each year is shown in the third row. The high proportion of this in the green leaves at their peak (row 4) further indicates the major demand of the production tissue for most of the macronutrient elements. We can use this temperate forest example to illustrate the further fate of the annual nutrient uptake. Earlier we suggested that about 20–40% of the NPP in such ecosystems provided the yearly increment in biomass. In the young Virelle forest this was rather higher, 62%, but the proportion of the nutrient uptake so stored in the plant biomass was in most cases only half as much. The major proportion of the annual

Table 1.8. Nutrient flux in a mixed temperate deciduous forest in Virelle, Belgium (re-calculated from Duvigneau & Denaeyer de Smet 1970).

		N	P	K	Ca	Mg
1. Total biomass	kg ha^{-1}	533	44	342	1248	102
2. Above-ground biomass	kg ha^{-1}	406	32	245	868	81
3. Uptake (U)	kg ha^{-1} yr^{-1}	81	5·2	54	191	18
4. Green leaf biomass (G)	kg ha^{-1}	73	4·7	36	54	4·6
5. Leaf biomass/uptake		0·90	0·90	0·67	0·78	0·26
6. Litter fall	kg ha^{-1} yr^{-1}	50	2·4	21	110	6
7. Leaching (L)	kg ha^{-1} yr^{-1}	1	0·6	17	7	6
8. Total released (R)	kg ha^{-1} yr^{-1}	51	3·0	38	117	12
9. Leaf litter fall (F)	kg ha^{-1} yr^{-1}	33	1·4	15	74	4·5
10. Withdrawal ($G-L-F/G$)		0·53	0·57	0·67	0·28	0·26
11. Return (R/U)		0·63	0·58	0·70	0·61	0·67
12. Increment ($U-R/U$)		0·37	0·42	0·30	0·39	0·33

uptake is thus released back to the soil during the year, even in a forest which shows a rapid increase in biomass. The details of the nutrient release are given in rows six to nine of Table 1.8. The major agency for N, P and Ca release is litter fall but K and Mg show a more mobile character and are released as readily by leaching as by litter fall.

The main component of litter fall by dry weight is leaf material. Combined with the very high concentration of nutrient in photosynthetic tissue already demonstrated this suggests that the major proportion of the nutrient uptake should be lost to the plant by this agency. The ninth row of the Table shows that this is a major pathway but that the proportions are very much lower— 41, 27 and 28 per cent of the uptake for N, P and K compared with the 90, 90 and 57 per cent allocation to leaves at peak photosynthetic activity. In the case of K this is largely due to the excess leaching loss from the senescing leaves but as already noted such losses are negligible for N and P. The discrepancy is indicative once again of nutrient conservation. During leaf senescence, and prior to abscission large proportions of the photosynthate and of some key nutrients such as N and P are withdrawn from the leaves into the twigs. The extent of this for the various elements is shown towards the

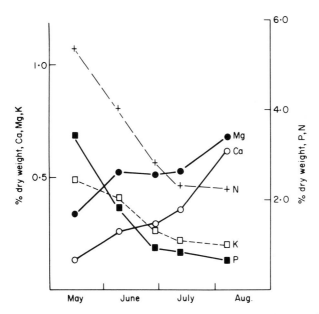

Fig. 1.13. The seasonal change in nutrient concentrations in leaves of *Rubus chamaemorus* on blanket bog in northern England. The withdrawal (N and P) or loss to leaching of three of the elements results in a marked decline in concentration prior to litter fall. In contrast Ca and Mg show significant increases in concentration during the same period (after Marks & Taylor 1972).

bottom of the Table (row 10). Again Ca shows a markedly different pattern, the falling leaves accounting for an excess of nutrient compared with the biomass at peak growth (cf. row 9 and row 4). This is a familiar pattern, the Ca concentration of leaves commonly increasing during senescence in contrast to the decrease characteristic of other nutrients (Brown 1974).

The withdrawn nutrient forms part of the incremental nutrient in the form of calculation adopted. It is probable however that much of this autumn-conserved fraction is stored in metabolically accessible form and is re-utilised in the next season's production in contrast to the fraction allocated to 'dead' perennial tissue. A dynamic view of the withdrawal process and the differing patterns of behaviour of the elements is shown in Fig. 1.13.

1.3.2   The decomposition subsystem

We have now established the context of the flow of nutrients through the decomposition subsystem, the inputs and outputs to the organic matter in soil. The main inputs, as shown in Fig. 1.10 are four in number; organic materials washed or blown in from other systems (1); leachate from the plant cover—including root exudation—(9); the corpses and faeces from the herbivore—carnivore food chain (11) (dealt with in the next section); and, the major component, the return of nutrient direct from the plant in litter fall (10). This comprises the sum of the resources available to the decomposers in the soil or elsewhere. There is only one major output (although there may be some loss of organic material to other ecosystems by leaching or blow)— the conversion of the organic nutrient to inorganic form, the process of mineralisation (17). This is the essential function of decomposition replenishing the soil solution equilibrium in relation to the demands of plant growth as previously illustrated in Fig. 1.12 and Table 1.5. We now turn to examining the mechanisms whereby that equilibrium is maintained.

The *immobilisation* of a nutrient element occurs when it is incorporated into or maintained in an organic form. It occurs when inorganic nutrients are taken up by plant roots; it also occurs when micro-organisms or decomposer animals take in inorganic nutrients (Fig. 1.12). *Mineralisation* occurs when inorganic forms of an element are released during catabolism of organic resources, e.g. CO_2 from carbohydrate, NH_3 from protein. The consequence of catabolism is the release of energy for anabolic activity which also involves the uptake and use of nutrient elements. Immobilisation thus inevitably accompanies mineralisation. This concept is very familiar for carbon; for instance the aerobic laboratory utilisation of a sugar by a microbe results in about 40% of the carbon being 'respired' to CO_2 (mineralised) and 60% going to growth yield (immobilised). It is not perhaps so commonly appreciated that the same balance sheets must be drawn for other elements. Thus the key to the availability of inorganic forms of nutrient elements lies in the *nett mineralisation*—the extent to which mineralisation exceeds immobilisation.

The extent of nett mineralisation of any element relates directly to the extent of availability of that element to the decomposer organisms. Thus C is usually in plentiful supply in the early stages of decomposition but other elements such as N or P may only be present in much lower amounts. Under these conditions one or more of these nutrients may limit decomposer growth. In the circumstances of a limiting nutrient, conditions of nett immobilisation of that element will tend to prevail. Similarly nett mineralisation will only occur where a nutrient occurs at non-limiting concentrations. An index of such availability could be given by the energy:nutrient ratio— indicative of the relative limiting power to growth of energy sources and nutrient concentration. Most commonly this is expressed as a C-nutrient ratio, C being easier to determine and broadly proportional to energy accessibility for a heterotroph.

Some of the implications of this for the activity of decomposer organisms can be seen by comparing the C:nutrient ratios of the organisms with that of their food resources. In Table 1.9 this is shown for wood decay—perhaps the most extreme example of nutrient limitation due to the very low nutrient contents of wood. The decomposer organisms have narrower ratios than the resources indicating a high demand for nutrients. This is demonstrated by calculating the extent of concentration of the nutrient, relative to carbon, which occurs during movement from resource to decomposer. It is not surprising therefore to learn that N and P availability may severely limit the rate of decay of wood by fungi (Findlay 1934).

As decomposition proceeds the C:nutrient ratio will decline. C is lost continuously but the limiting nutrient will be conserved in an immobilised form. Conditions may soon be approached when the nutrient is no longer

Table 1.9. Carbon: nutrient ratios for resource and decomposer organisms. Data for wood and wood decay fungi modified from Swift (1977b), data for 'typical Insecta' from Allen *et al.* (1974). Also given are the approximate extent of concentration for the nutrients during uptake from one level to the next. The carbon contents assumed for each level are Wood 47%, Fungus 45%, Insect 46%.

	C:N	C:P	C:K	C:Ca	C:Mg
Resource (wood)	157	1424	224	147	1022
Fungal decomposer	26	94	136	11	346
Concentration	6	15	2	13	3
Animal decomposer	6	51	66	157	235
Concentration (to wood)	26	28	3	1	4
(to fungus)	4	2	2	xs	2

Chapter 1

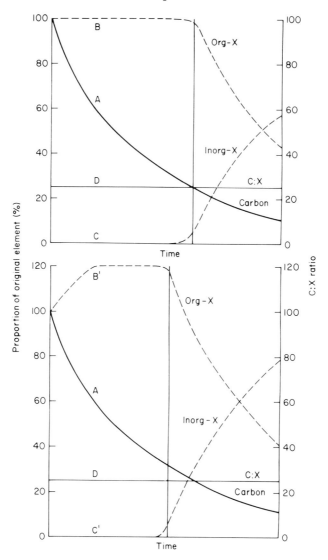

FIG. 1.14. A simple time-course model to illustrate the importance of C:nutrient ratios as indices of the equilibrium between immobilisation and mineralisation of an element X. Curve A = C:X ratio of an organic resource having an initial ratio of 100:1. Curve B = proportion of X in organic form (immobilised), (Curve B[1] = case where there is import of X from sources external to the resources being decomposed). Curve C = proportion of X in inorganic form (mineralised), (Curve C[1] = mineralisation in response to alteration of C:X ratio by import of X as in case B[1]). D = C:X ratio of decomposer organism. After this point nett mineralisation occurs and Curve A flattens off although carbon loss may continue on the same gradient.

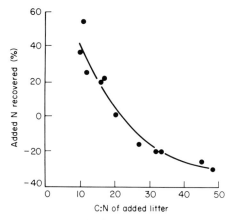

FIG. 1.15. Percentage recovery of nitrogen by tomato plants in four weeks from eleven different plant residues added to soil, versus ratio of carbon to nitrogen in the original plant residues. (From Black 1968, after Iritani & Arnold 1960).

limiting and nett mineralisation becomes a possibility. This can be illustrated by a simple time-course model (Fig. 1.14). The implication of this is that the release of inorganic forms of an element X will occur only when the C:X ratio of the resource drops to the level at which it is no longer limiting to the organisms decomposing it, i.e. to the C:X ratio of the decomposer organisms themselves. Prior to that all the nutrient is in organic form. Indeed, as curve B^1 suggests, if demand for the nutrient by decomposers is high enough then nutrient may be imported from the soil solution in contact with the decomposing resource, lowering the limitation threshold and altering the C:X equilibrium in favour of earlier release of mineral-X. The important influence of C:nutrient ratios has been long appreciated in agricultural practice. Fig. 1.15 shows the availability to tomato plants of N added to the soil in organic form as plant detritus over a four-week period. Only with plant materials which had C:N ratios below about 20 (close to that of microbial tissue) was any N mineralised. At higher C:N ratios mineral N availability was in fact decreased—due to immobilisation from the soil solution to the decomposing detritus. A number of similar examples can be found in the books by Black (1968) and Allison (1973).

The underlying principle of nutrient flux in decomposition is thus relatively simple. The time course of events may however be difficult to predict and we are a long way from being able to manipulate this important area of soil fertility at all precisely. A variety of factors other than simple C:nutrient ratios will affect the response of a decomposer to the nutrient composition of a resource. For instance the molecular form of the nutrient source and the ratio of one nutrient to others will also affect its limiting

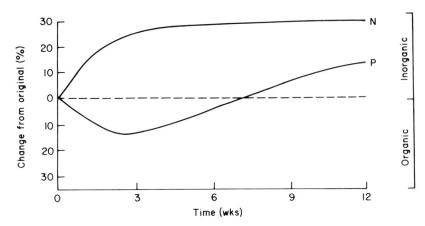

FIG. 1.16. Changes in nitrogen and phosphorus during decomposition of pasture grasses in the laboratory at 30°C. Increases in organic forms indicate phases of immobilisation, increases in inorganic forms of mineralisation (from Floate 1970a).

capacity. During decomposition particular nutrients may have periods of nett immobilisation and nett release, and will in turn be limiting and non-limiting. An example of this is shown in Fig. 16 which illustrates the mineralisation of N and P during decomposition of pasture grasses under laboratory conditions. As decomposition proceeded C was progressively lost but the patterns of P and N mineralisation differed markedly. P showed initial nett immobilisation but after about six weeks mineralisation exceeded immobilisation. In contrast a nett increase in mineral N was observed throughout consistent with the low initial C:N ratio of 17:1.

In Fig. 1.17 the decomposition of yellow birch leaves at Hubbard Brook Experimental Forest is shown. The initial C:N ratio of these leaves was 62:1. During the first eight months N was immobilised from external sources—probably leachates and other N-rich litter—but once the C:N had dropped to below 30:1 in June/July nett loss of N occurred. Although it is not possible to say with certainty that this corresponds to nett mineralisation it at least implies a change in the extent of N retention in the decomposing leaves. The patterns of S and P were similar to N indicating their limiting availability in fresh-fallen leaves; K, Mg and Ca behaved quite differently however, as is shown for K in the Figure. This mobile element was lost very rapidly from the leaves so that by April only 20% of the original content remained. This content remained constant however and even showed a slight increase suggesting that the critical C:K ratio for the decomposer organisms had been reached and that immobilisation exceeded mineralisation in the later phase.

It should be noted that the assumption that nett losses of an element from a decomposing resource are necessarily due to mineralisation and solubilisa-

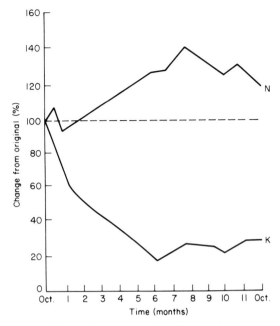

Fɪɢ. 1.17. Changes in nutrient content of yellow birch leaves over a twelve-month period after leaf-fall (from Gosz *et al.* 1973).

tion of the element may not be entirely justified. Just as import due to immobilisation can occur so can export in organic forms for instance due to emigration of animals, translocation in fungal hyphae or loss in fungal spores (Dowding 1976). It is not only C:nutrient ratios that affect the patterns of mineralisation during decomposition, factors of resource quality and physical environment and their effect on the decomposer organisms all act to determine the rate at which a nutrient will become available to plants. These factors will be explored in subsequent chapters.

1.3.3 The herbivore subsystem

The fraction of the NPP which enters the herbivore system annually is low compared to that directly entering the decomposition subsystem as plant detritus. In forest systems it is negligible; Carlisle *et al.* (1966a) estimated that 27% of the mean number of leaves were consumed by larvae of *Tortrix viridana* in a particularly severe year of defoliation, the remainder, plus branches, roots and reproductive materials passed directly into the decomposition subsystem; in other years less than 2% of the leaves were consumed. Whittaker (1975) estimates that in Temperate Deciduous Forest the average herbivore consumption of total NPP is less than 0·2% whereas in grasslands—temperate or tropical—which maintain the highest populations of herbivores among terrestrial ecosystems, the values range from

Table 1.10. Nutrient flux through the herbivore subsystem in a temperate montane grassland managed for sheep grazing, Snowdonia, UK. (Perkins *et al.* 1978). Above ground NPP only.

		OM	N	P	K	Ca	Mg
NPP	kg ha^{-1} yr^{-1}	1482	27·6	2·4	16·3	4·1	7·0
Litter input	% NPP	39	26	33	17	28	59
Herbivore consumption	% NPP	18	21	25	27	32	15
Return in excreta	% NPP	7	15	16	21	22	12
Herbivore harvest	% NPP	0·3	1	3	0·1	3	0·1

10 to 15%. The amount of nutrient removed may be proportionately greater however as illustrated for a managed grassland ecosystem in Table 1.10. Herbivores consume largely living tissue the material consumed has high levels of nutrient compared with that of detritus from which significant components may have been removed by leaching or withdrawal into the plant from the senescent tissues. Thus although the amount of dry matter consumed by sheep in the Snowdonia pasture is less than 50% of that directly entering the detritus pool, the proportions for N, P, K and Ca are 81, 76, 159 and 114% respectively.

The herbivore subsystem may therefore form a significant pathway in the cycling of nutrient elements within some terrestrial ecosystems. Theoretically all nutrient elements (other than carbon) entering the herbivore subsystem will pass in the form of carcasses and excreta, to the decomposition subsystem (see Fig. 1.3). What then is the main significance of this by-pass of the plant-decomposition-plant cycle that we have described in previous sections? In the first place it is clear that the herbivores may be responsible for some losses of nutrient to the decomposition subsystem. This can occur at a local or even an ecosystem scale by emigration by herbivore populations. Although this may appear dramatic in some cases—locust or lemming migrations for instance—it is doubtful if the amount of nutrient so removed

Table 1.11. Nutrient contents of herbivore products (% dry weight), after (1) Wilkinson & Lowrey (1973), (2) Henzell (1973), (3) Brasher & Perkins (1978), (4) Cooke (1967).

	N	P	K	Ca	Mg
Carcass (steer) (1, 2)	2·4	0·68	0·15	1·28	0·04
Milk (cow) (1, 2)	0·53	0·10	0·12	0·11	0·001
Carcass (sheep) (1, 2)	2·5	0·45	0·13	0·84	0·03
Unwashed wool (sheep) (1, 2)	11·4	0·03	4·66	0·13	0·02
Faeces (sheep) (3)	2·2	0·51	1·01	1·09	0·63
Urine (cattle) (4)	0·8	0·004	1·5	—	—

could substantially affect the productivity of the ecosystem. An analogous process is the harvesting of herbivore production by man. An example of this is given for sheep farming in Table 1.10; the proportion of nutrient taken for production is small in comparison to the return to the decomposition subsystem via excreta. The mineral element content of urine and faeces of herbivores is high (Table 1.11) and the excretion rate of the mature animal is such that the amount of mineral retention is negligible (Wilkinson & Lowrey 1973). Losses in milk from dairy herds (particularly phosphorus) may be larger than the losses removed in beef production for an equivalently stocked herd.

Other effects of the herbivores on nutrient cycling and decomposition processes are more difficult to assess quantitatively. They include for instance the nature of the nutrient return. The composition of animal products is markedly different from that of the plant tissues they consume (compare Table 1.11 with the nutrient compositions for grasses given in Table 4.9). Because of these differences in composition these products may have markedly different decomposition characteristics including the rate at which mineralisation takes place. This aspect is discussed in detail in Chapter 4. In addition to this composition factor the intervention of herbivores results in a relocation of significant amounts of nutrients. The input of nutrients to the decomposition subsystem in an ungrazed grassland will be relatively homogeneous over the whole area. Inputs from herbivores will be confined to specific areas of deposition of excreta or carcasses creating nutrient 'hot-spots'. Heterogeneous distribution patterns of this kind may have significance for the distribution of decomposer organisms (Usher 1976) and ultimately of the distribution of primary productivity over the area.

1.3.4 Nutrient turnover in ecosystems

Analysis of the varying patterns of nutrient flow through the decomposition subsystem may well turn out to be among the most revealing of insights to ecosystem function in differing climates. Such analysis requires detailed tabulation of all nutrients, micro- and macro-, the detailed partitioning of the inputs to decomposition and consideration of the differing molecular and ionic forms of the nutrients within the litter and soil. Currently it is only possible to attempt a few broad generalisations.

The concept of turnover, previously applied to organic matter (Table 1.1) may also be applied to individual nutrients. Table 1.12 shows data for a Beech forest in Sweden. In keeping with other temperate forests the litter layer is turned over in a period of between one and two years—as estimated on the basis of the C. Similar calculations based on the ratio of litter-fall to litter standing crop for individual nutrients are also shown in the Table (row 5). The rate of turnover varies for the nutrients—K > Ca > Mg >

Table 1.12. Nutrient turnover in a temperate deciduous forest (*Fagus sylvatica*) (after Nihlgard 1972).

		C	N	P	K	Ca	Mg
1. Litter fall (*F*)	kg ha^{-1} yr^{-1}	3000	69	5·0	14·4	31·7	4·3
2. Leaching (*L*)	kg ha^{-1} yr^{-1}	—	1	0·1	11·2	6·6	2·5
3. Released (*F* + *L*)	kg ha^{-1} yr^{-1}	3000	70	5·1	25·6	38·3	6·8
4. Litter standing crop (*x*)	kg ha^{-1}	2700	86	5·8	10·4	34·2	4·8
5. Turnover (*F/X*)	yr^{-1}		1·11	0·80	0·86	1·38	0·93	0·90
6. Turnover (*F* + *L/X*)	yr^{-1}		1·11	0·81	0·88	2·46	1·12	1·42
7. Soil exchange pool	kg ha^{-1}	—	—	84	56	175	38

P > N—in a pattern which is consistent with the previous observations made on the relative limiting effects of these elements on decomposer growth. The turnover can be recomputed to include the leaching input as well as litter-fall (row 6). This gives a more accurate picture of the much more rapid movement of K, Mg and Ca from immobilised form in plant materials to mineral form in the soil.

This picture of nutrient turnover is crude because no account is taken of inputs from the roots, nor is it possible to determine the rate of mineralisation directly. The size of the exchange pool seems in each case well in excess of the potentially mineralisable stock in the litter. If however it is remembered that the uptake demand each year is in excess of the release then the necessity of rapid replenishment through decomposition is re-emphasised. The mean residence time of K, Ca and Mg in the litter layer is somewhat lower than that of C; that of N and P however is somewhat longer—indicative presumably of 'lag' imposed by immobilisation in decomposers. This is a general pattern; Fig. 1.18 is derived from data of Rodin & Basilevic (1967) for fifteen vegetational types including our six ecosystem-types plus data for deserts and sub-tropical forests. The data plotted on a logarithmic basis show a linear relationship between the 95% turnover time for organic matter and that of the total mineral elements, including N. The line of best fit however shows that mineralisation of the other elements always lags behind C. The logarithmic scale of the graph tends to diminish the differences between the two lines. Two examples will serve to illustrate the difference however; at a mean 95% turnover time for organic materials of 1 year (e.g. in sub-tropical environments) the nutrient elements have a mean turnover time of 1·49 yr; in colder climes with a mean turnover time of 100 yr the nutrient value is 133·2 yr. This is suggestive of a general 'immobilisation' lag for the minerals of between fifty and seventy per cent of the organic turnover period. The data also suggest the hypothesis that immobilisation may be 'tighter' in those environments less conducive to decomposition.

Interbiome comparisons can also be made in terms of the distribution of

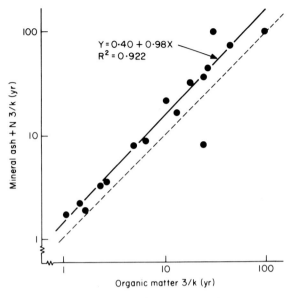

FIG. 1.18. Relationship between organic turnover and the turnover of macro-nutrient elements (other than C) in different ecosystems. Data calculated from Rodin & Basilevic (1967) as 95% turnover time (3/k) in the litter layer. The dotted line is the theoretical line of agreement if the rate of turnover of the two components were the same, the full line is the regression line of best fit. See text for further interpretation.

nutrients between the plant and decomposition subsystems. Again the data compiled by Rodin & Basilevic (1967, 1968) can be used to demonstrate a number of interesting similarities and contrasts. The amount of total nutrient (N plus ash elements) taken up by the vegetation very closely parallels the NPP in all but a few ecosystems. In twelve of the eighteen sites for which the calculation is possible the annual uptake only ranges between 38 and 52 kg nutrient per t ha^{-1} annual NPP. The six exceptions include the three coniferous Boreal Forest sites which have lower values (18 to 25) and two (out of four) desert ecosystems which have very high uptake values (81 and 135). The single Tropical Forest site also has a slightly higher than average value (62). If the same types are compared on the basis of the concentration of total nutrient in the plant biomass a different contrast emerges. The five Boreal and Temperate Deciduous Forest sites have concentrations ranging from 0·8 to 1·5% with the Tropical Forest lying just outside this at 2·2%. The grassland dominated systems on the other hand are all at a higher level, ranging from 2·5 to 4·2% for the three sites. Tundra systems resemble grasslands and deserts tend to have even higher nutrient concentrations in the plant biomass. The lower concentrations in the forests are clearly associated with the perennial woody tissue as is demonstrated when the concentrations in photosynthetic tissue are compared. On this basis Boreal Forest lies in the same range as grassland (2·6 to 4·0%) but Deciduous Forest and Tropical Forest have markedly higher concentrations (averages of 6·6 and 7·3% respectively). The productivity of Boreal Forest is thus maintained at a lower 'nutrient cost' than that of other ecosystems. That is, production in Boreal Forest is maintained at a lower unit nutrient uptake and with a lower unit maintenance of nutrient in photosynthetic tissue. This implies a high efficiency in relation to nutrients in this ecosystem-type.

When the concentration of total nutrients in the litter fall is compared then again Boreal Forest is the main exception to a general trend with concentrations ranging between two and three per cent (five sites, Rodin & Basilevic 1967), whereas the concentrations for another eleven sites ranging from Tundra to Tropical Forest all lie within the range of 3·7–5·6%. The data also show the extent to which forest systems conserve the nutrients in litter fall, as discussed earlier, for all the forest values are only about 65% of the concentrations in the living photosynthetic tissue.

The influence of climate in regulating availability of nutrients is probably one of the major factors determining the differences in NPP between the major biomes. One simple example illustrating this is the ratio of nutrients in the standing crop of undecomposed litter to that in the plant biomass. This gives an idea of the extent to which the vegetation can utilise the nutrient in the biosphere as opposed to its 'storage' in inaccessible form. In tropical systems such as Savannah or Tropical Forest the ratio is about 2%, in temperate forests or grasslands it is about 20% and in the Boreal Forest 80%. But in the Tundra the amount of nutrient in the litter is over three times as much as that in the plant biomass. This serves to emphasise that the climatic regulation of decomposer activity may be of as crucial importance in determining the primary production characteristics of an ecosystem as its direct effect on physiology of the vegetation. Nonetheless the arguments given above indicate that in Boreal Forest at least the characteristics of the vegetation compensate to some degree for the high extent of nutrient immobilisation in the litter.

1.4 Decomposition processes and ecosystem development

The foregoing account has detailed some aspects of the role of the decomposition subsystem in maintaining ecosystem structure and function. We have also compared the decomposition characteristics of ecosystems of widely differing type. It should be apparent however that a very considerable level of variation in decomposition characteristics may be found between ecosystems which differ in only relatively minor ways. Local variations in climate, topography, soil type and vegetational cover can produce markedly different rates of organic turnover, mineralisation, and soil organic matter accumulation. In the same way that variations may occur in space so the decomposition characteristics of an ecosystem may vary with time. The most fundamental aspect of this is the changes that occur during the establishment of a climax ecosystem through its successional stages.

An interesting example of this is given by the studies of Dickson & Crocker (1953a, b) on a forest ecosystem at Mount Shasta in Northern California. They identified five soils which had been deposited by mudflows at different times over a period of about 1500 years so that the vegetation on them could be identified as representing a primary succession on parent material of identical origin and under constant climatic conditions. Measurements were concentrated on stands of the dominant *Pinus ponderosa*; on the youngest soil (27 years) this formed virtually a monospecific canopy but in the mature forests (500 years plus) the canopy was more diverse with oaks, firs and cedars co-dominant. The development of the organic horizons of litter and soil is shown in Fig. 1.19A. The amount of carbon in these layers may be taken as representing the equilibrium between input due to litterfall (F) and output due to decomposition (k). The litter layer was characterised in the earliest stages by a phase of nett accumulation ($F > k$). No measurements of production were made but this would be in accord with the general theory that the NPP of an ecosystem increases during the early stages of succession as biomass is accumulated (Odum 1969). The litter fall can also be expected to increase and if the decomposition rate is relatively constant (being mainly regulated by climatic factors) then nett accumulation of organic matter should ensue. The subsequent decline in the standing crop of litter may be due to a number of factors; NPP is probably declining during this phase; there is also a redistribution of organic matter downwards in the soil profile as is shown by the comparable stability of the soil curve in Fig. 1.19A and in more detail in the soil profile diagram (Fig. 1.19B). This may imply that it is at this stage that steady state equilibrium between the formation and decomposition of humus colloids is established. From the 566-year-old soil the organic status as a whole seems to be in steady state equilibrium.

The nitrogen dynamics are broadly similar, and show the same shifts downwards presumably of humus-bound N. An interesting illustration of the establishment of equilibrium in decomposition processes is given by plotting the C:N ratio for the different horizons (Fig. 1.19C). If the C:N ratio is taken as expressive of an equilibrium between immobilisation and mineralisation then comparison in terms of horizon and time indicates the stage in the decomposition cascade which each equilibrium represents. Thus the litter layers are always showing nett immobilisation while the soil shows stable, mineralising conditions with C:N ratios constantly below 20:1. In the early stages of succession the extent of equilibrium immobilisation is at a higher

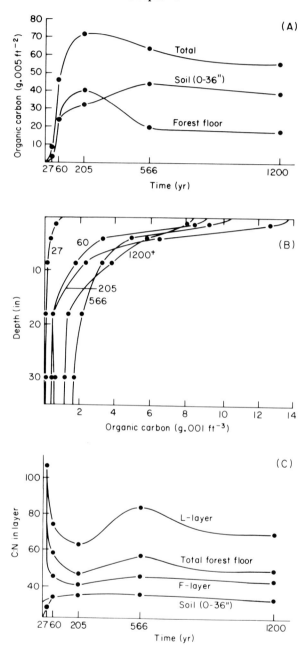

FIG. 1.19. Establishment of soil and litter characteristics during the primary succession on a virgin soil of a conifer forest at Mount Shasta, California: (A) Accumulation of C in the litter layer and upper soil horizons; (B) Changes in the distribution of C in the soil profile; (C) Changes in the C:N ratio of litter and soil horizons (Dickson & Crocker 1953b).

level than later. This suggests a less critical demand for mineralisation in the early stages which is in accord with Odum's (1969) theory that detritus becomes increasingly important for the regeneration of nutrients as ecosystems mature and the proportion of nutrients in biological circulation increases.

This example provides a graphic illustration of the stabilisation of the decomposition subsystem during succession. Accompanying these energetic and nutrient aspects is an increase in species diversity and of community homeostasis. Olson (1963) has reflected on these considerations in some detail and derived models for a number of varying circumstances. There is however too little data currently available to enable us to test such hypotheses in greater detail. This is of some consequence for the stable ecosystem structure is fragile to the manipulations of man. The strong link between the organic status of soil and its fertility has promoted a large amount of attention to the stability of SOM in agricultural soils but regrettably little attempt has been made to derive useful information from the study of natural ecosystems. The conversion of forest ecosystems to arable culture is tantamount to reversing the successional process and re-introducing an immature seral stage. Predictably then one of the major changes that occurs is a decline in the content of organic matter in the soil. Thus, for instance, in continuous cropping of wheat over a period of thirty years Salter & Green (1933) found that the soil organic carbon content declined to 63% of its original level. Under maize the decline was even greater over the same period and only 36% of the original content remained. Such changes undoubtedly reflect the decline in the qualities of primary resources entering the soil. Changes in the composition of the detritus input may also be significant and cultivation may also have the effect of increasing the rate of decomposition of humus fractions.

Fortunately the position can be reversed by allowing land to return to a more natural vegetative state. In 1881 at Rothamsted Agricultural Station in England a wheat field on which cereal cropping had been carried out for many years was allowed to revert to natural woodland. Legumes were early in evidence but within fifteen years trees and shrubs became dominant and the area, Broadbalk Wilderness, is now a well established deciduous woodland. The soil carbon in the upper 23 cm has risen from 0·98% in 1881 to 2·7% in 1964 (Russell 1973). Decreases in organic carbon are accompanied by other changes of great significance to soil fertility—decline in nutrient levels, particularly N, decline in CEC, etc. The management of the decomposition processes in soil may therefore be regarded as a matter of some significance. Attempts to correct for the imbalances caused by the cropping of herbaceous or gramineous plants have accompanied the development of all agricultural systems. In the intensive agriculture of temperate regions manuring with organic material to provide both nutrient and organic matter has largely been replaced by short circuiting of the decomposition processes by the provision of inorganic fertilisers. In the tropics a more ecological solution was traditionally used. The system of shifting cultivation replaces short periods of intensive cropping by fallow periods of forest regeneration. During these periods the organic and nutrient status of soil is re-established and the land can be brought into cultivation again. High levels of population growth in recent decades have created the demand for more intensive types of agriculture in tropical lands and continuous cropping is steadily being introduced, with consequent problems in conserving soil fertility. These important questions we shall take up in more detail in Chapter 7.

2

THE DECOMPOSITION SUBSYSTEM

In the first chapter we established and examined some aspects of the role of decomposition processes in ecosystems. We also contrasted the patterns of these ecosystem processes in ecosystems of differing structure and composition. It is now our purpose to analyse the decomposition subsystem in greater detail; to identify and define its components and describe the ways in which they interact to determine the functioning of the subsystem.

In this chapter we shall look at the structure of the subsystem as a whole. In the first stage of analysis we shall identify and define some major components of the subsystem and try to build up a conceptual model of how the subsystem functions. That is to say at this stage we shall confine our description to hypotheses of the mode of function; the evidence that validates or contradicts this model is developed in the following five chapters. In the later part of this chapter we initiate the analytical process at a coarse level of resolution by describing two systems models of C flux in the decomposition subsystem that have been developed as part of the ecosystem-analysis studies in the International Biological Programme. Models of this kind, with explicit mathematical statements describing complex processes and structures, provide a valuable method of testing ecological hypotheses. We shall examine the extent to which the various hypotheses of decomposition processes are borne out, and we shall see the extent to which aspects which other evidence suggests are important, can be ignored or remain only implicit in an apparently satisfactory model.

2.1 The structure of the decomposition subsystem

We shall use the term *resource* for any identifiable component of detritus entering the decomposition subsystem. We prefer this to the commonly used term *substrate* because the latter has a more specialised meaning than the present context. In biochemistry it is used to refer to a molecule entering into a chemical reaction with a specific enzyme; often resulting in the formation of an enzyme—substrate complex. It is thus confusing to describe also the complex mixture of molecules which constitute a piece of organic detritus as the 'substrates' of decomposition. The term 'resource' does not have this ambiguity and also has the advantage of implying that it is a food source

49

for the decomposer organisms. Later in this chapter we shall consider the range of different *resource-types* that it is useful to recognise.

Decomposition essentially results in a change of state of a resource under the influence of a number of biological and abiotic factors. This bald view of the process is pictured in Fig. 2.1. During the time period t_1 to t_2 the state of the resource changes from R_1 to R_2. The 'valve' symbol indicates the regulating influences and the arrow the process of change or reaction rate. We shall use this simple descriptive 'module' to build up our concept of the structure of the decomposition subsystem.

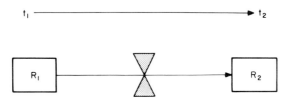

FIG. 2.1. The basic decomposition module. This is a diagrammatic statement of the progress of decomposition between time t_1 and time t_2 resulting in the change of a resource from state R_1 to R_2. The rate of state change (as indicated by the arrow) is subject to regulation (as indicated by the valve symbol) by biological and abiotic factors.

The simplest expression of the state change is as a decrease in mass of the resource. More detailed analysis shows that this includes a loss of matter from the resource and a change in the chemical composition of the remainder, which may or may not be accompanied by a process of fragmentation, i.e. a reduction in the particle size of the resource. These are the chemical and physical changes which we recognise as decomposition. They may be attributed to the effect of three distinct processes; those of leaching, catabolism and comminution.

Leaching is the abiotic process whereby soluble matter is removed from the resource by the action of water. Leaching thus causes weight loss and change in chemical composition. It should be noted however that its consequence is to transfer the soluble resource material to a different site— usually lower down the vertical profile of the ecosystem—where it may subsequently be acted upon by further decomposition processes.

Catabolism is the biochemical term which describes energy-yielding enzymatic reaction, or chains of reactions, usually involving the transformation of complex organic compounds to smaller and simpler molecules. The most familiar example is the aerobic respiration of glucose to carbon dioxide and water, $C_6H_{12}O_6 + 6O_2 \rightarrow 6CO_2 + 6H_2O + \text{energy}$. In this case the oxidation is complete and the products are all inorganic. Over a given time period however the catabolism of a given substrate or mixture of substrates

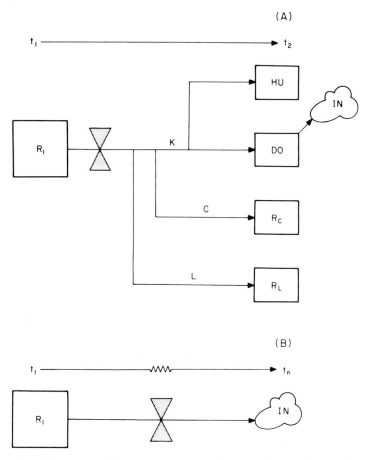

FIG. 2.2 (A) Decomposition of a resource (R) over a short period of time
(t_1 to t_2). The three component processes, catabolism (K), comminution (C) and
leaching (L) result in chemical changes, e.g. mineralisation giving rise to inorganic
forms (IN) and the resynthesis of decomposer tissues (DO) and humus (HU),
physical changes, e.g. reduction of particle size of chemically unchanged litter
(RC) and the removal of soluble resource materials in unchanged form to other
sites (R_L). (B) Decomposition of a resource over a longer period of time
(t_1 to t_n), resulting in complete mineralisation.

may be incomplete. Some products will be inorganic (CO_2, NH_4, PO_4 etc.);
other intermediates will have entered the metabolic pool of the decomposer
organisms and been *resynthesised* into the polysaccharides and proteins of the
decomposer tissues; others may have been incorporated in non-cellular
organic matter such as humus. Matter is lost from the original resource by
catabolic formation of volatile or soluble inorganic forms, or by creation of

soluble organic intermediates which are subsequently leached out, or by transport out as components of the decomposer tissues. The chemical composition is changed as a result of the production of intermediates and the synthesis of decomposer tissues and humus *in situ*.

Comminution is the reduction in particle size of the organic resource. Comminution differs from catabolism in being a physical rather than chemical process and is largely brought about by the feeding activity, both ingestion and digestion, of decomposer animals. During passage through the digestive system comminution is accompanied by catabolic changes; the residue from comminutive and catabolic activities is excreted as faeces characteristically of smaller particle size and different chemical composition to the ingested food material. Some organic matter may be comminuted without ingestion and is reduced in size without change in chemical composition. A similar effect occurs when comminution is brought about by abiotic factors such as freezing and thawing or wetting and drying cycles.

In Fig. 2.2A we illustrate these changes in modular form. Over a relatively short time period t_1 to t_2, usually weeks, the original resource is acted upon by decomposer organisms to comminute and catabolise it to inorganic products, decomposer tissues, humus and comminuted particles of chemically unchanged residue. Leaching also removes a soluble component from the resource. Over a much longer period decomposition may be complete (Fig. 2.2B). This time scale may be hundreds or even thousands of years, as has already been intimated, because of the very slow turnover of the humus residues (Section 1.2.2). In some ways then decomposition is synonymous with mineralisation—but mineralisation is always incomplete in the short term. In some literature decomposition and catabolism are regarded as synonymous—but we shall use the broader definition for the former to describe the overall change in state and specify the component processes where necessary.

In practice the three processes act simultaneously on the same resource and it may be impossible to distinguish the three effects. The processes also interact in their overall effects on the resource. Catabolic activity may soften plant materials and render them more readily comminuted; catabolic processes also result in the release of soluble components which may be removed in leachate; reduction in particle size may improve the access of catabolic enzymes and increase the ease with which soluble compounds may be leached.

The organic products of decomposition processes will themselves at some stage re-enter the detritus pool. We have already identified this aspect of decomposition processes in two previous models. In Figs. 1.3 and 1.5 the *re-cycling* of organic matter between decomposers and detritus was emphasised. What this means in simple terms is that after the death of the decomposers their corpses become detrital resources available to other decomposers.

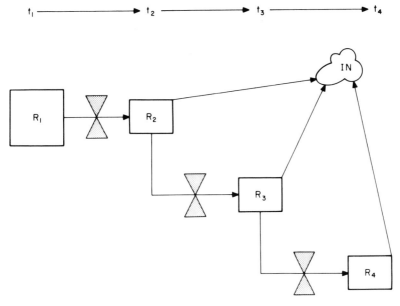

FIG. 2.3. The cascade structure of decomposition processes. The progressive change of the primary resource R_1 through states R_2, R_3 and R_4 is shown during the time period t_1 to t_4. This time has been divided into a number of discrete periods (t_1 to t_2, t_2 to t_3 etc.) to show different stages in the decomposition process (e.g. change of state from R_1 to R_2 and from R_2 to R_3). Thus for any intermediate state (e.g. t_2 to t_3) the products of the preceding module (e.g. R_2) is the starting resource state. During each of these stages matter may be lost from the system in the form of inorganic molecules but note that there may be some recovery due to re-immobilisation—this is not shown in the modular structure. The diagram recognises that different rates of change may occur during each stage and that the nature of the regulatory factors may also differ.

Another way to represent this in module form is to show the decomposition subsystem as having a *cascade structure* (Fig. 2.3). Over any given time period some of the products of decomposition become the starting resources of the next module in the cascade. As the cascade proceeds so the proportion of matter entering the inorganic pool increases. But note that inorganic materials may also recycle in that matter released (mineralised) during a primary module may become re-immobilised during a succeeding module. We shall define resources entering the decomposition subsystem directly from primary production as *primary resources* (i.e. equals plant detritus) and those formed by secondary production as *secondary resources* (i.e. equals corpses of microbes and animals). Other major categories of resource are *faeces* which are a mixture of microbial cells and comminuted primary resource materials, and *humus*, newly synthesised extracellular organic

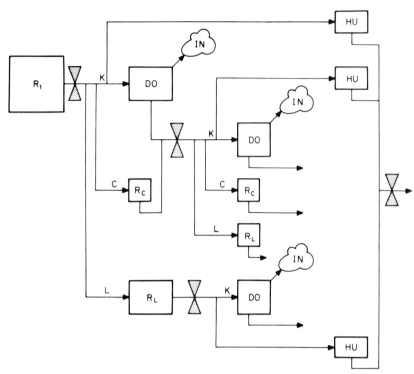

FIG. 2.4. A resource model of the decomposition subsystem. See text and legends to Figs. 2.2 and 2.3 for explanation.

matter. We may thus see that secondary resources plus faeces constitute the *cellular organic matter* of soil which we distinguished from humus in Chapter 1.

A simple model of the structure of the decomposition subsystem (Fig. 2.4) may be obtained by combining the two concepts of process differences (leading to branching of the module, Fig. 2.2A) and re-cycling (leading to a cascade structure, Fig. 2.3). The products of the primary module will decompose at different rates and by different processes. Thus, taking in turn the products of decomposition of a primary resource (R_1) as pictured in Fig. 2.2A; the organic leachate (R_L), consists largely of carbohydrates and polyphenolic compounds. The carbohydrates will be catabolised very rapidly to form CO_2 and microbial tissue but most of the polyphenols are resistant to catabolism and may become incorporated in humus. Although humus (HU) has an extremely slow rate of decomposition the products may be much the same, i.e. inorganic molecules and microbial tissue. The cellular fraction ($DO + R_C$) will have an intermediate rate of decomposition and a pattern of breakdown similar to that of the primary module (Fig. 2.4). This emphasises another aspect of the structure of the decomposition subsystem—that of *convergence*. Whilst the effects of different processes may lead to branching of the resource chain at an early stage, the subsequent processes may lead to the formation of similar products and the convergence of the later cascades.

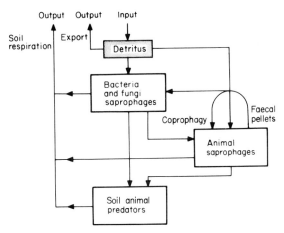

FIG. 2.5. An integrative model of the functioning of the organisms of the decomposition subsystem (From Wiegert, Coleman & Odum 1970).

Convergence is a feature which makes the analysis of the community of decomposer organisms particularly difficult. Analysis of the herbivore subsystem has been greatly facilitated by recognition of the differing trophic levels of herbivore, carnivore and top carnivore. Thus while the food webs within a given community may be relatively complex they can often be simplified by assignment of the organisms to distinct trophic positions. Relatively simple quantitative relationships between the organisms occupying different trophic levels on the basis of numbers, biomass or production can be observed fairly consistently. The trophic type of analysis has not proved possible with decomposer organisms. This question is analysed in detail in Chapter 3 but it is appropriate to make some general points at this stage.

The decomposer community is taxonomically diverse embracing a wide range of Bacteria, Fungi, Protista and Invertebrates. Many of the representatives of the groups have specific attributes which enable them to play widely differing roles in decomposition processes. These abilities are however often exhibited without much apparent limitation in relation to the resources consumed. Simple scenarios of decomposer food webs can be constructed, e.g. fungus A decomposes plant litter—animal B eats fungus A—bacterium C decomposes animal B—protozoan D eats bacterium C. Wiegert, Coleman & Odum (1970) constructed a general scheme of this kind (Fig. 2.5). However, the same organisms may exhibit markedly differing roles. The bacterium, the protozoan and the fungivorous animal may all also participate in consumption of the primary resource for instance, and the fungus may assist in the decomposition of the secondary resources. One solution to this problem is to assign only broad functional roles to decomposer populations without designating any hierarchical status to their involvement. This is the policy adopted in the model of Heal & MacLean (1975) and described in detail earlier (Fig. 1.4). Here the trophic levels embrace organisms differing in decomposer function and the re-cycling system simulates the cascade of our model but without distinguishing between the involvement of specific groups of organisms at the different levels of the cascade.

Another solution which we shall see adopted in the models discussed in the second part of this chapter is to omit the organisms from quantitative models and to view them only as a 'driving variable'. *Driving variables* are the factors which influence the rate of change of state. They are the third major component of our model system (with the state variables—the boxes, and the fluxes—the arrows) shown as the valve symbol in Figs. 2.1 to 2.4. Organisms can be sensibly placed as a driving variable as the rate of the biotic decomposition processes will clearly be related to the composition and size of the decomposer populations. It should nonetheless be emphasised that, while useful for some purposes, this is clearly an inadequate way in which to describe the structure of the decomposition subsystem. Definition of the roles of different groups or species of decomposer organisms will require different models closer to that of Wiegert, Coleman & Odum (1970) (see Chapter 3).

The rate of decomposition is also regulated by two other categories of driving variable, the physico-chemical environment and the resource quality. The *physico-chemical environment* influences all three of the decomposition processes. The organisms and processes respond to the interactive effects of a wide range of factors. At different times and under different circumstances particular features of the environment may be more or less significant. The climatic variables of moisture availability and temperature have a fundamental influence however; the availability of free-water is essential to the maintenance of decomposer activity and is modified by the nature of the micro-habitats. Availability may be partially determined by the size and shape of the resource particles, indicating the interactive nature of environmental and resource factors. In addition to the climatic variables, features of the soil environment, such as the composition of the gaseous atmosphere and the pH of the liquid phase, also affect the activity of decomposer organisms; the chemical and physical nature of soil is largely determined by the nature of the parent materials from which the soil is derived (Fig. 1.8) and is also an important regulating factor. The physico-chemical environment of decomposition processes can therefore be conveniently subdivided into *climatic* and *edaphic* complexes. Physical factors may directly intervene in decomposition as well as regulating biological activity. Thus the leaching rate is directly related to the extent and intensity of rainfall. Comminutive action may be brought about by freezing and thawing or drying and wetting cycles as a result of the interaction of temperature and moisture regimes.

The pattern and rate of decomposition of a resource are also determined by various aspects of its physical and chemical composition which we term the *resource quality*. This may operate to determine its palatability to animals; some animals can ingest only soft materials whereas others possess the ability to chew their way through wood. The presence of relatively minor chemical components such as polyphenols may stimulate or inhibit microbial activity or feeding by detritivore animals. The main sources of energy and carbon in plant materials are the polysaccharides and lignin of their cell walls; relatively few organisms possess the enzymes capable of degrading all the molecular types present and the substrate composition of the resource may impose a selective effect on the composition of the decomposer community. The substrate molecules also vary in the rates at which they are catabolised—for instance lignin is degraded much more slowly than cellulose. Essential nutrient elements may occur within resources at low concentrations which limit the rate of decomposer activity. These aspects of the 'quality' of the resource will thus determine not only the composition of the effective decomposer community but also the rate of the processes for which they are responsible.

The introduction of *resource specificity* into our concept of the decomposer community suggests the idea of resource specificity among organisms. For this reason it may be important to define the type of resource for any particular decomposition module. The definitions employed may be fairly broad, e.g. primary resources can be subdivided into roots, stems, leaves and reproductive structures and secondary resources into muscle, bone, microbial cells, etc. Each category can be further subdivided in terms of the species of origin. The main features of variation are differences in the chemical composition of these resources and it is important to determine these. Physical features may be of importance however and not least in significance is size. Detrital resources are distributed as discrete 'particles'—individual leaves, branches, faecal pellets, etc. The size of these particles varies considerably from the massive tree stem to the microscopic microbial cell. The size can be a factor in determining what organisms can colonise or consume the resource. The size and shape also determine to a considerable extent the micro-environment around and within the detrital particle, emphasising again the interactive nature of the driving variables. The type of resource may influence the organisms occupying it directly through resource quality factors and indirectly by influencing the micro-environments established. In turn the organism by its activity changes the quality of the resource and the nature

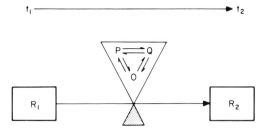

Fig. 2.6. Regulation of decomposition processes by the driving variables, the physico-chemical environment (P), the resource quality (Q) and the decomposer organisms (O).

of the micro-environment. This interactive nature is pictured diagrammatically in modular form in Fig. 2.6.

In the foregoing discussion we have established a concept of the structure of the decomposition subsystem and how it operates. The proposals we have made should be viewed as hypotheses and in the remainder of the book we shall examine the evidence for and against them. Before starting this analysis there is one further structural feature that must be mentioned; that of scale. It is axiomatic in science that the scale in time or space over which a measurement is made will determine the phenomena that are observed. In Table 2.1 we have listed some of the features of the decomposition subsystem relevant to different spatial scales of observation. Chapter 1 was largely about the features at the ecosystem level. The smaller scales are given closer attention in the later chapters of the book. Similar variations could be listed for temporal scales of observation—for instance the climatic environment shows marked variations of differing intensity by the hour, the day, the month, the year and the millenium.

2.2 Models of the decomposition subsystem

Mathematical models have become increasingly employed to describe complex ecological processes or systems. The value of such simulations lies in their lack of ambiguity. When the structure of the system is described in mathematical expressions then the author's concepts become fully explicit. This enables hypotheses concerning the structure and function of an ecosystem to be explored and tested. All such models are necessarily simplifications however and must be recognised as such. The limitations placed upon them commonly stem from practical considerations; it is of little value, for instance, building a model which produces predictions which cannot be evaluated. Mathematical models then are often more limited in scope than conceptual ones.

Table 2.1. The regulation of decomposition rate by organisms, physical environment and resource quality at different scales of operation.

Scale of resolution	Resource type	Organisms	Scale of operation — Environment	Resource quality
Ecosystem	Total detritus	Decomposer community	Macroclimate Edaphic complex	Relative proportions of main resource types
Ecosystem component	Main resource types (primary, secondary etc.) down to taxonomic origins of components (leaf species etc.)	Resource specific selection within community	Local variation in environment (down to level of resource size, i.e. micro-environment)	Chemical and physical composition of resources
Resource components (substrates)	Cellulose, lignin, keratin, etc.	Enzymatic capability of species	Micro-environment down to molecular environment of enzymes, etc.	Substrate specificity of enzyme systems

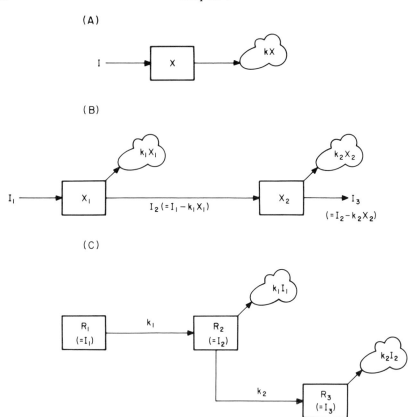

FIG. 2.7(A). The 'general paradigm' of simulation models of decomposition processes. I = input of detritus to a given standing crop, X, k = the decomposition rate, the loss due to decomposition is shown in the 'cloud' compartment. (B) A compartment model in which part of the standing crop in X_1 (the first module) is transferred and becomes the input to the second. This can be manipulated so as to maintain the standing crops at steady state (as shown) but transferring the residue of decomposition; or the transfer component can be based on a different mathematical simulation, e.g. a simulated rate of leaching or comminution. Note that there may be any number of such transfers originating from a compartment and compartments may also receive more than one input. (C) Structural relationship between this type of dynamic model and our conceptual model. The latter omits permanent cumulative compartments and only traces the fate of discrete resource particles.

Bunnell (1973) and Bunnell & Tait (1974) have given good general descriptions of the processes and purposes of such mathematical simulation models with reference to the decomposition subsystem. It is not our intention to reiterate model theory but some general statement may be helpful in interpreting the discussion which follows. The general structure of a compartment model of decomposition may be developed from a simple illustration of the organic turnover equation of Jenny, Gessell & Bingham (1949) which we introduced in Chapter 1. In Fig. 2.7A the compartment X represents the standing crop of detritus, the input I the litter fall and the output kX the loss to decomposition. The decomposition rate k as we have seen is regarded as proportional to the size of the standing crop. At steady state the inputs and the outputs balance—the case considered in Chapter 1. If non-steady-state conditions prevail or more particularly if time intervals shorter than that necessary for steady-state are considered then input may exceed decomposition and X, the standing crop will accumulate. Thus the value of X after any time period can be predicted from a given starting value if I and k are known. In more complex models which incorporate some idea of what we have termed the cascade concept of decomposition a proportion of the residue in X, after a given period, may be transferred to a second compartment and decomposed at different rate (Fig. 2.7B). This basic structure has been termed the 'general paradigm' by Bunnell & Tait (1974) and is fundamental to most simulation models of decomposition processes. It differs slightly in structure from our conceptual module but the relationship can be seen by comparing diagrams B and C in Fig. 2.7.

In Fig. 2.8 the general paradigm is shown built into a fourteen-compartment model of C flux in a tundra ecosystem. The model, entitled ABISKO II after the place in Sweden where it was conceived, was designed to allow

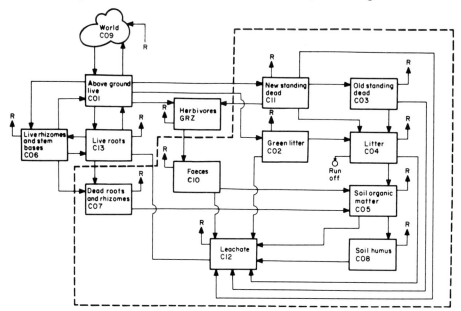

FIG. 2.8. ABISKO II—a compartment model describing the carbon flux in Tundra ecosystems. The decomposition subsystem is enclosed within the broken line. For details see text (after Bunnell & Scoullar 1975).

comparison of ecosystem processes in the National Sites contributing to the Tundra Biome studies in the IBP, but has been developed particularly for a wet meadow site at Point Barrow, Alaska.

The main structural features are the fourteen biomass C compartments which simulate the plant, herbivore and decomposition subsystems, the latter being the most complex, embracing nine of the compartments. One of the constraints on the system is that only a single species of primary producer can be simulated but primary production is allocated separately to above-ground components (C01) and two below-ground components, roots (C13), and rhizomes plus stem bases (C06). Herbivore biomass is also shown as a separate compartment (GRZ). There are four inputs to the decomposition subsystem; the death of plant tissues produces dead standing resources (C11) and dead roots and rhizomes below ground (C07). Herbivore activity produces faeces (C10) or green litter (C02), i.e. undecomposed leaf litter reaching the surface directly as a result of feeding without ingestion by the herbivores. Successive stages in the decomposition of these inputs are shown in some cases—for instance new standing litter (C11) decomposes in the canopy for one year, one component falls to form litter at the soil surface (C04) and another remains attached for a second year (C03). The designation of these different types of resource is indicative of the recognition of resource quality, i.e. that different types of resource have different decay rates and that these rates change as decomposition proceeds; it also acknowledges the different local environments—canopy, litter surface, soil—in which de-composition takes place.

Three categories of flux are recognised. Microbial respiration is the main pathway of carbon flux—the loss of CO_2 from each biomass compartment by aerobic respiration is indicated by R in Fig. 2.8. This then is a simulation of catabolism. It is limited in scope however for it is confined to aerobic respiration by micro-organisms; there is no anaerobic C dissimilation, and no animal respiration. The respiratory loss is simulated daily on the basis of driving variables of temperature and moisture availability. The mathematical function used to derive these values is a sophisticated submodel named GRESP, the basis for which will be discussed in Chapter 6. Suffice it to say that it also has an implicit resource quality limitation to the optimum rate.

The second type of flux from the resource compartments is by leaching, simulated in a crude way compared to respiration. Leaching is regarded as proportional to the availability of free flowing water, apart from a narrow region of temperature dependence close to freezing point. It operates largely within a maximum imposed by the content of soluble materials in the resource but a small allowance is made for leachate produced by microbial activity although the authors recognise that this has not been effectively simulated. The third loss from each compartment is the transfer of the residue to the next compartment which is always a resource compartment at a more decomposed state. This type of transfer is termed 'comminution and ageing' and recognises transport by biotic or physical agencies to environmentally different sites or degradation of material to a different quality level. The two most important sets of transfers in this respect are to soil organic matter (C05) and soil humus (C08) which may be taken as equivalent to our distinction of the cellular and humic fractions of SOM.

We can thus see that ABISKO II makes explicit or implicit reference to many of the structural features of our conceptual model of the decomposition subsystem. The practical constraints impose considerable simplifications which the authors of the model readily acknowledge, most striking perhaps is the absence of any decomposer biomass from the model. As suggested earlier the practical difficulties of obtaining good estimates (eg. of microbial biomass) to test the model has led to the organisms being reduced to the level of driving variables. Even further the only biotic agency recognised is that of microbial respiration; although intercompartment transfers are termed 'comminution' there is no explicit simulation of the feeding activity of animals, nor are animals allowed to play any regulatory role. ABISKO II does incorporate the cascade concept of decomposition processes which we referred to earlier—this may be more apparent when part of the model is redrawn in our modular form (Fig. 2.9). The convergence of decomposition processes in the later stages is also shown by the large number of inputs to the soluble organic carbon compartment (leachate, C12, Fig. 2.8), and soil organic matter (C05, Fig. 2.8). Detailed accounts of ABISKO II and its development are given by Bunnell & Dowding (1974) and Bunnell & Scoullar, (1975).

In Fig. 2.10 the C flux from the below-ground system (root respiration plus CO_2 released by decomposition) is shown as predicted by the model and as measured at Point Barrow over an eighty-five day period in 1973. The agreement between prediction and observation is good—the cumulative value for the predicted CO_2 output was only $3 \cdot 7\%$ higher than the measured. It should be remembered however that a single value validation tells us little, for agreement can arise as readily from errors cancelling one another out as from accurate summation.

ELM (Hunt 1978) is a model designed to simulate various aspects of grassland ecosystems in North America. It is more comprehensive than ABISKO II with regard to state variables but these can be subdivided into eight submodels. Four of these—primary producer, insect and mammalian consumer, and decomposer—are used to simulate C flow through the ecosystem. The inputs to the decomposition submodel are diverse, coming from the variety of primary and secondary resources represented in the other three submodels. As with ABISKO II however this diversity has not been maintained within the decomposition submodel. Instead the inputs have been subdivided into two components termed 'labile' or 'resistant'. This is an

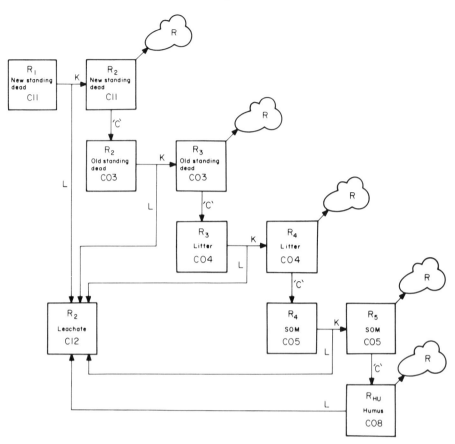

Fig. 2.9. Part of the ABISKO II compartment model (Fig. 2.8) redrawn in modular form. The compartment numbers of the original model are given with the prefix C and are also designated a resource number (R). Catabolic changes are shown by K, leaching by L and comminutive transfers by C. This lay-out of the model shows its intrinsic cascade structure and the convergence of the decomposition process to leachate.

explicit recognition of resource quality factors—the decomposition rates of the two components within the model are set at defined proportions. As with the Tundra model above- and below-ground resources are treated differently.

The only decomposition process recognised in the model is microbial catabolism. This is again driven by moisture and temperature functions which were derived from published data on the effects of temperature and soil moisture potential (see Chapter 6) on microbial activity. There is thus a good deal of parallelism with the ABISKO model but a major difference

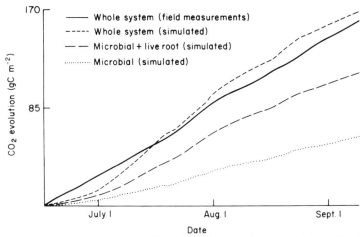

FIG. 2.10. A comparison of predicted and observed decomposition in Tundra. The predicted cumulative CO_2 output for various compartments of the ABISKO II model are shown in broken lines. These predictions were based on simulations using abiotic data taken during 1973 at Point Barrow, Alaska. The observed CO_2 output at the same site is shown by the solid line (from Bunnell & Scoullar 1975).

lies in the inclusion within ELM of compartments for Decomposer (microbial) Biomass. These are partitioned into Active and Inactive components. This distinction is based on the recognition that active growth of micro-organisms probably only occurs for very limited periods of time and that for much of the year most decomposers exist at only a maintenance level of metabolic activity (see Chapter 5). The microbial biomass is derived from the decomposition rate on the basis of assumptions concerning the efficiency of microbial growth yield in relation to substrate consumed. Thus the apparent sophistication of including two decomposer states is nonetheless still subject to marked constraints on their manner of handling.

In summary we can see that at the first level of analysis it is possible to define decomposition in relatively simple terms—a cohort of resource types distinguished in simple chemical terms with their rate of decomposition regulated largely by variation in the temperature and moisture environment. These simple criteria of resource quality and physical environment suffice to explain much of the observed seasonal trends in carbon pools. The details of the mechanisms of transformation of the resource variables are left unexplained—in particular with regard to the role of the decomposer organisms. It is now our task to explore this aspect in detail.

3

THE DECOMPOSER ORGANISMS

3.1 The diversity of decomposer organisms

From the brief account given in Chapter 1 it should be apparent that a major limitation to investigations of the structure and functioning of decomposer communities is their complexity. A temperate woodland soil with a well-developed litter and humus layer, for example, may contain more than a thousand species of soil animals, with total populations of several millions per square metre in just the top 5 cm of the profile. Virtually every class or order of invertebrate is represented within these communities (Table 3.1) including many forms, such as copepods, nemerteans, turbellarians, leeches, rotifers and protozoa generally thought of as 'aquatic' rather than 'terrestrial' in their mode of life. The micro-organisms are equally diverse in taxonomy (Table 3.2) with several hundred species of fungi and a wide range of bacterial types at population levels of 10–1000 million cells occupying each cubic centimetre of soil. The extent of this diversity at the taxonomic level emphasises the problems faced in analysing the decomposer community. It also imposes an urgent necessity on the ecologist to devise simplified schemes for classifying and quantifying the roles of these organisms in an easily assimilable way. We have approached this problem from a number of different points of view.

We first consider the decomposer organisms in terms of their functional ecology. In doing this we take up again the difficult question of the trophic structure of decomposer communities but we also consider the practical ways in which decomposer functions can be described. In the later sections of the chapter we consider the patterns of distribution of decomposer organisms and the quantitative role they play in different ecosystems. More detailed information on the dominant taxonomic groups of terrestrial decomposer communities can be found in general texts such as Burges & Raw (1967), Dickinson & Pugh (1974), Alexander (1971a) and Wallwork (1970). The biology and ecology of termites is reviewed by Harris (1971) and Lee & Wood (1971).

3.2 The functional ecology of decomposer organisms

Soil ecologists investigating decomposition processes in a particular habitat have two basic approaches open to them. Either they can establish the broad

66

Table 3.1. Taxonomic diversity of decomposer animals. Invertebrate members of terrestrial decomposer food webs are listed in approximate rank order from those groups which maintain an essentially aquatic mode of life in the soil to wholly terrestrial arthropods.

Phylum	Class	Sub-class or order	Common name
Protozoa	Flagellata		flagellates
	Sarcodina		protozoa { amoebae (naked and testate)
	Ciliata		ciliates
Rotifera			rotifers
Tardigrada			tardigrades ('water bears')
Nematoda			nematodes
Gastrotricha			gastrotrichs
Platyhelminthes	Turbellaria	Tricladida	planaria or flatworms
Nemertinea		Metanemertini	nemerteans or ribbon worms
Annelida	Oligochaeta		earthworms and white worms (enchytraeids)
	Hirudinea		leeches
Mollusca	Gastropoda	Pulmonata	slugs and snails
Arthropoda	Crustacea	Ostracoda	
		Copepoda	
		Amphipoda	'sand hoppers'
		Decapoda	crabs
		Isopoda	woodlice
	Diplopoda		millipedes
	Chilopoda		centipedes
	Pauropoda		
	Symphyla		
	Insecta	Collembola	spring tails
		Diplura	
		Protura	
		Thysanura	bristle tails
		Isoptera	termites
		Coleoptera	beetles
		Diptera	flies
		Lepidoptera	moths (and butterflies)
		Hymenoptera	ants (etc.)
		Orthoptera	grasshoppers and crickets
		Dermaptera	earwigs
		Dictyoptera	cockroaches
	Arachnida	Scorpionida	scorpions
		Pseudoscorpionida	pseudoscorpions
		Solpugida	sun spiders
		Uropygi	whip scorpions
		Amblipygi	
		Ricinulei	
		Opiliones	harvestmen
		Acari	mites
		Araneae	spiders

Table 3.2. Taxonomic diversity of decomposer micro-organisms.

Kingdom	Division	Class	Selected orders[1]	Common name
Prokaryota[2]	Eubacteria		*	'true' bacteria
			Myxobacteriales	fruiting bacteria
			Cytophagales	gliding bacteria
			Spirochaetales	spirochetes
			Actinomycetales	actinomycetes
	Cynobacteria			blue-green 'algae'
Fungi[3]	Myxomycota	Acrasiomycetes		cellular slime moulds
		Myxomycetes		slime moulds
		Plasmodiophoromycetes		'club root' organisms
	Eumycota Mastigomycotina	Chytridiomycetes	(4)	
		Hypochytridiomycetes	(1)	zoosporic fungi
		Oomycetes	(4)	water moulds
	Zygomycotina	Zygomycetes	Mucorales	pin moulds
			Entomophthorales	entomogeous fungi
			Zoopagales (4)	nematode trapping fungi
		Trichomycetes		

Ascomycotina	Hemiascomycetes		(3)	yeasts
	Loculoascomycetes		(5)	
	Pyrenomycetes		(4)	flask fungi
	Plectomycetes		(1)	
	Discomycetes		(7)	cup fungi
	Laboulbeniomycetes		(1)	
Basidiomycotina	Teliomycetes		(2)	rusts and smuts
	Hymenomycetes	Agaricales		mushrooms and toadstools
		Aphyllophorales		bracket fungi
			(7)	
	Gasteromycetes		(9)	puff balls and allies
Deuteromycotina	Hyphomycetes		(1)	moulds, imperfect fungi
	Coelomycetes		(2)	pycnidial fungi and allies

1 Orders are entered where the ecological distinction is sufficient for them to be mentioned as a group elsewhere in the text otherwise only the number of commonly recognised taxa is given.

2 After Buchanan and Gibbs (1974)—note that the hierarchical classification of bacteria above the family level is controversial. For this reason no distinction of orders for the vast majority of the 'true' bacteria (*) (which contain the majority of the heterotrophic decomposers as well as photo- and chemo-autotrophs) has been given.

3 After Ainsworth, Sparrow and Sussman (1973).

Sweet chestnut (*Castanea sativa*) leaf discs

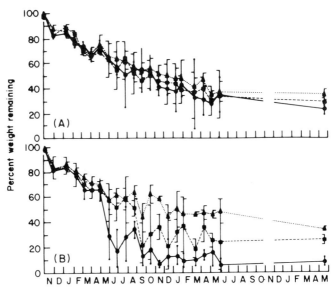

Beech (*Fagus sylvatica*) leaf discs

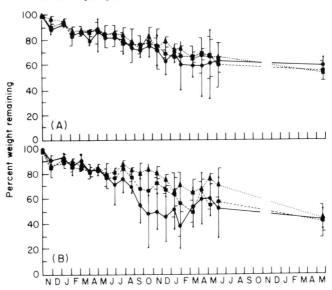

functional roles of the decomposer community irrespective of the taxonomic identity of the organisms, or the taxonomic structure of the food web can be unravelled and a detailed understanding of decomposition processes built up from a knowledge of the functional biology of the organisms so identified. The two approaches are complementary in any thorough investigation of the decomposer subsystem but the latter approach involves a very high degree of research commitment. In this section we consider a number of functional attributes of decomposer food web organisms and try to synthesise information obtained by both approaches. The function of the decomposers may be considered under groupings made on body size, or on various physiological aspects of trophic function.

3.2.1 Body size

The simplest initial approach to analysing the functioning of a decomposer community is to investigate by means of the litter-bag technique the contribution of different organisms, grouped in terms of their body size, to decomposition (see Appendix). The results of a typical litter-bag study are shown in Fig. 3.1 (Anderson 1973a). Two species of leaf litter, chestnut (*Castanea sativa*) and beech (*Fagus sylvatica*) were placed in two contrasting moder woodland soils in South East England. The litter-bag mesh sizes chosen were 45 μm, 1 mm and 5 mm. The coarsest mesh size allowed access to the entire soil and litter community including the largest animals present at these sites and was theoretically a measure of the total weight losses attributable to leaching, catabolism and loss of particles of less than 5 mm diameter, following comminution. The medium sized mesh excluded the larger soil animals such as millipedes, isopods and earthworms but allowed entry of mites, collembola, nematodes and enchytraeids. The fine mesh excluded all organisms except the fungi, bacteria, nematodes, protozoa and other very small members of the soil fauna.

Weight losses from the fine mesh bags were similar for each species in the two sites, but consistently greater from chestnut than from beech. This could be taken to indicate that levels of microbial activity were similar in the

---

FIG. 3.1. Decomposition of leaf litter in two woodland soils with contrasting humus forms: (A) a mor-like moder under sweet chestnut (*Castanea sativa*); (B) a mull-like moder under beech (*Fagus sylvatica*). The soil fauna activities on the *Castanea* site are dominated by the micro- and meso-fauna while earthworms and macroarthropods are important fauna elements in the *Fagus* site. Mean percentages ($\pm 95\%$ limits) of initial weights are shown for coarse ($\bullet$), medium ($\blacksquare$) and fine ($\blacktriangle$) mesh litter-bags containing chestnut and beech leaf discs. The mean and confidence limits for the final set of samples are not comparable with the other data because of larger sample sizes (Anderson 1973a).

two sites but that chestnut was more biodegradable than beech. While this is generally true, weight losses from similar litter-bags suspended above the soil where microbial activity was greatly reduced by desiccation, revealed that a significant proportion of the weight losses from the fine mesh bags was attributable to physical leaching processes (this aspect of the experiment is considered in greater detail in Section 6.6). Losses from the medium and coarse mesh bags showed marked differences in pattern between the two sites.

The *Castanea* site has a mor-like humus form and like most acid soils the fauna was dominated by large populations of enchytraeids, Collembola, mites and Protozoa whilst the larger fauna, woodlice, millipedes and earthworms, were scarce or absent. The negligible weight loss differences between the three mesh sizes of bag for both litter types in that soil indicate that decomposition was dominated by microbial catabolism and leaching, and that litter comminution by soil fauna was negligible. In the *Fagus* site the litter breakdown pattern was rather different. The soil had mull-like characteristics due to a higher pH and burrowing earthworms (*Lumbricus terrestris*), large millipedes and woodlice were present. Weight losses from the coarse mesh bags were much higher than in the *Castanea* site due to the feeding activities of these groups of soil animals and only small amounts of litter remained in the bags after a period of a few months in the field. Weight losses from the medium mesh bags were also higher than from comparable bags in the more acid soil. This was not due to larger Collembola or mite populations but to juvenile earthworms and millipedes entering the 1 mm mesh bags.

From this experiment a number of features of the functioning of the decomposer community emerge; the effect of resource quality factors can be seen from the differences in rates of decay of the two leaf species in the same site; the effect of edaphic environment is visible in the differences in rate between the same species located at different sites. But both these factors are seen to operate by their selective effects on the decomposer organisms. This is shown up by the distinction of the size of the organisms made by use of the litter-bag technique.

Division of the decomposer community on the basis of size has been made ever since the introduction of mesh litter bags by Edwards and Heath (1963), but the total soil fauna have been traditionally subdivided on the basis of *body length* (Fig. 3.2). Under this scheme the *microfauna* is almost entirely composed of the Protozoa, the *macrofauna* has a significant size overlap with the *meso-* or *meio-fauna* for a number of invertebrate orders, while the mesofauna contains almost the entire soil animal community. It is difficult to classify the bacteria and fungi in a scheme based on length considerations since the thallus of bacteria occupies a range between $0 \cdot 1$ and $2\,\mu$m while the mycelium of fungi may extend for several metres.

The division of the decomposer community using litter-bags of different mesh sizes separates the organisms according to their *body diameters*. This provides a more natural functional classification with respect to litter breakdown and decomposition processes (Fig. 3.3). The fungi and bacteria can now be incorporated in the scheme as the *microflora*. The *microfauna* under this classification is composed of the Protozoa, nematodes, rotifers, tardigrades and the smallest Collembola and Acari included at the upper limits. Functionally this is a coherent group as none of these animals are involved in litter comminution. The Collembola and other apterygotes, the Acari, Enchytraeidae, most Diptera larvae and some of the genera of smaller Coleoptera are included in the *mesofauna*. Most members of this group can attack intact plant litter but their sum contribution to litter breakdown is generally insignificant, as was shown in the litter-bag study. Their major role in decomposition processes is in regulating microbial populations and reworking the faeces of the macrofauna. The termites are an exception and are probably the only members of the *mesofauna* whose presence or absence can directly determine energy and nutrient flux pathways in terrestrial ecosystems. The *macrofauna* consists of the large litter-feeding arthropods such as the millipedes, isopods, amphipods and insects, as well as the molluscs and the larger earthworms. These animals are responsible for the initial shredding of plant remains and its redistribution within decomposer habitats. Their presence can significantly affect decomposition pathways and con- tributes directly to the structure of the soil. Van der Drift (1951) proposed the term *megafauna* for animals over 20 mm in length. This term could usefully be used for members of the decomposer food web with body diameters over 20 mm in diameter, and would include the largest earthworms, some molluscs, tropical millipedes, dung beetles and vertebrate scavengers.

These size divisions are essentially practical as many organisms may move through two or even three of the size classes during the course of their development, as was demonstrated in the litter-bag study described earlier. The comments above indicate however that there is a useful degree of correlation between the size classes and the decomposer function of the organisms. In any litter-bag study however the choice of mesh sizes will be determined by some knowledge of the community present and with the experimental objectives of the study in mind.

3.2.2 Trophic function

In Chapters 1 and 2 we considered some of the difficulties encountered in proposing any fixed structure to the trophic relationships within decomposer communities. Whilst the organisms of the herbivore subsystem can generally be readily ascribed a trophic position within clearly defined food chains, species of decomposers rarely occupy such clearly defined roles. The food web of a decomposer community has a more fluid, interactive structure with

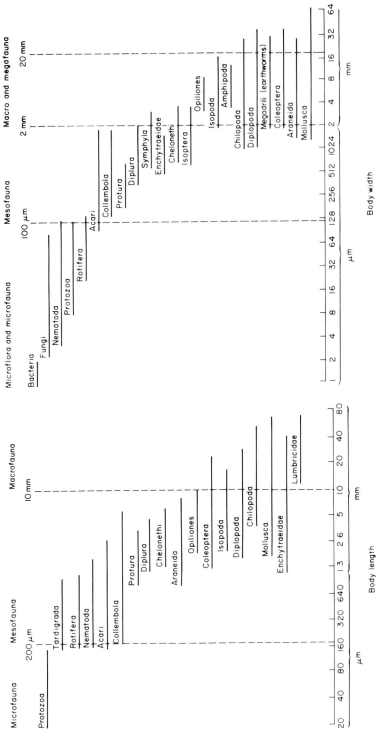

FIG. 3.3. Size classification of organisms in decomposer food webs by body width.

FIG. 3.2. Size classification of the temperate soil fauna by body length (after Wallwork 1970).

74

individual species operating on several levels which might be distinguished as trophically different. In this section we shall analyse this situation in more detail by considering the problem from the standpoint of the functional attributes of the decomposer organisms. In a later section (3.2.8) we shall come back to the question of the structure and organisation of the decomposer community in trophic terms.

Let us consider first whether we can readily recognise any distinct groupings of function among decomposer organisms. Decomposer communities have frequently been subdivided into the *carnivores* (feeding on animals), *microbivores* (feeding on micro-organisms) and the *saprovores* (feeding on dead plant and animal remains)—e.g. in the Heal & MacLean (1975) model described in Chapter 1. This classification is not equally applicable to animals and micro-organisms since fungi and bacteria utilise their resources by external digestion while the suffix '-*vores*' refers to food ingestion by animals. We propose therefore to use the suffix '-*troph*' which does not imply a particular method of resource utilisation. We also propose the use of the trophic level classification into *necrotrophs*, *biotrophs* and *saprotrophs*, devised by Lewis (1973) for fungi. This classification defines the trophic role of decomposers in terms of the dynamic relationship between the organism and its food resource.

The *necrotrophs* have a short-term exploitation of living organisms which results in the rapid death of the food resource. This trophic group includes some herbivores and many plant parasitic microbes (which feed on and kill plant tissues), predators (animals and micro-organisms such as the nematode trapping fungi, which kill animals) and the microtrophs (animals and micro-organisms feeding on living fungi and bacteria). The *biotrophs* have a long-term exploitation of their living food source which is dependent upon the continued existence of the host. Some herbivores (e.g. root-feeding nematodes and aphids), obligate plant parasites (such as rusts or mildews), and organisms forming mutualistic symbioses with their hosts (mycorrhiza, legume nodules) are included in this category. *Saprotrophs* are organisms utilising food already dead and the majority of decomposers therefore fall into this category. As a step in the direction of further defining trophic structure in the decomposer community we may distinguish the trophic groups in terms of the type of resource which they exploit, *i.e. primary* or *secondary*. We shall consider later to what extent this particular distinction can be maintained.

The boundaries between these trophic categories we have defined are not rigid. Some organisms may occupy different trophic roles at different stages of their life cycle or under differing ecological conditions. The interrelations of the three groups are indicated in Fig. 3.4. The basis and utility of the classification may be illustrated by considering examples of some of the interactions, utilising the numbers given in the diagram. One of the commonest examples of flexibility in trophic position, and one of some considerable significance in decompositions is that of a necrotroph showing a saprotrophic

the mode of apical extension of the hypha, and its ability to adhere to the surface of the host or detrital particle.

Penetration is followed by the production of a frond-like mycelium which is able to exploit planes of weakness between cells and within the molecular lattice of the structural polymers within cell walls. Strong bonds exist between molecules within the same chain, which can only be enzymatically attacked, but only relatively weak hydrogen bonds link the molecules of adjacent chains in many structural polymers which have a crystalline or semi-crystalline structure, such as cellulose or keratin. Penetrating hyphae appear to be able to exploit these planes of weakness. For instance in the growth of keratinophilic fungi such as *Curvularia*, there appears to be a clear-cut division of labour into penetrating hyphae and the surface mycelium that absorbs nutrients (Griffin 1972). These two hyphal growth forms are also produced by cellulolytic fungi growing on cellophane but are not produced if the cellulose substrate is in the form of amorphous cellulose dispersed in agar (Moore, unpublished data, see Fig. 3.5). The form of the fungus is thus a response to the physical configuration of the substrate.

The formation of an extensive mycelial growth form is also an important adaptation for microbial growth in a physically heterogeneous habitat such as the soil. The mycelium allows the translocation of nutrients between microsites so that resources which are potentially limiting for hyphal metabolism in one microsite can be supplied from elsewhere in the mycelial complex. Translocation probably occurs over distances of a few microns in most situations but Basidiomycetes which form rhizomorphs may translocate materials over several metres.

FIG. 3.5. Growth forms of decomposer micro-organisms. (A) Colony of bacteria on the surface of a mineral particle in the soil ($\times 5100$ from Gray & Williams 1971). (B) Mycelial growth of fungi—shown here ramifying over the surface of cellulose film. Note the occurrence of conidia ($\times 255$). (C) Mycelial growth of fungi shown here as discrete mycelia within organic resources—in this case as 'penetrating' mycelium within cellulose film. The points of entry of the hyphae lie at the centre of the mycelium in each case and the mycelium grows radially away from this point of entry along planes of weakness with the film (stained with phenolanaline blue, $\times 255$). (D) Cellulolytic activity of penetrating hyphae. A higher magnification ($\times 680$) view of a penetrating mycelium within cellulose film that has been dyed so that the site of cellulolysis is shown by cleared zones (Moore, Basset & Swift 1978). Note that activity is confined to the immediate vicinity of the hyphae. The thicker hyphae at the centre are the point of juncture with surface hyphae. (E) Penetration of organic resources by fungi. A cattle hair is here shown penetrated by hyphal growth into the keratinous matter largely from the hollow centre. Penetration follows surface growth both inside and outside the hair. Note sporulation on the outside. Note also how vertical penetration is followed by horizontal growth within the keratin. The latter forms the equivalent penetrating mycelia to that shown in (C) for cellulolytic fungi. (From English 1965.)

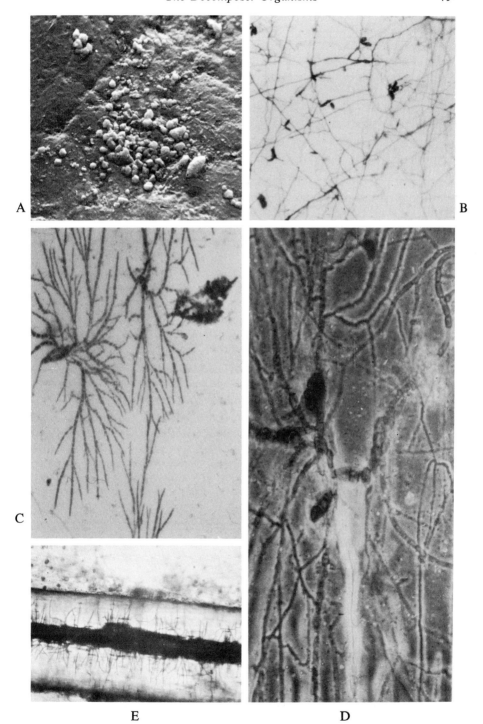

The mycelium may not remain intact for very long however. Studies in pure culture have shown that active metabolism and growth is generally only maintained in apical regions and that death and autolysis of older compartments quickly follows (Trinci 1971). This is probably an adaptation of considerable significance in decomposition for it enables the re-cycling to the actively growing regions of nutrients that might become otherwise limiting (Swift 1978). In more diverse situations lysis is speeded by the action of microtrophic animals or saprotrophic bacteria. The consequence of this is to produce a large number of hyphal tips acting independently. This is one means of dispersing fungal mycelia; widespread dispersal results from the copious production of asexual or sexual spores. These are dispersed, either actively in the case of 'aquatic forms' such as the chytrids, but more generally by carriage in wind, water or by small animals. Spores germinate and colonise new resources under favourable circumstances. Under unfavourable circumstances spores remain dormant. The cycle of exploitation, reproduction, dispersal (dormancy), colonisation, exploitation, etc. etc. is fundamental to the ecology of fungi and different taxonomic groups have different ecological strategies which partly explain their roles in decomposition processes. Reviews of these aspects can be found in Gray & Williams (1971), Swift (1976), and Smith (1976).

Whereas mycelial organisms are adapted to penetration and exploitation of relatively massive detrital particles over relatively long time scales the unicellular microbes are adapted to surface habitats which are rapidly exploited largely in the short term. The unicellular mode of life does not facilitate penetration of hard materials but the very small size of the bacterial cell does enable the colonisation of all available crevices and pores. Bacterial colonies are then formed on the surfaces by binary fission of the individual cells. Thus bacteria are adapted to occupation of particulate detritus where the surface to volume ratio is relatively high. This was demonstrated in experiments by Parr, Parkinson & Norman (1967). Using an artificial system with glass beads of varying size they showed that whereas fungal activity was related to pore space that of the bacteria was to the surface area of the solid phase.

The size of colonies formed at any one site may be small (see Fig. 3.5) but dispersal may be rapid. Many bacteria are motile due to a variety of arrangements of flagella. Their small size also enables efficient and easy dispersal through water films and porous systems (see Chapter 6). The exponential pattern of growth by cellular reproduction and the small body size clearly adapt the bacteria both to rapid exploitation of available resources and response to heterogeneous environments. The genetic mechanism of these prokaryotic organisms also confers a considerable capacity for adaptation particularly at the nutritional level. Under unfavourable circumstances bacterial cells may die, adopt a low level of maintenance activity, or form dormant spores. Bacteria seem particularly susceptible to antibiotic influences and are probably more adapted to the avoidance of competition rather than for particular competitive strategies.

Nutrition. The general pattern of nutrition in fungi and bacteria is similar and is different to all other organisms—it may be given the general term 'absorptive'. Saprotrophic micro-organisms function by secreting extra-cellular enzymes which break down complex polymers to smaller molecules which can then be absorbed and utilised intracellularly. In contrast with animals there is no energy expenditure or morphological adaptation for the purposes of ingestion, although there is of course loss of energy and nutrient in the enzyme secretions.

Among the micro-organisms the bacteria are the most nutritionally diverse and it may be useful to digress slightly to consider the range of nutritional capacities of which these organisms are capable. A purely saprotrophic function is found in the majority of bacteria which fall into the physiological group designated as *chemoheterotrophs*. This name implies that both energy and carbon are obtained from organic sources. There is one family of bacteria, the purple non-sulphur bacteria (Rhodospirillaceae) which obtain energy from sunlight but utilise organic compounds for their carbon. These are therefore known as *photoheterotrophs* and because of their dependence on organic sources of carbon can be strictly regarded as decomposers. This is in contrast to the autotrophic bacteria whose sole source of carbon is CO_2. This includes both the photosynthetic bacteria, including the blue-green bacteria—the *photoautotrophs*—and the *chemoautotrophs* which obtain their energy from the oxidation of inorganic compounds rather than from sunlight. Despite the non-saprotrophic role of these latter organisms, because of their independence from organic sources of C or energy, they are nonetheless of great significance in the decomposition subsystem. Included in this group for instance are the nitrifying bacteria which oxidise ammonium to nitrite and nitrate and the *Thiobacilli* which oxidise reduced forms of sulphur to give sulphate. These autotrophs are thus responsible for the conversions of nutrient elements into mineral forms which increase their availability to plants—conversions which have already been described as highly significant in the linking of the decomposition subsystem to the plant subsystem.

If this classification is extended beyond the bacteria then the fungi can be seen as totally chemoheterotrophic in function. They are generally regarded as the main decomposers of primary resources because it is within this group that there is the most widespread distribution of extracellular depolymerising enzymes such as cellulases or lignin degrading enzymes. Only a limited number of bacteria possess these capacities but they are probably the prime agents of such processes in extreme environments. Thus fungi are largely restricted to aerobic environments whilst anaerobic cellulo-lytic bacteria such as species of *Clostridium* are fairly ubiquitous in de-composer habitats. Cellulolysis by thermophilic actinomycetes may also be important in high temperature environments. Whereas fungi may be particularly adapted for degradation of primary resources bacteria appear to be better suited to secondary resources, chitinolytic ability being for instance widespread in the Actinomycetales.

Although there may be significant differences between the fungi and the bacteria in their depolymerising abilities there is probably little distinction in terms of the range of metabolites—sugars, amino-acids, aromatic derivatives etc.—that can be utilised; both groups are probably ubiquitous although marked differences may exist between species. The main distinction at the kingdom level is again the wider range of environments under which bacteria can operate than fungi. This leads to bacteria being the dominant saprotrophic microbes of anaerobic environments, including those of animal digestive systems.

Many attempts have been made to classify the nutritional abilities of decomposer fungi and bacteria and thus assign more specific ecological roles. Winogradsky (1924) first proposed a biological classification of bacteria into *zymogenous* species (opportunistic species usually in a dormant condition but capable of rapid growth to exploit readily metabolised resources) and *autochthonous* species (those resident in the soil as vegetative populations and utilising complex polymeric substances). This classification was based on the observation that when plant materials reach the soil there is usually a rapid initial increase in microbial populations which then dies away as the metabolism of simple sugars, pectins, and amino acids is completed. The remaining materials, particularly cellulose and lignin, are more intractable and are decomposed more slowly by micro-organisms which do not show the intense activity of the earlier stages.

Garrett (1951, 1963) proposed a similar scheme for fungi which also postulated certain taxonomic correlations. The '*sugar fungi*' were pictured as primary colonisers of plant resources which were dependent on simple non-polymeric substrates. This group included most of the zoosporic fungi, pine moulds and most yeasts and many imperfect fungi. All these fungi are capable of rapid germination and growth in the presence of suitable resources. As the simple organic compounds are rapidly exploited, the fungi sporulate and disperse, the mycelium dies off and the sugar fungi then remain dormant until similar resources become available again. Garrett also recognised two further distinct groups of *cellulolytic* and *ligninolytic* fungi. The latter consisted largely of Basidiomycetes and the former of Ascomycetes and imperfect fungi. These species have slower growth rates than the sugar fungi, were supposed not to compete successfully with them for simple organic compounds, and thus to appear later in the decomposition process. In later modifications Garrett (1963) recognised that *secondary sugar fungi* might occur in the later stages, in association with the cellulolytic fungi. A good example of this is given by Frankland (1966) in a study of the decomposition of bracken (*Pteridium*) petioles. Most of the primary colonisers of the dead material reaching the soil were cellulose or lignin decomposers and the sugar fungi did not reach a maximum until the fifth and sixth year after the initiation of the experiment. None of the original soluble carbohydrate was present at this time and the source of these sugars was the hydrolysis of cellulose by Basidiomycetes.

These classifications have been of great value in promoting a study of decomposer organisms but their reality is now sharply questioned. It is becoming clear that our knowledge of the enzymatic capacities of groups or species of microbes is incomplete and it is thus dangerous to assume what their biochemical role may be. A number of examples have emerged recently for instance of cellulolytic capability among fungi previously believed to lack it as a group (Park 1975). Moreover, a species may utilise a particular enzymatic capacity in one set of conditions but not in another. In the face of such flexibility, groupings of decomposers based on nutritional capacity are not very instructive. Schemes like Garrett's do serve to emphasise however the close relationships between fungi and primary resources. One grouping based on nutritional characters that does seem to work is that of the wood-decomposing Basidiomycetes. These have been distinguished into brown-rot and white-rot forms on the basis of having either polysaccharide-digesting ability alone or combining it with ligninolytic ability.

Direct observation of microbial habitats has shown that many bacterial species are closely associated with fungal hyphae. These bacteria are probably utilising fungal catabolic products in the same way as the secondary sugar fungi are dependent upon the Basidiomycetes. The extent of these relationships is not yet known, but it indicates the possibility that many bacteria occupy a secondary associative role to fungi in decomposition processes.

One further biochemical ability that is restricted to prokaryotic organisms should be mentioned before completing this section—that of nitrogen fixation. Nitrogen is particularly important for plant growth as well as for the decomposition of organic remains as it contributes to the formation of proteins. It is an abundant element, forming about 80% of the atmosphere, but it is an extremely inert gas which is not generally biologically available. Some bacteria are able however to convert it to organic compounds which can thus enter the biological cycle. The bacteria may be biotrophs in symbiotic association in root nodules of legumes (*Rhizobium* species) and some trees and shrubs, autotrophs (some blue-green bacteria) or free-living saprotrophs (*Azotobacter, Beijerinkia*). Nitrogen fixation requires energy expenditure by the micro-organism. In the biotrophic and autotrophic forms this is supplied by plant or microbial photosynthate. The saprotrophic forms appear to be dependent on the presence of readily metabolisable resources as no N-fixing bacteria have yet been shown to be capable of cellulolysis. Further research may reveal mutualistic symbioses between bacteria and fungi on this basis,

opens into the more spacious pharynx (Fig. 3.8F). The pharynx is muscular and is used in feeding on fragmented plant materials or soil. The mouth is applied to the food material and the pharynx then undergoes a series of contractions which pump the food into the mouth. Large leaf fragments may be torn up by gripping the leaf between the prostomium and the lower border of the mouth and then contracting the body. This type of feeding behaviour restricts the worms to feeding on soft plant materials since tough litter cannot be broken down in this way. This is illustrated by the characteristic appearance of European forest soils during the summer months containing large earthworm populations; the lamina have been stripped from the leaves and the soil surface is bare except for a covering of leaf petioles and small twigs.

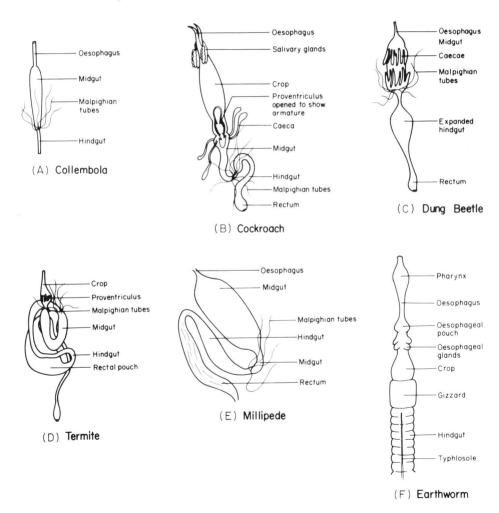

(A) Collembola

(B) Cockroach

(C) Dung Beetle

(D) Termite

(E) Millipede

(F) Earthworm

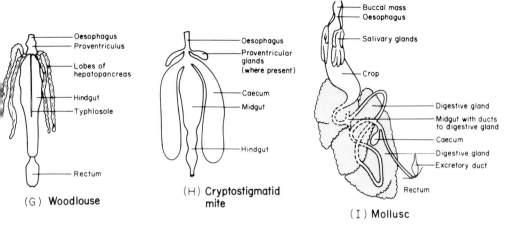

FIG. 3.8. Gut morphology of saprotrophic animals (not to scale). For details see text. (A) Collembola (*Tomocerus*); (B) Cockroach; (C) Dung beetle (*Oryctes*); (D) Lower termite (*Eutermes*); (E) Millipede (*Glomeris*); (F) Earthworm; (G) Woodlouse (*Oniscus*); (H) Cryptostigmatid mite (*Platynothrus*); (I) Mollusc: slug (*Agriolimax*).

The slugs and snails feed on plant materials by rasping the tissues away with the grater-like radula. The movements of this structure are complex and transfer the particles detached from the food to the ciliary lining of the roof of the buccal cavity and thence back to the oesophagus (Fig. 3.8I). These molluscs mainly feed on leaf materials but Mason (1970) has recorded that two small temperate woodland snails, *Discus rotundus* and *Marpessa laminata*, also feed on rotten wood. Other species are carrion feeders or facultative microtrophs but the structure and functioning of the mouthparts is basically the same.

The arthropoda have sclerotised mouthparts though these may be less well-developed in the larval stages of insects. The Diptera for example contain three sub-orders, the Nematocera, Brachycera and Cyclorrhapha which represent a series showing progressive reduction of the head complex (Fig. 3.9). Some of the Nematocera have well-developed heads and biting mouthparts. This group includes the Bibionidae and Tipulidae which may be an extremely important component of the invertebrate fauna in some temperate soils (Szabo *et al.* 1967); other tipulids are wood boring. The Mycetophilidae and Chironomidae have small biting mouthparts and are predominantly fungivorous; a major proportion of their gut contents is composed of short lengths of fungal hyphae though fine plant detritus may also be ingested (Anderson 1975). There is a transition through the Brachycera to the typical maggot of Cyclorrhapha which is virtually headless with vestigial mouthparts. Little is known of the trophic ecology of the Brachycera though

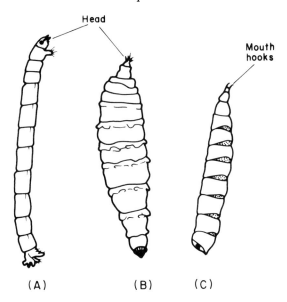

FIG. 3.9. Reduction of the cephalic complex in Diptera larvae; (A) Chironomid larva (*Nematocera*); (B) Tabanid (*Brachycera*); (C) Muscid larva (*Cyclorrhapha*).

Healey & Russell-Smith (1971) found that the gut contents of Phoridae and Longchopteridae contained fungal hyphae, algal cells, pollen, spores, testate amoebae and fine humic material suggesting that these fly larvae feed by scraping leaf surfaces. The Dolichopodidae and Empidae are apparently predators feeding on other fly larvae and enchytraeids.

The mouthparts of the Cyclorrhapha are reduced to a pair of chitinous hooks. These larvae are poorly adapted to feeding on solid materials but the Muscinae (Muscidae) are important members of decomposer communities in compost and dung while the blowflies (Calliphoridae) are major agents in carrion decomposition. Both these groups of flies have mouthparts adapted to feeding on fine particles (Dowding 1967). The ventral aspect of the pharynx is divided into a series of longitudinal ridges which are roofed over by lateral lamellae to form a sieve which retains particles of about $0 \cdot 6 \, \mu$m in size. The food of these larvae appears to be bacteria rather than detritus and the sieve reduces the amount of non-nutritious matter ingested. A similar filter structure is found in the proboscis of the adult Coelopidae (kelp flies). The proboscis is used to 'puddle' sand so that the interstitial water is brought to the surface and the suspension of algae, diatoms and bacteria is then filtered out and ingested (Cheng & Lewin 1974).

The mouthparts of the Collembola and Acari are varied in structure and reflect the adaptive radiation in the feeding biology within these groups. The

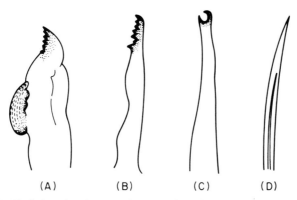

(A)   (B)   (C)   (D)

FIG. 3.10. Variation in the mouthparts of Collembola. Chewing type of mandible (A), biting type of mandible (B), sucking type of mandible (C) and maxilla (D).

majority of Collembola chew their food and have strongly developed mandibles and maxillae (Fig. 3.10A). The mandibles have a roughened plate in species such as *Tomocerus* which feed on hard plant litter; the size of the plate is reduced in species feeding on soft food materials. Typical soil-inhabiting species such as *Friesia* have biting or rasping mouthparts with no molar plate whilst piercing and sucking mouthparts are present in certain genera like *Neanura* (Fig. 3.10E and D). A comprehensive account of the feeding biology of Collembola is given by Christiansen (1964). The cryptostigmatid mites also exhibit variation in the structure of their mouthparts which can be broadly related to their feeding habits (Fig. 3.11). The xylophagous Phthiracaridae have massive chelicerae while those of the Pelopsidae are elongated structures for feeding on fungal hyphae; the mouthparts of most Cryptostigmata, which are general feeders, are intermediate between these extremes.

Other arthropod groups such as the Crustacea (Isopoda and Amphipoda), Orthoptera and worker castes of the Isoptera have generalised mouthparts with strong mandibles for biting and tearing and the muscles of the pharynx are adapted for swallowing large chunks of food. Most Diplopoda have well-developed biting mouthparts but families of the order Colobognatha exhibit a progressive development of suctorial mouthparts with a corresponding degeneration of the mandibles. These changes culminate in the tropical Siphonophoridae in which the mouthparts are developed into a long piercing beak for feeding on plant juices. In the Polyzonidae, represented in temperate regions by *Polyzonium*, the mouthparts are semi-suctorial.

Digestion. After this initial comminution the food particles pass into the gut where further comminution may take place. The gut is also the site of

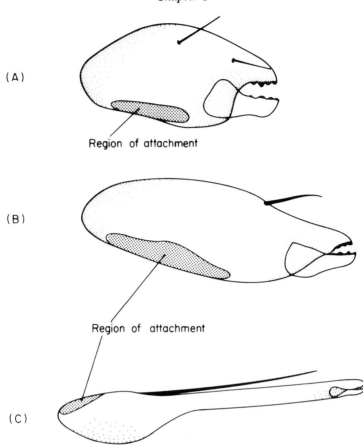

(A)

Region of attachment

(B)

Region of attachment

(C)

FIG. 3.11. Chelicerae of Cryptostigmata showing broad adaptations to diet. (A) Wood boring species (*Phthiracarus*) with strengthened biting surfaces and a massive fixed digit for the insertion of the powerful adductor muscles, (B) A general browser (*Liacarus*), (C) The elongated chelicera of a fungal feeder (*Pelops*).

catabolic action and the various morphological regions represent sites of differing significance in these two respects. Fig. 3.8 illustrates a variety of gut morphologies representative of many decomposer types.

Ingested material passes into the foregut where it may be stored in a thin-walled crop formed by a posterior enlargement of the oesophagus (B, D, F, H and I). In the crustacea, terrestrial oligochaetes and some insects (e.g. cockroaches and most beetles), the distal end of the crop is developed into a proventriculus or gizzard which is highly muscular and armed with cuticular processes, which crush and triturate the food (B, E, F). In some molluscs the crop is lined with chitinous processes and functions as a gizzard. This complex structure is absent from the gut of Cryptostigmata (Acari), Diplopoda and Collembola (A, G and H). There does not appear to be a close relationship between the presence of a well-developed proventriculus and diet in insects. In the xylophagous cerambycid beetle larvae *Xystrocerca globosa* the proventriculus is absent and the particles of wood in the midgut are comparatively large, whereas in *Macrotoma palmata*, which has a well-developed proventriculus, the contents of the gut are finely pulverised. The earthworms, however, have one (Lumbricidae) or several gizzards (Megascolecidae) which apparently represent an adaptation to feeding on tough terrestrial vegetation as this structure is considerably reduced in size in species which have returned to aquatic or semiaquatic habitats.

In the Cryptostigmata, Diplopoda, Collembola, Diptera, Dermaptera and Isopoda, the oesophagus projects into the stomach to form the oesophageal invagination. The cells of this structure secrete a peritrophic membrane which encloses the food bolus during its passage through the remainder of the gut. The peritrophic membrane is secreted by epithelial cells throughout the length of the midgut in Coleoptera, Lepidoptera larvae and Orthoptera. It is generally regarded as protecting the midgut cells from abrasion by hard fragments of food, particularly in the rectum where the faeces become consolidated due to water absorption.

The main site of digestion is usually the stomach or midgut but in Orthoptera and some Diplopoda the secretion from the midgut is passed forward and the crop is probably the main region of activity. This may also be true of most animals (e.g. blowfly maggots and some primary saprotrophs) which re-ingest their own faeces in which digestive enzymes are still active. In the Isopoda, Amphipoda and Mollusca finely comminuted particles from the foregut pass to the hepatopancreas. This organ is equivalent to the midgut of other invertebrates and is the site of most digestive processes.

Assimilation of low molecular weight products of digestion takes place in the mid and hind regions of the gut, though the whole alimentary canal is involved in absorption to a greater or lesser degree. In the Isopoda and Oligochaeta the hindgut forms a major portion of the alimentary canal and is convoluted on the dorsal surface to form a typhlosole. This structure is believed to increase the absorbtive surface area of the gut and is of massive size in the deep-soil living Megascolecidae which ingest soil and plant remains of extremely poor nutritional status. The hindgut also represents a major part of the alimentary canal of termites. In the primitive wood-eating termites the hindgut is dilated to form the rectal pouch containing cellulolytic protozoan symbionts (Fig. 3.8D). The Macrotermitinae have a similar hindgut structure but much of the catabolism of plant structural carbohydrates is carried out in the fungus combs. The soil feeding Termitidae have a long coiled hindgut in which the large intestine is divided into two voluminous chambers. These appear to be the sites of extensive catabolic activity by

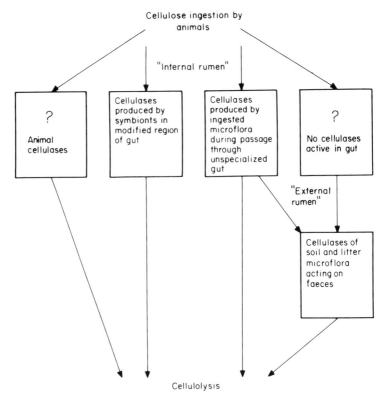

Fig. 3.12. The action and interaction of saprotrophic animals and micro-
organisms in cellulose decomposition.

bacterial symbionts. The rectum is also capacious; the excreta being used
for nest construction. A further important function of the hindgut in many
terrestrial animals is the reabsorption of water from unassimilated materials
before egestion.

Broadly speaking the enzyme complements of saprotroph guts reflect the
materials on which the animals predominantly feed. Most primary resource
feeders possess a rather similar basic enzyme complement, exhibiting only
limited protease and lipase activity, but strong carbohydrase activity particularly of enzymes which are active against disaccharides and oligosaccharides.
Whilst the distribution of the α- and β-glycosidases is widespread and
most saprotrophs can digest starch, glycogen, sugar and hemicellulose
components such as the pectin and xylan of higher plant cell walls, the
occurrence of cellulases and lignin-degrading enzymes is more limited
(Nielsen 1962). This provides an interesting contrast with the saprotrophic

micro-organisms. Indeed in many cases where cellulolytic activity has been detected in animal guts it has been shown to be due to the enzymes secreted by microbial inhabitants. This can take place in three ways (Fig. 3.12) representing a series ranging from obligate symbiosis, through facultative symbiosis to the 'external rumen' phenomenon where the animals assimilate or digest the metabolic products of the normal microflora associated with decomposing plant remain.

Certain species of cockroach, dung beetles and termites (with the exception of the Macrotermitidae) rely upon symbiotic bacteria or protozoa for the digestion of structural plant polysaccharides. In Lower Termites the symbiotic protozoa can make up two-thirds of the body weight of the insects. The gut flagellates ingest fine particles of wood (Fig. 3.6) and are responsible for extensive cellulolytic activity though bacteria are also implicated. Termites feeding on wood containing $54 \cdot 6\%$ cellulose, $18 \cdot 0\%$ pentosans and $27 \cdot 4\%$ lignin produced faeces containing 18% cellulose, $8 \cdot 5\%$ pentosans and $75 \cdot 5\%$ lignin indicating extensive polysaccharase activity but little lignin breakdown (Lee & Wood 1971). This latter seems to be the case for most termites, though the same authors quote values exceeding 80% lignin digestion for a species of *Reticulitermes*.

Cellulases are secreted in the guts of some xylophagous beetle larvae. Cellulose and pentosans are present in the frass of *Xestobium* in much smaller amounts than in the original wood: the cellulose to lignin ratio varied from $2 \cdot 38$ to $2 \cdot 81$ in the sound wood on which the larvae were feeding in comparison with ratios of $0 \cdot 86$ to $1 \cdot 24$ in the frass. Assuming no lignin was digested, 80% of the total weight of cellulose was hydrolysed. The wood adjacent to the borings showed similar cellulose to lignin ratios to the frass suggesting that the enzyme was still active in the excretion (Campbell 1929). There has been considerable debate as to the origin of these enzymes; in most cases it is assumed that they are microbial, but not necessarily obligate symbionts. Cellulose decomposing bacteria are always present in rotting wood and vegetation and may commonly form an important component of the diet of insects. These organisms may act as an inoculum for a fermentation chamber in some wood-boring Diptera and Coleoptera larvae. The hindgut of cerambycid larvae is enlarged for this purpose (Fig. 3.8c). The woody particles may be held for several weeks within this sac and are fermented by the bacteria. The enzymes secreted in the midgut are also active and may ultimately digest the bacterial cells as well as the flagellate Protozoa which feed on them.

This type of facultative symbiotic association is also common in litter-feeding animals. The digestive secretions of isopod guts are alleged to contain cellulase but Hassall & Jennings (1975) have demonstrated that this enzymatic activity is attributable to micro-organisms ingested with the food. Excised guts were incubated under a film of carboxy-methyl-cellulose and then stained to demonstrate localised cellulolysis around the gut. When anti-microbial agents were added to the food the cellulolytic activity was absent but was regained when the animal fed on leaf litter with the associated micro-flora. This clearly demonstrates the role of ingested micro-organisms in the digestive processes of isopods though the extent to which they contribute to total carbohydrate metabolism is unknown.

The case for cellulases of actual animal origin has been established in a few cases. Cellulases have been reported from a range of gastropod molluscs but there has been contention as to whether they originate from bacteria or are produced by the molluscs themselves. Recently however Koopmans (1970) has demonstrated conclusively for the herbivorous species *Helix pomatia* that a range of cellulase enzymes are secreted by the digestive gland. The extent to which similar enzyme complements are present in saprotrophic gastropods is unknown. Soedigdo *et al.* (1970) demonstrated cellulase activity in crude hepatopancreas extracts from the giant land snail *Achatina fulica* but showed that cellulolytic gut bacteria were able to induce cellulases in sterilised eggs and larvae which were unable to develop without these symbionts.

In a very wide range of primary saprotrophs no endogenous cellulolytic activity is detectable and they must rely on the saprotrophic micro-flora to digest most of the plant cell-wall polymers. This has been described as the 'external rumen' system and probably reaches its highest level of sophistication in the fungus combs of Macrotermitinae and fungus gardens of leaf-cutting ants (see Section 7.1).

It is interesting to speculate why a capacity of such apparent selective value is lacking from all but a minority of saprotrophic animals. We would hypothesise that this is because selection is more in favour of the ability to withdraw N, P or other nutrient elements which may be limiting to sapro-trophs (Chapter 1, Table 1.7). This imposes a necessity for the rapid processing of large volumes of material. The energy yields from cellulolysis, if this was the means employed, would create serious 'waste disposal' problems. A corollary to this hypothesis is the question of how some animals such as molluscs, worms, termites and xylophagous invertebrates do digest cellulose efficiently without encountering this problem. In the former two cases this may be related to the production of large quantities of mucopolysaccharides for locomotion and, in the case of molluscs, for feeding. A particularly interesting phenomenon has been described by Breznack *et al.* (1973) who found that N-fixing bacteria were present in the hindgut of all four termite species investigated in their study. The termites and the symbiotic cellulolytic protozoa were both dependent for survival on the bacteria as N sources when the termites were fed wood or filter paper. Nitrogen fixation requires considerable energy expenditure by bacteria which is in this case provided by extensive cellulolysis by the flagellate symbionts. Further investigation may reveal that this symbiotic relationship is widespread in invertebrates which maintain anaerobic conditions in their guts.

If we turn now to the secondary saprotrophs their diet is different and their enzyme complements show corresponding differences in composition. These organisms (in common with predators) ingest highly proteinacious foods. Carbohydrase activity is weak or absent but protease and lipase activity is more vigorous, particularly in carrion-feeding animals. In the blow-fly *Lucilia* a collagenase is also present, separable from the tryptase of other secondary saprotrophs which can digest collagen and elastin (tendons and soft bones). Keratin, the chief constituent of hair and feathers, forms the basis of the diet in tineid moths and dermestid beetles which are characteristic of the later stages of carrion decomposition. These structural proteins cannot be attacked by hydrolytic enzymes until the covalent S–S linkages binding the peptide chains are broken. This is effected by a strong reducing agent secreted by the midgut of these insects (Linderstrom-Lang & Duspiva 1935).

An adaptation to the low nutrient content of their food shown by primary saprotrophs which it is appropriate to mention is the habit of re-ingesting faecal material. McBrayer (1973) found that the litter feeding millipede *Apheloria montana* showed small weight increases if it was solely fed on leaf litter ($1 \cdot 1\%$ in 30 days). A second group of animals was allowed access to their faeces and increased their ingestion rates by more than 60% and their growth rate by 16% over the controls. Apparently the conversion of the unavailable substrate to digestible microbial protoplasm and particularly degraded plant compounds was essential to the diet of this species. Mason & Odum (1969) have also demonstrated that a wood-eating beetle, *Popilius disjunctus*, died if it was excluded from its faeces. The coprophagy enhanced mineral nutrient retention and the tunnel acted as a fermentation chamber or 'external rumen' for cellulolytic bacteria.

We may therefore conclude that whilst saprotrophic micro-organisms are highly adapted to metabolise primary resources, saprotrophic animals have developed feeding and digestive systems capable of exploiting the secondary levels of decomposition and utilising the products of the microbial activity.

3.2.5 Saprotrophic vertebrates

No natural vertebrate populations feed entirely on dead plant materials though many herbivores, particularly the game herds of the African savannahs, feed on dead vegetation at certain times of the year. The leaf litter from woody perennials has a low nutritional value due to nutrient withdrawal before leaf abscission. The savannah grasses however become rapidly sun-dried at the end of the wet season and have a comparatively high protein content. The game animals utilising these grasses can be technically regarded as saprotrophs during these periods. The functioning of the vertebrate rumen is then directly analogous to decomposition processes in the decomposition subsystem.

There are also a number of vertebrates which feed on dung and carrion but as the distribution of these resources is unpredictable in space and time most of the obligate coprophages and carrion-feeders are flying insects. These invertebrates efficiently exploit these resources by virtue of their high population densities and by their sensitivity to gradients of faecal gases or putrifaction odours.

Herbivore dung contains unassimilated simple sugars, undigested polysaccharides and proteins of primary and secondary origin in quite high proportions (see Table 4.6). It is not extensively utilised by vertebrate saprotrophs however although birds and mammals may scavenge undigested seeds from herbivore dung. The faeces of omnivorous animals and predators have a higher protein content. Some vultures and kites feed on human faeces ('shit hawks') and the arctic fox is known to scavenge for polar bear faeces during the winter months when prey or carrion is scarce. The hyenas and most of the Canidae (foxes, jackals, etc.) are predators as well as carrion feeders. Indeed it has been shown in the Serengeti in East Africa that the hyenas are major predators and that the lions often scavenge from hyena kills. The jaws of hyenas are immensely powerful and break up and ingest large quantities of bone. The hyenas may therefore be important agents in the redistribution of calcium and phosphorus which have been concentrated in the bones of herbivores.

The energy returns for the effort of searching for carrion are so low that few vertebrates are obligate carrion feeders. Some vultures are able to overcome these limitations by low energy expenditure during their soaring flight on thermals high above the ground. The downward spiralling of a vulture is a signal to others in the area that a dead or dying animal has been spotted and the vultures are thus collectively able to procure food over a very large area of land.

3.2.6 Necrotrophs

In the strict sense—that they take living food rather than detritus—the necrotrophs do not have a decomposer function. This is clearly recognised by distinguishing between the herbivore and decomposition subsystems. There are many necrotrophic organisms however whose activity is an essential part of the functioning of the decomposition subsystem. These include the necrotrophic plant parasites, the microtrophic animals feeding on saprotrophic micro-organisms and the predators feeding on saprotrophic invertebrates. These latter two groups of organisms can play an important regulatory role within the decomposition subsystem in the way that the carnivores do within the herbivore subsystem. Furthermore we see again examples of flexible trophic roles with many microtrophs and even some predators capable of purely saprotrophic behaviour and the plant parasites also persisting in a saprotrophic role.

Plant parasites. The input of plant detritus into the decomposition subsystem may be due to the natural senescence of the tissue or it may be a consequence of the activity of necrotrophs such as the plant parasitic fungi. The essential distinction between the necrotrophic and the saprotrophic fungi is the ability of the former to overcome or avoid the natural resistance mechanisms of the plants and invade living tissues which are then killed, often by the action of toxins. These organisms share with the saprotrophs the extracellular enzymes capable of degrading cell wall polysaccharides and it is by this means that they obtain energy and nutrients from dead host tissues. Classic examples of such behaviour are the facultative parasites of fruits and tubers including that fungal genera *Penicillium*, *Rhizopus* and *Botrytis* and bacteria such as *Erwinia*. Indeed it is interesting to note that necrotrophy is more characteristic of fruit, root or stem invading parasites; the more metabolically active photosynthetic tissues have a greater proportion of biotrophic associates.

The necrotrophs may persist as active decomposers for some considerable time after the death of the host tissue. In a review of fungal successions on plant detritus, Hudson (1968) pointed out that most primary colonisers were parasites and that many of them were still present to fairly late stages in the progress of decay. Necrotrophs vary in their pathogenicity and some show specific host relationships whilst others are very general in their ability to attack plants. The same species may even include 'races' which are non-pathogenic and thus exhibit the characteristics of true saprotrophs. An interesting example of the specialisation between parasitic and saprotrophic strains has recently been shown in *Lophodermium* species on pine needles. Forms of this fungus have been found both as necrotrophic parasites of seedlings, colonisers of senescent needles on standing trees and in dead needles in the litter. Initially these were attributed to a single species, *L. pinastri*, which was therefore thought to show a spectrum of activity, initially as a necrotroph then persisting as a saprotroph after needle fall. More recent work however suggests that three distinct species may be involved in these niches (Minter & Miller personal communication).

It is interesting to speculate as to how widespread this phenomenon may be and whether the apparent persistence of necrotrophs from the parasitic phase into the decomposition phase may involve the replacement of one race or even species of the fungus by another adapted to the saprotrophic mode of life. Garrett (1956, 1970) has suggested that the parasitic and saprotrophic abilities of root-infecting fungi occur in inverse proportion.

This proposition may be a good one to bear in mind in examining the niche relationships of apparently opportunistic necrotrophic fungi.

Microtrophs

As previously explained, primary resources represent a low-quality food for animals both because of the low mineral nutrient contents, which are potentially limiting to their populations, and because of the inability of most animals to digest the main sources of energy and carbon—the polysaccharides. The micro-organisms are generally more efficient than the animals in annexing these resources and as a result of their activities the low-quality plant resources are converted into microbial tissues with much narrower C:nutrient ratios. Microbial tissue thus represents a higher-quality food for animals providing a rich source of N and P at low metabolic cost to the consumer. We have earlier seen an example of this (Table 1.9) where the original low accessibility of nutrient in the primary resource (wood) was concentrated by the activity of the primary decay fungi so that the microtrophic, wood-invading animal was feeding on material with a C:N ratio of 15–30 rather than 100–500. This probably explains the widespread distribution of micro-trophy among the decomposer animal groups. A dependence on microbial food does pose problems of availability and selection however and many microtrophic organisms display a good deal of flexibility in their feeding habits.

Within the mesofauna (Fig. 3.3), the ability to select microbial food resources varies in both space and time. Anderson (1975) has investigated the inter- and intra-specific variation in the gut contents of woodland Cryptostigmata. He found that most species selected fungal hyphae among the moist, freshly fallen litter during the autumn months when microbial activity was generally confined to exterior surfaces of the leaves. As the plant material became more decomposed and moved down the soil profile, the larger species initially and finally the smallest species of mite apparently lost their ability to select microbial tissues from the matrix of leaf fragments and frass. Species which were considered as microtrophs in autumn were therefore 'generalised' feeders in summer and 'saprotrophs' by late summer when the standing crop of fungal hyphae in plant tissues was at a minimal level. Each species also showed a high degree of intra-specific variation in gut contents. The question of definition as saprotrophic or microtrophic species thus becomes purely academic in such circumstances. A similar pattern of inter- and intra-specific variation in feeding biology also occurs in the Collembola (Anderson & Healey 1972) and the Enchytraeids (O'Connor 1967).

Little is known of the enzyme complement of microtrophs, although the Cryptostigmatid mites and Collembola appear to have the same protease, lipase and carbohydrase complement as most saprotrophs. Trehalose activity has also been shown (Luxton 1972; Zinkler 1971) which is significant for whilst trehalose is uncommon in higher plants it is the only disaccharide which is formed in appreciable amounts by fungi. Luxton (1972) also recorded weak chitinolytic activity in Cryptostigmata. This may assist in the digestion of fungal cell walls. The enzyme activity of these organisms may also depend on relationships with micro-organisms of a highly complex kind. Von Törne (1967a, b) has demonstrated examples of this in the Collembola where different species vary in their ability to digest different carbohydrate components of their food by virtue of their specific gut microflora which is selected from the species complement of ingested microflora. Two examples of microbivory, in which the animals actively culture the fungi involved in processing plant remains, are worth considering in detail.

We have described how the termites, like most other animals, apparently lack cellulases. The endo-symbiotic associations with bacteria and protozoa have resulted in a highly efficient utilisation of plant structural polysaccharides by the six families of Lower Termites. A very different type of association—in this case with fungi—is found in the Macrotermitinae. These termites build special structures in their nests known as 'fungus combs' which support various fungi of the genus *Termitomyces* in what may be a highly specific association. The 'combs' are constructed from faecal material and their function seems partly nutritional and partly concerned with the maintenance of a humid environment in the nest by the production of metabolic water. There is some controversy about the nutritional importance of the 'combs' to the termites but Lee & Wood (1971) suggest that undigestible plant remains, voided as excreta, are placed on the combs, the fungi then break down the resistant materials such as lignin and the older comb material, including fungal tissues, is re-ingested by the termites for further digestion. The end products of this cycle are water, carbon dioxide, termites and in some cases a small resistant remnant of comb. This is an example of a reduction of the normally complex decomposer community to a single animal and a single fungal species (Fig. 3.13). As fungal tissue is probably a significant component of the food of the termites this behaviour qualifies as microtrophy.

Another important mutualistic association between fungi and insects is found in the bark beetle families Scolytidae and Platypodidae which do considerable damage to forest trees. The beetles feed on fungi growing on the walls of tunnels which they excavate in the wood. The fungi are transported by the beetles from tree to tree, or overwinter in special flask-shaped receptacles derived from modified oil glands, on tufts of hair on the abdomen under the elytra or in invaginations of the exoskeleton which open near the base of the coxae. The fungi principally involved in these associations are commonly species of the Ascomycete *Ceratocystis*, and its imperfect stages.

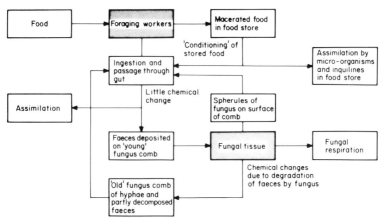

FIG. 3.13. Utilisation of food by Macrotermitinae. Decomposition of the detritus gathered by the foraging workers largely takes place in the nest due to the combined activity of termite and symbiotic comb fungus. This model may be considered as a reduced or specialised case of the functioning of the decomposer community, here dominated by two organisms as compared with the diversity of 'free-living' decomposer food webs. Note that food stores are constructed only by certain species (from Wood 1976).

Predators. These include both the predatory soil fungi, which snare nematodes by means of contractile loops or sticky projections on their hyphae, and the predatory soil animals. The predatory animals show few taxonomic affinities. The spiders and pseudoscorpions of the litter and soil are all predators but some species within the other groups of predators such as the staphylinid and carabid beetles and the prostigmatid and mesostigmatid mites, are facultatively omnivorous. The centipedes (Chilopoda) are generally thought to be predatory, but Lewis (1965) has shown that two *Lithobius* species may also eat leaf litter during the winter. Many of the predominantly saprotrophic and microtrophic groups of soil animals show adaptive radiation with predatory or facultative predatory members. The Collembola, Crypto-stigmata (Acari), Diptera, Isopoda and Diplopoda all contain species which feed on live animals.

Predators are an important component of the decomposer community and may have a regulatory influence on other animal populations. Ultimately this influence affects the rate and pattern of decomposition processes. For example it has been suggested that the predatory macrochelid mites (Meso-stigmata) found in phoretic association with dung beetles may feed on fly larvae in the dung and hence reduce competition between the coprophilous beetles and fly larvae. Predators do not generally however have any direct involvement in decomposition processes and are not considered in detail in this book.

3.2.7 Biotrophs

These are organisms characterised by their dependence on a living host-organism for energy and nutrient requirements. The clear-cut examples are parasitic or mutualistic associates of plants. Other symbionts, such as the micro-organisms in animal guts and the microbial components of such mutualistic associations as the termite and bark beetle symbioses mentioned above do not qualify for this definition. Although food is provided for them by the activity of their animal associates, and they may in some cases be dependent on their hosts for some component of their diet, their activities are essentially those of saprotrophs and they act as intermediates in energy flow and nutrient cycling between primary resources and decomposer animals. The biotrophs which form mutualistic associations with plants link decomposition and plant productivity in different ways. The mycorrhizal fungi act to transfer mineral nutrients from soil to plant and perhaps as an agency for movement of C and energy from plant to soil. The N-fixing bacteria of plant root nodules are important in nutrient cycles but have no direct link with decomposition.

The biotrophic fungal parasites of plants are a large group which have little direct connection with decomposition processes. Although they may be the agents of plant death and thus responsible for entry of plant detritus to the decomposition subsystem they do not persist and participate in the decomposition process. One characteristic of the adoption of the biotrophic habit (in contrast to that of necrotrophy) in such fungi as rusts and powdery mildews is the loss of extracellular polysaccharase activity. This may be either totally absent or may be so sensitive to catabolite repression (see Section 5.3.1) as to be totally insignificant. These organisms thus rely totally on plant photosynthate for energy and C and once the plant metabolism is so disrupted as to terminate the supply of photosynthate the parasites cease to function. These aspects have been discussed in detail by Lewis (1973, 1974) and Cook (1977).

Examples of animals which have a biotrophic habit for at least part of their life cycle are aphids and some root-feeding nematodes, both of which feed directly from plant photosynthate by means of stylets. The production of honeydew by the aphids acts as a unique pathway from primary production to the decomposer subsystem, largely in the form of the sugar melizitose. Several aspects of this pathway and its effect on the soil fauna and flora have been investigated by Dighton (1977).

3.2.8 Conclusions: the structure of the decomposer community

The difficulties involved in constructing models of the functioning of decomposer communities have already been mentioned. They arise perhaps from three main sources. Firstly the sheer diversity of organisms involved in decomposition in a relatively small area creates problems in the attribution of niche differences between them (Anderson 1975; Swift 1976). This difficulty is further aggravated by the functional flexibility of many of the individual species which cannot therefore be easily categorised in terms of trophic function (e.g. organisms which consume plant, faecal and microbial food indiscriminately). Thirdly the range of scales at which these processes have to be described, which ranges from the microscopic bacterial colony to broad activity range of the macrofauna and megafauna, increases the risk of attributing generalisations to one scale which are not applicable at another.

Nevertheless there are a number of consistent patterns which make it possible to derive generalisations whilst retaining the picture of a very fluid system which the above considerations imply. These patterns derive from some aspects of the functional ecology of the organisms which clearly adapt them for particular roles. Thus the possession of the wide spectrum of extracellular enzymes and the form of the thallus enabling them to penetrate into hard plant tissues clearly fits certain of the saprotrophic fungi to act as primary decomposers. Many of the necrotrophic parasites may be included among this group. Plant detritus is also subjected to comminution by the activities of saprotrophic animals such as millipedes, beetle larvae and (in the tropics) termites. The activities of these primary decomposers creates a wide spectrum of secondary resources and faeces which may then be decomposed by a diverse community of secondary decomposers. Important members of the active community at this level are the microtrophic animals, both micro- and mesofauna, the bacteria and coprophilic fungi. The corpses and excreta of these organisms in turn provide resources for a third cascade of decomposition.

This description implies an open, branching structure to the food web which could be readily simulated by our cascade model (Figs. 2.3 and 2.4). This pattern is broken however by the variations in decomposer function mentioned earlier. Whilst some organisms have very specific roles and thus a fixed position within the cascade (e.g. necrotrophic fungi), many others have variable roles and are active at several levels of the cascade (e.g. microtrophs which consume primary resources, saprotrophs which also consume microbial tissue, saprotrophic fungi active both on plant detritus and faeces). This is the reason for cyclical forms to the model such as that used by Wiegert, Coleman & Odum (1970) (Fig. 2.5). Whilst this incorporates the trophic flexibility concept it also cuts across the cascade concept of decomposition and obscures the extent of specialisation and the enormous diversity of function that does exist. Models of the decomposer community are thus fairly easy to conceptualise but very difficult to make explicit beyond a very superficial level of generalisation. It is probably better at this stage to avoid the temptation to produce a generalised model comparable with the trophic level model of the herbivore subsystem. Relatively simple statements that retain some validity can be made in specific circumstances (e.g. see Fig. 3.13) and as research proceeds it may prove possible to derive these for a sufficiently wide range of circumstances to serve as a general description.

Let us now consider in more detail the interrelationships between the different functional groups of the decomposer community which we have so far considered largely in isolation. Microbivory and saprophagy may be seen as occupying opposite ends of a spectrum of trophic relationships which are difficult to define except at their extremes. At one end are invertebrate saprotrophs such as the Lower Termites and blow-fly larvae which have a high assimilation efficiency of ingested organic matter, and at the other microtrophic Diptera and Coleoptera larvae which develop inside large fungal fruiting bodies. In between these extremes there is a gradient of trophic relationships broadly related to the body size of the animal. The largest animals of decomposer communities are predominantly saprotrophic, most of the mesofauna and microfauna are microtrophic, but here some specialisation may occur with the Collembola and Acari feeding mainly on fungi whilst the Protozoa and Nematoda utilise bacteria.

If we recognise that bacteria are largely associated with decomposition of particulate detritus and the lysis of fungal mycelia this implies a secondary role to them. The bacterial role does extend into the primary cascade however in the utilisation of readily assimilated sugars, and by the activity of the gut symbionts of primary saprotrophs. Under conditions of anaerobiosis cellulolytic bacteria may occupy the primary module to the almost total exclusion of all other organisms. Conversely there is a significant component of the fungal community which shares the secondary role with the bacteria— for instance a wide range of coprophilic fungi (Webster 1970) which includes the Mucorales, a group that is probably adapted to surface rather than penetrative growth. Much interest is currently being paid to the regulatory roles that microtrophic or saprotrophic animals may play in relation to the activities of the microflora. The possible significance of comminution in this respect has been commented on already and is discussed in detail as a resource quality factor in Chapter 4. The feeding activities of animals may have other significant interactive effects—for instance in improving the release of nutrients immobilised in fungal tissues. We have previously stressed the importance of immobilised nutrients in microbial tissues for the feeding and nutrition of soil animals. The complementary process of increased mobilisation of nutrients resulting from the feeding activities of animals on micro-organisms is equally important to the functioning of the decomposer subsystem. Crossley & Witkamp (1964) compared losses of weight and Cs^{134} from leaves in litter-bags placed in control sites and sites defaunated by adding 100 g naphtha per square metre. After approximately one year only small differences in weight losses between the two sites were detected but there were major increases in Cs^{134} mobilisation in the presence of soil

Chapter 3

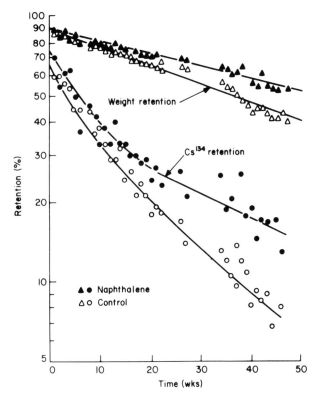

FIG. 3.14. The effect of the decomposer fauna on the rate of weight loss and the mobility of a mineral element. Weight retention and Cs^{134} retention by bagged leaf litter are shown for control plots and in plots treated with naphthalene (100 gm^{-2}), in deciduous forest. Lines fitted by eye (after Crossley & Witkamp 1964).

animals (Fig. 3.14). The processes involved in the caesium losses were not determined but it was suggested that microbe stripping by microarthropods was involved.

Clear demonstrations of the role of microtrophs in releasing biologically significant nutrients from microbial immobilisation have yet to be made. Swift & Moore (unpublished) have shown that the turnover of fungal mycelium in the presence of grazing mesofauna such as Collembola is orders of magnitude faster than in their absence but the significance of this to nutrient recycling has yet to be ascertained. Swift (1977a, b) has shown in a different context that the activity of wood-boring animals, entering wood after considerable decay by fungi, prompts a rapid release of nutrients previously immobilised in the fungal mycelium. Such release is of interest as there must presumably be some considerable demand for the nutrient on the

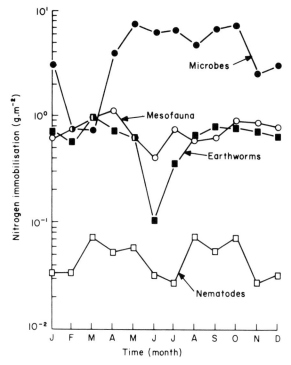

FIG. 3.15. Comparison of microbial nitrogen immobilisation with that of mesofaunal arthropods, earthworms, and nematodes. While microbial immobilisation is much greater than that of fauna, faunal immobilisation is much more constant through the year (after Ausmus, Edwards & Witkamp 1976).

part of the microtroph, but the losses of nutrient from comminuted wood are suggestive of increased solubilisation of non-utilised nutrient.

In view of the uncertainties that continue to dominate the picture, the significance of these interactive processes to ecosystem function is difficult to determine. The most detailed work comes from the long-term ecosystem studies at the Oak Ridge Laboratories in the U.S.A. of a *Liriodendron* forest summarised in a paper by Ausmus *et al.* (1976). They reported a detailed analysis of nutrient fluxes through soil fauna, microflora and tree roots. A major proportion of the total nitrogen, phosphorus and potassium in the soil and litter subsystem was located in microbial biomass particularly during the summer months and early autumn (Fig. 3.15). During the early spring (February and March) arthropods and earthworms immobilise as much as the fungal and bacterial populations which are then at their annual minimum. Maximum root growth and uptake of nutrients from solution occurred over the same period. Even though lysimeters recorded considerable soil water percolation during spring no significant leaching of N, P or K occurred. Thus despite considerable seasonal shifts in the location of the nutrient pool the soil and litter subsystem maintained a high degree of nutrient homeostasis. The soil fauna, while representing a smaller pool than the micro-organisms, is less susceptible to climatic perturbations than bacteria and fungi and faunal immobilisation is more or less constant throughout the year. Thus the importance of the soil fauna—the mesofauna in particular—to nutrient cycling goes beyond the quantitative role indicated by measurements of biomass or production.

3.3   Distribution patterns of decomposer organisms

The primary determinants of decomposer niches are factors associated with resource quality and the physico-chemical environment. In Chapters 4 and 6 respectively we discuss in detail the ways in which the distribution and activity of decomposers is regulated by these two complexes. In this section however we shall briefly consider whether any generalisations can be made concerning the patterns of distribution exhibited by decomposers as diverse in their biology as bacteria and earthworms.

We can conveniently subdivide the general spectrum of effect of the physico-chemical environment and resource quality into operations expressed at a number of different levels of scale—the biome or biogeographical scale, the ecosystem, the habitat and the micro-habitat (see Table 2.1). Each of these may represent discontinuities in the various factors which determine the distribution of decomposer organisms. Indeed we find that there is a certain degree of correlation between the scale of operation of these determinants and both the hierarchic level of taxon and the size of the decomposer organisms.

In general terms, the higher the taxonomic category, the longer the time that the group has had to become widely distributed. Hence, most classes and orders of soil organisms are cosmopolitan. Families are more restricted while genera and species show a higher degree of specific endemism in isolated habitats or land masses (Fig. 3.16). The size of an organism is also an important determinant of its distribution pattern. The microfauna and microflora may be widely dispersed by biotic and abiotic agencies, whereas most of the larger saprotrophic animals are rather sedentary and are rarely dispersed except by their own locomotive efforts. An organism's niche, in the sense of MacArthur (1968) is defined by the 'n' environmental variables required for survival. The habitable space of the 'n' dimensional hypervolume which the organism can utilise in the absence of competitors is termed the fundamental niche. The actual amount of this niche space which is utilised in the presence of competitors is the realised niche. Thus the fundamental niche space of an organism will be related to its body size: a large animal will

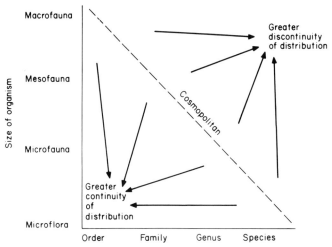

FIG. 3.16. Distribution of soil organisms according to size and taxonomic resolution. Some macro-fauna orders, e.g. Diplopoda, have almost global distribution but a high degree of endemism at the species level. Many bacteria and fungi species, however, are cosmopolitan (after Anderson 1977).

define its living space in terms of large units while small organisms will define the same habitat as a complex mosaic of small units. For example, a millipede will define a woodland as a series of habitats from several centimetres to perhaps a metre in size (logs, leaf litter pockets, loose bark, etc.) but it is unlikely to be affected by the structure of these habitats below a scale of a few millimetres. The mites and Collembola, however, will define the same woodland on a structural scale of millimetres (Anderson 1978) while the bacteria and fungi can define resources down to a macro-molecular level. An example of the importance of determinants at the micro-scale is the influence of resource factors on fungi.

Swift (1976) divided the decomposer microflora into two components: *Resource-specific* (RS) species are those with a distribution limited to only specific categories of resource types, i.e. occurring on only deciduous leaves but not on conifer leaves, herbivorous dung, or wood, etc. In some cases the degree of specificity may be even higher and only a single species of a resource type may be occupied (Table 3.3). The basis of this specificity is largely unknown but in some cases it may be related to the distribution of specific 'modifier' chemicals (see Chapter 4). Many other micro-organisms have a wider type of distribution in relation to resources; these were termed *Resource-non-specific* (RN) forms. These affinities with the type of organic matter occupied account for many patterns of distribution observed at a micro-scale. At a larger scale, resource types are more homogeneously distributed and discontinuities are less obvious at a habitat level although

Table 3.3. Distribution of Resource-specific and Resource-non-specific fungi. The occurrence of fungal species on four species of leaf (= four different primary resources) in the litter layer of an oak wood is shown in data calculated from Hering (1965) (after Swift 1976).

		Numbers of species			
1.	Total species		50		
		Ash	Birch	Hazel	Oak
2.	Distribution between resources	28	28	23	31
3.	Resource specific (RS)	8	6	2	10
4.	Total RS		26		
5.	Resource non-specific (RN)				
	(a) All resources		15		
	(b) Three resources		6*		

* A further three species were found on two of the leaf types.

they clearly become important at the ecosystem level where the differences between the RS microflora of a grassland and a woodland are very marked. Nonetheless there are many RN forms which are common to both, and this is particularly so in the soil flora.

Thus it is probably true to say that above the level of the genus and perhaps even down to the species, all the taxa mentioned in Table 3.2 are global in distribution. Peyronel (1956) analysed the distribution of eight taxonomically unequal groups of soil fungi and distinguished a number of general trends which are related more to the abiotic environment than resource factors. For instance Phycomycetes (Mastigomycotina plus Zygomycotina) formed a greater proportion of the isolated fungi in the cold wet soils of higher latitudes in contrast to the genus *Aspergillus* which was more predominant in the hot desiccated soils of the tropics. An interesting distribution pattern is found in the free-living nitrogen-fixing bacteria. Whilst the genus *Azotobacter* is cosmopolitan in distribution, species of *Beijerinckia* appear to be confined to tropical soils although global in their distribution within this region. Differences at the level of soil type (i.e. habitat level) are frequent and have been summarised for the fungi by Christensen (1969) and for the bacteria by Sundman (1970).

The importance of the edaphic environment is also apparent when the microfauna and mesofauna are considered. Here again the higher taxa are cosmopolitan and the major discontinuities occur at the level of genus and species. The Enchytraeidae have been recorded from every continent, from the poles to the equator, but because of their sensitivity to drought they reach their greatest abundance in moist temperate climates. The Collembola and Acari have a global distribution at the family level and some species such as the cryptostigmatid mites *Tectocepheus velatus* and *Oppia nova* and the Collembola *Isotomurus palustris* and various *Onychiurus* species have been identified on worldwide basis in many habitat types. Many species of Protozoa and Nematoda (e.g. *Plectus parientinus*) may also be truly cosmopolitan as far as can be judged from the limited numbers of studies on these animals.

On a local scale the soil fauna is often heterogeneously distributed between different habitats enabling the recognition of rather loosely defined species assemblages which are usually associated with certain habitat types. The basis for this association is unclear. Blackith & Blackith (1975) showed that within the West European province an overriding ecological factor determining the species distribution of Collembola was the vegetation cover. Other workers, however, conclude that this type of association is not a general phenomenon. Studies on the distribution patterns of Cryptostigmata along ecological gradients have repeatedly shown that the species populations overlap and that the populations must be considered as a mosaic rather than a sequence of communities (Wallwork 1976).

There seems to be little evidence that the distribution patterns of soil animals are determined by specific food resources (Anderson 1977). Deciduous tree leaves ranked in terms of their palatability to soil animals remain in approximately the same order irrespective of whether the feeding trials are carried out with earthworms, millipedes or micro-arthropods at the class, family, genus or species level. The herbivorous insects feeding on living leaves of the same tree species exhibit a far higher degree of trophic specialisation and monophagy. Most of the smaller soil animals are obligate or facultative microbial feeders (Anderson 1975). The selection of individual species of fungi by mites and Collembola has been widely demonstrated in the laboratory (see reviews by Luxton 1972, Harding & Stuttard 1974) but Dash & Cragg (1972) found, by placing fungal baits in the field, that there was little evidence of selectivity in natural populations of a wide variety of soil animals.

These problems of interpretation of the distribution patterns of small soil animals may be the result of failing to define the habitat on a scale which is meaningful in relation to the organism concerned. Bulk measurements of soil conditions (pH, soil organic matter, moisture, etc.) may only obscure the relationship between small organisms and their environment (see Chapter 6). The same criticisms apply to the interpretation of microbial distribution patterns. Large animals, however, are distributed in relation to macro-environmental parameters (pH, rainfall, soil organic matter content, base status, etc.) which can be easily measured at a scale relevant to the organism and discontinuities on their distribution can be readily demonstrated in relation to these macrofactors. Their contribution to decomposition processes in different biomes and in habitats of differing soil type is thus extremely variable. Millipedes for instance are absent from cold temperate and montane soils and reach maximum biomass and species diversity in humid tropical areas.

The terrestrial Oligochaeta include four main families of earthworms, the Lumbricidae, Glossoscolecidae, Megascolecidae and the Eudrilidae (in addition to the Enchytraeidae). The taxonomic status of the earthworm families, and their subdivisions, is complex and further information is available in Edwards & Lofty (1977). The Lumbricidae are dominant in temperate areas of Europe and Asia, while the remaining families are predominantly tropical or subtropical in their distribution. The Megascolecidae, which have the largest family with over eighty genera, are particularly found in Australasia and East and South East Asia. The Glossoscolicidae occur predominantly in Central and South America, but are also represented in North Africa and Southern Europe. Several lumbricids, mostly of European origin, have been reported as having a wider distribution as a result of the activities of man in distributing them.

In America, north of Mexico, the endemic species for instance were exterminated over large areas of the continent during the Quaternary glaciation and early settlers found no earthworms over the glaciated regions of the United States and Canada as well as large areas of the Great Basin and the Great Plains. The present-day impoverished earthworm fauna of these regions have been introduced by man from the southern areas of North America and from Europe (Gates 1966). Consequently the rich mull-soils of Europe are not typically found in North American forests. Local distribution

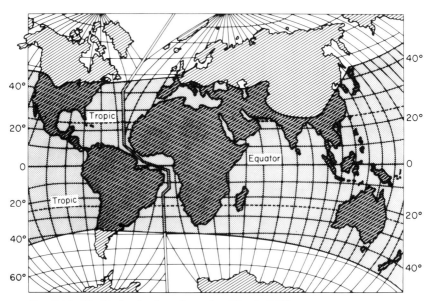

FIG. 3.17. World distribution of termites (Harris 1971).

patterns of earthworms are closely related to soil types. One of the most striking patterns is the differences between soils of mull and mor type. Satchell (1967) found eight species of lumbricid in mull soils in the English Lake District but only two in neighbouring mor sites.

The termites are distributed over about two-thirds of the land surface from latitude 45° N to 45° S (Fig. 3.17). The greatest diversity and activity of termites occurs in the tropics and declines rapidly with low minimum temperatures; though *Reticulitermes* has adapted to cold winters in the North Temperate zone. The geographical distribution of the termite families is complex (Harris 1971) and often obscured by the spread of termites by man. For example, many *Coptotermes* species, which are economically significant pests, have spread throughout the tropical zone in commercial timber, floating logs and ships.

3.4 The quantitative ecology of decomposer organisms

Our original intentions for this section were to classify decomposer communities in contrasting biomes according to the usual ecological parameters of number and size of organisms, taxonomy, biomass and productivity. It has become apparent that at the present time it would be impossible to carry out this exercise with any degree of completeness. There are two principal reasons for this problem: firstly the very considerable variation in microbial and invertebrate populations in space and time within the same habitat, and secondly the lack of standardisation and the variety of techniques used in population sampling.

It is worth digressing a little on this latter point as it is one of considerable importance. An enormous amount of research effort has been devoted, particularly over the ten years of the IBP, to the description of the quantitative role of the decomposer organisms described in Section 3.2. It is nevertheless difficult to draw any general conclusions on the basis of the data obtained. This is largely because of the practical problems encountered in the study of decomposer organisms. Even the most basic estimates of quantitative ecology, those of population size, have provided almost insuperable problems. For instance, a quick search of the literature will reveal numerous estimates of bacterial population size for practically all known soil types under all conceivable conditions. Yet few of these data can be validly compared because the numbers and species list obtained are as much a product of the method employed as of the prevailing ecological conditions.

The same strictures apply to other groups of organisms and when other ecological parameters such as biomass or production are considered the problems of method are even more formidable. Nonetheless ecologists have shown a high degree of ingenuity in the face of these difficulties and methods abound for tackling one or more of these particular problems. These are reviewed in the Appendix where references to more detailed information are also given. Lest this seems too gloomy a picture let us re-emphasise that the major difficulty lies not in the accuracy, precision or repeatability of individual measurements or studies but in the comparability between them. This point must be borne in mind when any comparison is made between soil types, ecosystems or biomes, in terms of the quantitative roles of decomposer organisms.

There is also an inverse latitudinal gradient in our knowledge of the structure and functioning of decomposer communities. Biomes exhibit a pattern of increasing complexity and variety from the poles to the equator and while it is possible to make tentative generalisations about the ecosystems of high latitudes the amount of information available from the much more complex tropical systems is very much less. It is cautionary for temperate ecologists to reflect that there are more contrasting forest habitat types in the Matto Grasso of South America than in the whole of the temperate region. The 'simplicity' of tundra ecosystems has become axiomatic in the ecological literature but a wide variety of habitats are present even in this region, ranging from mountain slopes to plateaux and from dry lichen heath to waterlogged moss tundra soils. The soil organism populations are correspondingly variable: bacterial plate counts range between 16×10^3 cells g^{-1} soil and $3 \cdot 7 \times 10^8$ cells g^{-1} soil (Parinkina 1974); mite and Collembola populations in the same region may be lower than ten thousand per square metre and as high as 200–$300 \times 10^3 \, m^{-2}$ (estimated from Whittaker 1974). In fact, estimates of fungal hyphal lengths, bacterial cell numbers and soil invertebrate populations (Protozoa, Nematoda, Acari and Enchytraeidae) are as variable in space and time within the tundra biome as between contrasting biome types from the poles to the equator! With these cautionary strictures in mind we have attempted to make a comparison between biome types on very general grounds only.

3.4.1 Decomposer communities in contrasting biomes

In Table 3.4 we have synthesised data from a wide variety of sources which broadly illustrate the major regional differences in structure of decomposer communities. Individual ecosystems which have been subject to extensive study and for which more or less complete data are available for the decomposer sub-system will be considered as case studies in Chapter 7.

A general interbiome pattern emerges of increasing bacterial cell counts and hyphal lengths in the soil with decreasing latitude. Little information is available on the proportion of the fungal material which is active or on the ratio of bacterial plate counts to direct counts of cells in these soils but it is likely that the turnover time of microbial populations is far shorter in the tropics than sub-arctic and cool temperate biomes. This will result in higher

Table 3.4. Soil organism populations in major ecosystem-types.

Soil organism groups		Tundra = Arctic alpine	Mor forest soil = Boreal forest	Mull forest soil (M) = Warm temperate forest (WTF)	Temperate grassland (TG) = prairie (P)	Tropical savannah	Tropical forest
microflora	bacteria (no. g⁻¹ soil)	7×10^6	5×10^6	6×10^6	10×10^6	55×10^6	50×10^6
	fungi (m hyphae g⁻¹ soil)	1,200	4,000	3,000	3,000	?	6,000
microfauna	Nematoda Protozoa (no. m⁻²)	28×10^6	270×10^6	200×10^6	500×10^6	30,000	65,000
mesofauna	Collembola Acari	100,000	400,000	40,000	25,000	2,000	15,000
	Enchytraeidae	150,000	750,000	30,000	10,000	0–400	0–1000
	Isoptera	0	0	M 0 / WTF 1,000	TG 0 / P 1,000	4,000	5,000
	Diplura Symphyla Protura Diptera	1,000	1,000	500	1,000	500	1,000
	Total (no. m⁻²)	251,000	1,151,000	71,500	37,000	6,900	22,000
macrofauna	Earthworms	<10	20	200	750	1–100	0–250
	Diplopoda Isopoda	0	500	1,000	500	<1	400
	Coleoptera Mollusca Orthoptera Dermaptera etc.	0	300	200	200	100	1,000
	Total (no. m⁻²)	<10	820	1,400	1,400	200	1,650

rates of catabolism and less immobilisation of potentially limiting nutrients in microbial tissues. The high soil temperatures are mainly responsible for these latitudinal differences in decomposition rates. As pointed out in Chapter 1 the general trend is for most plant litter to decay over a period of several years at higher latitudes, with a concomitant accumulation of organic matter in the soil, while the net primary production of warm temperate and tropical biomes decays in a few months, when water is not limiting, and organic soils rarely develop.

The amount of organic matter in the soil, and the climatic conditions which promote the accumulation of this material, are the main determinants of microfauna population sizes. The Nematoda and Protozoa inhabit interstitial water films and organic matter stabilises the moisture regimes of soils. In the tundra soil microfauna populations are closely related to the lichen crust, moss cover and the development of turf not only because of the increase in food resources, but also as a consequence of the heat insulative properties of the organic material. The hot, dry mineral soils of the tropics have low nematode and protozoan populations. The population sizes of most mesofauna, like those of the microfauna are generally proportional to soil organic matter content. The deep organic soils of temperate forests maintain the highest populations and the numbers of Collembola, Acari, Enchytraeidae and most other soil animals decline towards the tropics where they are replaced by the termites in predominantly mineral soils.

Decomposition processes in temperate soils are intermediate between these extremes. The formation of mull and mor humus forms in European woodlands under the same climatic conditions, and even under the same vegetation type, are an example of the interaction of the organisms and the physico-chemical environment in the determination of characteristic decomposer communities and the development of different soil types. Deciduous tree vegetation producing acid litter or conifer litter with a high resin content tend to inhibit bacteria and promote the growth of fungi. The fungal metabolites further decrease litter pH and burrowing earthworms, millipedes and to some extent Isopoda, which have higher pH optima, are excluded from the soil organism community. The deep, organic mor soils formed under these conditions particularly favour Protozoa, Nematoda, and enchytraeid worms, but Acari and Collembola are also present in large numbers. Decomposition rates may be inhibited by the tanning of cellulose by protein phenol complexes (Handley 1954) and by the water-holding capacity of the organic materials. Less acidic vegetation, or base rich parent soils which neutralise the phenolic acids, allow the macrofauna to become established and their feeding and burrowing activities, particularly those of earthworms and iulid millipedes, rapidly break down the litter on the soil surface and mix it with the underlying mineral material. Decomposition rates are higher than in mor soils as a result of litter comminution and the maintenance of well drained aerobic conditions in the soil. The standing crop of soil organic matter is lower in mull than in mor soils under equilibrium conditions with the same litter input (see Section 1.2.2 and Table 1.3). Bornebusch (1930) has suggested that the total animal biomass in the two soil types is similar: the high micro- and meso-fauna populations in mor soils compensating for the absence of the macrofauna (particularly earthworms) and *vice versa*. There are few detailed studies of soil animal communities to support this hypothesis at the present time. Another hypothesis is that bacteria are dominant in mull soils while fungi predominate in mor soils. There is more evidence to support this general principle. Bacteria are favoured by more basic conditions than fungi but other factors than soil pH are probably involved. Clarke (1967) has shown that bacterial counts are highest in arable land, lower in mull soils and lowest in mor humus. This could indicate that bacteria are better adapted to physical disturbance or the finely divided nature of plant material in the soil because of the unicellular structure of the thallus. On the other hand whilst the fungal biomass is probably highest in the acid mor soils and lowest in the arable soils where primary resources are low, it is uncertain as to what proportion of the biomass in each soil is active.

Mesofauna populations are generally inversely correlated with decomposition rates but the termites are a major exception. Their activities determine major energy and nutrient flux pathways in many tropical ecosystems. The principal reasons for this are the symbiotic associations between termites and protozoa or microflora, which result in far higher cellulolytic efficiencies than are characteristic of other soil animals. In addition, some species of termites (particularly the Macrotermitinae) are not limited to the passive exploitation of food materials located in humid microclimates like most members of the mesofauna, but can construct runways and sheets of soil cemented with saliva over plant materials on or above the soil surface. This strategy enables them to annexe food resources during dry seasons when they are unavailable to other organisms of the decomposer community including microflora.

The macroarthropods are generally more abundant in tropical ecosystems than at temperate and sub-arctic latitudes while earthworm population densities reach a peak in most temperate soils. There is a general pattern, however, for tropical saprotrophic animals to have a larger individual body size so that the macroarthropod biomass is highest in the tropics while the earthworm biomass is probably comparable to that in temperate ecosystems. Reliable data on the tropical macrofauna are extremely few, however, principally because of the sampling difficulties. The decomposer fauna in tropical forests is not limited to the soil and earthworms, millipedes, woodlice and cockroaches may be found in standing dead wood and among epiphytic growths on tree trunks and even in the forest canopy.

The contribution of the micro-, meso- and macrofauna to total animal

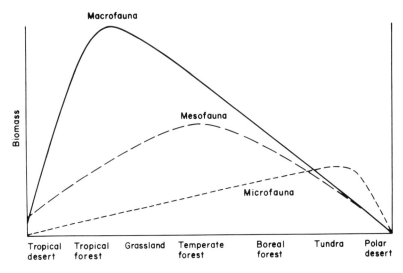

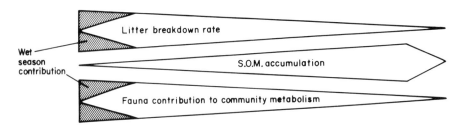

FIG. 3.18. Hypothetical patterns of latitudinal variation in the contribution of the macro-, meso- and micro-fauna to total soil fauna biomass. The effects on litter breakdown rates of changes in the relative importance of the three fauna size groups are represented as a gradient together with the fauna contribution to soil community metabolism. The favourability of the soil environment for microbial decomposition is represented by the cline of soil organic matter accumulation from the poles to the Equator; SOM accumulation is promoted by low temperatures and waterlogging where microbial activity is impeded.

biomass in decomposer communities at different latitudes is represented in Fig. 3.18. From the biomass and body size/food consumption relationships described by Reichle (1968) we can predict from the decomposer population data that litter breakdown rates will be much higher in the tropics than at temperate and polar latitudes. This effect will be accentuated by the influence of temperature on invertebrate metabolic rates. The tropical invertebrate faunas are therefore important components of the decomposer subsystem and their activities can be more significant determinants of decomposition rates and nutrient cycling pathways than in temperate regions. At the other extreme the microfauna in north temperate and tundra soils have little influence on decomposition processes and their populations are determined by the physico-chemical environment which governs soil organic matter accumulation.

Some examples where more detailed energy and nutrient budgets have been worked out on intensively studied sites are given in Chapter 7.

4

THE INFLUENCE OF RESOURCE QUALITY ON DECOMPOSITION PROCESSES

In Chapter 2 we identified two major groups of factors which influence the rate and pattern of decomposition processes—those of the abiotic environment and those of resource quality. In our decomposition module (Fig. 2.6) resource quality is pictured as a factor which regulates the action of decomposer organisms in a 'feed-back' manner—that is to say that as the nature of the resource is changed by the activity of the organisms so this change itself modifies the future activity of the decomposers. In this chapter we analyse the attributes that we have grouped as resource quality, examine the mode of their regulatory action on the decomposer organisms and describe in more detail the changes that occur in their influence as decomposition proceeds. Before embarking on this detailed analysis we must consider the validity of the basic hypothesis—that decomposition is affected by differences in the quality of resources.

4.1. The effect of resource quality on decomposition rate

In Chapter 1 we distinguished between a wide variety of resource types. Primary resources, originating from plants, comprise a variety of tissues differing in chemical and physical properties. Among these we may distinguish the major plant organs—leaves, reproductive organs, stems, roots. Each of these categories show characteristically different rates of decomposition. This is illustrated in Table 4.1. In both tropical and deciduous forest the rate of turnover of reproductive material is greater than that of leaves which is greater than that of wood. These differences are consistent whatever the climatic conditions and the rate at which decomposition proceeds.

The resource categories could be subdivided further to distinguish different species or forms of leaf, fruits from seeds and pods, twigs from branches and stems, different size classes or different age classes. In all groupings of these kind group-specific decomposition rates are found. The same applies to secondary resources such as microbial tissues, faeces or corpses of herbivore, carnivore or decomposer animals. Some further examples will serve to establish how specific these differences are.

118

Table 4.1. Turnover of main types of primary resource in forests of differing climates. Wood includes branches (> 2 cm diameter) and twigs (< 2 cm diameter) but not main stems. Data from Satchell (unpublished) for Meathop Wood, U.K. and Healey & Swift (unpublished) for Barro Colorado Island, Panama.

		Temperate Deciduous Forest			Tropical Rain Forest		
		Leaf	Wood	Reproductive	Leaf	Wood	Reproductive
Fall Standing	kg ha^{-1} yr^{-1}	3240	1580	600	7040	3020	3280
crops	kg ha^{-1}	2010	4810	270	2760	8090	340
k	yr^{-1}	1·61	0·33	2·22	2·55	0·37	9·65

Leaf litters from a range of deciduous trees were placed on the soil at three stations in a Belgian oak forest on calcareous soil (Mommaerts-Billiet 1971). The same pattern of decay rates was found at each of three stations differing in soil type; low loss rates for *Fagus sylvatica* and *Quercus robur* and high loss rates for *Prunus avium, Carpinus betulus, Tilia platyphyllos* and *Acer pseudoplatanus* (Fig. 4.1), the difference between lowest and highest rates being in each case tenfold. Mikola (1955) used a range of primary resources including mosses, ferns, grasses, heaths and deciduous and coniferous tree leaves. His data from controlled laboratory conditions (Fig. 4.2) show a threefold range in decomposition, with mosses at the low and deciduous tree litters at the high end of the range.

In these two studies total weight loss was used as the measurement of decomposition. As defined in Chapter 1 this includes losses due to catabolism, leaching and removal or export following comminution. These processes may be estimated separately and in such experiments resource-specific rates are still apparent. For example, the rate of aerobic catabolism of litter by heterotrophs may be estimated from the rate of uptake of oxygen during decomposition. In a study of primary resource decomposition on a blanket bog litter-bags were retrieved from the field after one year and measured for

Table 4.2. Differences between resources in rate of catabolism and total weight loss. Respiration rate at 10°C and field weight loss after one year for four types of litter on blanket bog (from Heal, Latter & Howson 1978).

Litter type	Respiration μl O_2 g^{-1} hr^{-1}	% loss in weight
Rubus chamaemorus leaves	66 ± 4·3	38 ± 1·6
Eriophorum vaginatum leaves	50 ± 4·2	26 ± 2·1
Calluna vulgaris shoots	33 ± 4.0	15 ± 1·4
Calluna vulgaris stems	11 ± 2·2	8 ± 0·6

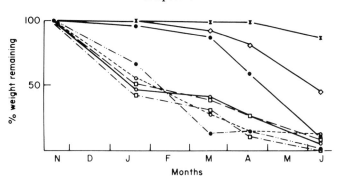

FIG. 4.1. Variation in decomposition of deciduous tree leaf litters in a woodland site on calcareous soil in Belgium. The species are □———□ *Carpinus betulus*, ○————○ *Acer pseudoplatanus*, ●————● *Acer campestre*, ●—·—●*Prunus avium*, ○------○ *Fraxinus excelsior*, □—·—□ *Tilia platyphyllos*, ◇————◇ *Quercus robur*, *————* *Fagus sylvatica* (from Mommaerts-Billiet 1971).

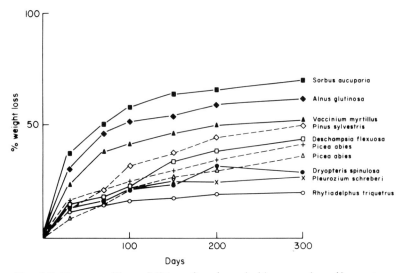

FIG. 4.2. Decomposition of litter of various deciduous and coniferous tree leaves, grasses, dwarf shrubs and cryptogams under laboratory conditions (from Mikola 1955).

the dry weight loss that had occurred and the current oxygen uptake (Heal, Latter & Howson 1978). The results (Table 4.2) show marked differences in respiration rate between resources, which strikingly correlate with the differences in first year decomposition which had already occurred.

Leaching is a component of weight loss which is rarely measured as a separate component of decomposition. The amount of leachable material is

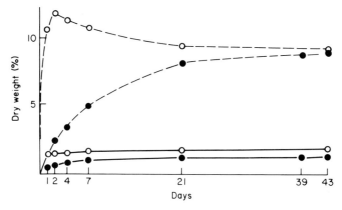

FIG. 4.3. Losses of water soluble organic (– – –) and inorganic (——) material from *Betula* (○) and *Picea* (●) leaf litter during leaching under anaerobic conditions at 25°C. Results are the dry weight of leachate expressed as a percentage of litter dry weight, each point being a mean of three samples (from Nykvist 1961a).

related to the content of water soluble material within the litter and also to its accessibility to water. In a series of studies Nykvist (1959a, b; 1961a, b) showed that the amounts of organic and inorganic substances leached from leaves of *Fraxinus excelsior*, *Pinus sylvestris*, *Betula verrucosa* and *Picea abies* under standard laboratory conditions varied considerably. Even when the amounts were similar, the pattern of leaching loss varied between species (Fig. 4.3). The difference in pattern probably reflected structural differences in the leaves. When the litter of spruce (*P. abies*) was ground, the leaching loss in the first day was increased by about ten times, indicative of the

Table 4.3. Variation in resistance to decay of stem and root wood of a variety of conifers to *Fomes annosus* a common root-rot necrotroph. The fungus was cultured on wood blocks under standard conditions in the laboratory for a period of twelve weeks (from Platt, Cowling & Hodges 1965).

	Extent of weight loss (%)	
	Stem wood	Root wood
Juniperus virginiana	3·5	9·5
Pinus virginiana	7·8	20·0
P. palustris	10·8	20·3
P. resinosa	11·6	12·7
P. echinata	11·8	24·8
P. taeda	14·3	25·8
P. ellioltii	17·1	23·0
P. strobus	20·6	21·1

interaction that may occur between comminution and leaching and of the way that resource-specific differences may be altered by the action of decomposer organisms.

The examples given so far showing the variation in rate of decomposition of different types of litter have involved mixed populations of heterotrophic organisms, but there are also many examples which show these differences in 'decomposability' are maintained when only individual species of microflora or fauna are involved. Table 4.3 illustrates this for wood decay fungi under laboratory conditions.

The examples quoted here serve to establish the validity of the resource quality hypothesis; different resource types do consistently decompose at different rates, even under conditions when all other factors, such as those of the physical environment, are controlled. The categorisation of resource differences simply in terms of the tissue or organ of origin or in taxonomic terms does not however reveal much concerning the factors which determine the differences in resource quality. These are to be found in the chemical and physical attributes of the resources. Before considering these in detail however we must look a little more closely at some of the characteristics of the progression of decomposition through time.

4.2.   The time-course of decomposition processes

We have emphasised the feed-back nature of resource quality regulation. That is, as the decomposition of a resource progresses through time, so the physical and chemical structure changes and modifies the factors contributing to resource quality. A number of examples of 'decomposition-curves' have already been presented in graphical form (Figs. 3.1, 4.1, 4.2). In these cases as in most decomposition studies the data are presented in a standard fashion with only minor variations; the fraction (or percentage) of the original weight lost or, more commonly, remaining, is plotted against time. Conversion of the original weights to fractions or percentages is a means of comparing the rate of decomposition of materials of discrepant initial weight. Mathematical analysis, usually in the form of regression equations then provides means of stating an overall decomposition rate. This is clearly a useful step in the analysis of decomposition for it provides a relatively simple summary statistic of a complex pattern of processes and also facilitates comparison of decomposition curves subject to different regulatory effects. Curve fitting may also be an instructive process in that it enables theories about the time course of decay to be tested against data. In this respect disagreement may be more instructive than agreement in that the anomalies so revealed stimulate further investigation with a consequent expansion in theory (see Bunnell & Tait 1974 for a discussion of this).

The simplest form of decomposition curve is one which shows a simple linear relationship between the weight remaining and time, e.g.

$$M_t = M_0 - kt$$

where M_0 and M_t are the resource weights at the start of the experiment or after time t respectively. This expression would fit well to curves like those of *Carpinus* and other fast decomposing tissues in Fig. 4.1. In the majority of cases however the plot of weight against time is curvilinear and a different form of equation is required. Thus for the data of Fig. 4.2 a much closer fit is achieved by use of the formula

$$M_t = M_0 e^{-kt}.$$

This negative exponential form of curve implies the loss of a constant fraction of weight rather than loss of constant increments of weight over successive equal intervals of time. The linear regression equation commonly fitted is thus

$$\ln \frac{M_t}{M_0} = M_0 - kt.$$

Although good fits to data can be achieved with the negative exponential model in a great many cases, the large number of exceptions stimulate more detailed analysis of the situation. Perusal of the curves in Figs. 3.1 and 4.1 shows that in a number of cases the curves are more complex than can be expressed either by a simple linear model or even a more complex curvilinear form such as asymptotic, quadratic or power (Howard & Howard 1974). The reason for this is that the overall weight loss curve is the product of the interaction of the three processes of decomposition (K, C and L) with the regulating variables (P and Q). Differential variation of one or more of these produces interactions which can clearly result in a family of curves differing quite considerably in both quantitative and qualitative respects. We shall consider the impact of a varying physical environment on the decomposition curve in Chapter 6. The interaction of the three decomposition processes with the different chemical components of plant resources is a matter of considerable significance to the understanding of resource quality.

4.2.1. Changes in resource composition during decomposition

Close fits to the negative exponential model are commonly obtained when a single substrate molecule such as cellulose is decomposed under relatively stable environmental conditions. In a natural complex resource, the total weight loss may therefore be expected to reflect the summation of the decay

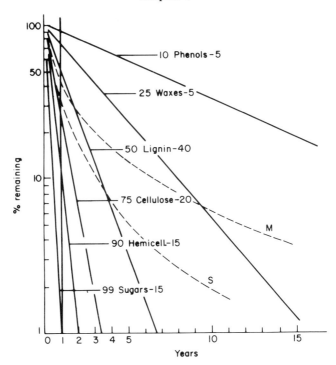

FIG. 4.4. The decomposition curves of the various groups of constituents, if their decomposition could be represented by a logarithmic function (the straight lines from the point 100%). The number in front of the name of the constituent indicates the loss after one year. The number after the constituent represents its percentage in weight of the original litter (these values are rough averages and they do not represent a specific analysis). The line *S* shows the summation curve obtained by annual summation of the residual values of the separate components. The line *M* gives an approximation, based on some analyses, of the probable course of the decomposition of similar resources in the mor-type at Hackfort (from Minderman 1968).

curves of the individual substrate fractions. Minderman (1968) analysed this hypothesis and plotted the loss rates of the main constituents of woodland litter (Fig. 4.4) representing them as logarithmic functions with annual loss rates varying from 10% for phenols to 99% for sugars. The summation of the residual values of the separate components gives a theoretical decay curve for the total resource (curve S in Fig. 4.4). This summation curve does not conform to a simple logarithmic function, reflecting the increase in proportion of resistant components and subsequent slowing of the decomposition rate as the resource decomposes. Minderman also plotted the course of decomposition of litter based on data for a deciduous wood in Holland. The curve (*M* in Fig. 4.4) showed a slower loss rate than predicted from the

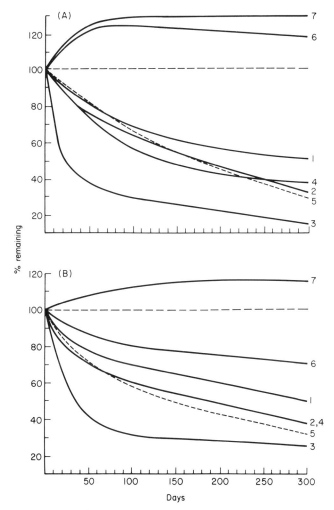

FIG. 4.5. Decomposition of the component substrates of leaf litters of (A) *Pinus* and (B) *Betula* under laboratory conditions (Mikola 1955). (1) Total weight; (2) Ether and alcohol soluble; (3) Water soluble; (4) Hemicelluloses; (5) Cellulose; (6) Lignin; (7) Crude protein.

summation of the component curves, indicating that interaction between the components and/or the production of new components results in greater resistance to decomposition than is predicted by consideration of the components individually. Clark & Paul (1970) suggested that 'if Minderman's data on the individual components (were) corrected for re-synthesis of secondary metabolites which break down at a slower rate than the primary components, the summation curve (S) closely approximates to that actually found (M)'. This theoretical analysis by Minderman indicates that for complex resources the representation of the time course of decomposition as a negative exponential function, i.e. a constant fraction loss, does not provide a complete description of the process although it is a convenient first approximation.

The evaluation of this hypothesis against data is difficult because detailed analyses of the decomposition rate of constituent substrates as well as the whole resource are rare because of the considerable labour involved. Mikola (1955) made a comprehensive analysis of the decay of a number of leaf litters. Results for litter of *Pinus* and *Betula* (Fig. 4.5) show a decline in weight of the total litter and of most of the components in a pattern which approximates to a negative exponential. There is however a marked increase in the absolute amount of crude protein in both litters and of lignin in the pine litter. It should be noted that this is not an artefact of the method of calculation as the data have been corrected for the loss in weight. Confusion can arise if this is not done, for the concentration of immobilised nutrients such as N necessarily increase as mineralisation of C proceeds. Increase in absolute amounts of N and other mineral elements is relatively common and is probably due to import from external sources or to N-fixation (see Chapter 1). The apparent increase in lignin may be due to the alteration in efficiency of the extraction procedures as the resource changes in composition or to the synthesis by micro-organisms of substances which are extracted in the lignin fraction. This is clearly a feature which imposes some caution in interpreting time courses in compositional change.

The immobilisation and accumulation of N is an indicator of the concomitant growth of decomposer micro-organisms as decomposition proceeds. There are few quantitative data to show this in more detail because of the practical difficulty of separating microbial tissue from the resource material. Swift (1973) was able to make an estimate of the growth of a fungus during decomposition of wood sawdust by measuring the change in chitin content which is restricted to the fungal cell walls (Table 4.6). By the end of the experiment a 39% weight loss was recorded for the sawdust, but the biomass estimate showed that 58% of the remaining material was of secondary (i.e. fungal) origin (Fig. 4.6.). This shows the probable validity of Clark & Paul's (1970) suggestion that re-synthesis (i.e. decomposer growth) may account for the discrepancy between a fast growth rate estimated on the basis of the substrate components and the slower observed rate for the total. The apparent rate of decomposition (k) of the sawdust was 0.04 wk^{-1} but when re-calculated by exclusion of the mycelial content the k value rose to 0.09 wk^{-1}.

Returning now to consider the component substrates in Mikola's data it is apparent from Fig. 4.5 that their rate of decomposition varies considerably. The most rapid loss is of the soluble sugar fractions, followed by the poly-

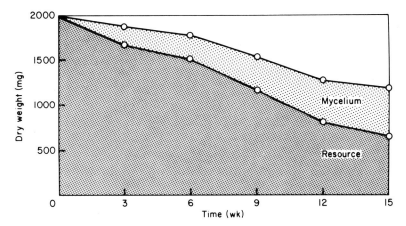

FIG. 4.6. Accumulation of microbial biomass during decomposition. Growth of the Basidiomycete fungus *Coriolus versicolor* on sawdust of *Castanea sativa*. The regression equations for the apparent (i.e. including mycelium) and actual decomposition of the sawdust are $\ln M_t/M_0 = 1.09 - 0.04t$ and $\ln M_t/M_0 = 1.02 - 0.09t$ respectively (Redrawn from Swift 1973).

saccharides, cellulose and hemicellulose, and finally lignin. After 300 days, the remaining 50% of the litter consists of microbial tissues plus an increased content of the more resistant substrates. This agrees well with the suggestion of Minderman that there will be latterly a lower decay rate than in the initial stages of decomposition. This may be illustrated by another example (Fig. 4.7) showing the decomposition of four leaf litters in a Hungarian woodland (Toth, Papp & Lenkey 1975). The accumulation of hemicellulose and lignin fractions in the terminal stages at the expense of sugars and cellulose is quite apparent as is also some marked differences in pattern between the litter species. Another point of interest emerges from these data by comparing the cellulose with the hemicellulose and sugar curves. The cellulose has apparently the most rapid rate of breakdown of these three carbohydrate fractions. This may not in fact be the case, for the products of cellulose decomposition include short chain oligosaccharides, which are alkali soluble and sugars which are water soluble. If these substances accumulate to any extent they will be estimated as components of the hemicellulose and sugar fractions respectively thus diminishing the apparent decomposition rate of these fractions. This type of analytical anomaly, similar to that observed for lignin in Mikola's data (Fig. 4.5), can in some cases lead to apparent increases in the hemicellulose fraction.

The assessment of changes in decay rate with time using weight loss data is difficult because of sampling and seasonal errors and because it is a cumulative measure so that conclusions must be based on comparisons of different

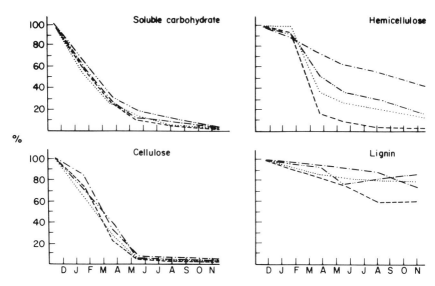

F<small>IG</small>. 4.7. Changes of watersoluble carbohydrates, hemicelluloses, cellulose and lignin in % of the starting quantities during litter decomposition. *Quercus cerris* (–··–), *Quercus petraea* (–····–), *Cornus mas* (– – –), *Acer campestre* (····).

parts of the same curve. A more sensitive measure is gained from measuring the 'instantaneous' respiration of the decomposing resource. Litter of known age can be retrieved from the field and placed in a respirometer at constant temperature and field moisture to give a measure of oxygen uptake or carbon dioxide output from the microflora and fauna involved in catabolic processes. Data for four types of litter from a blanket bog in northern England show a significant decline in rate of respiration with increasing decomposition in the two leaf litters which had high initial rates of respiration and large soluble sugar contents (*Rubus chamaemorus* and *Eriophorum vaginatum*, Table 4.4). The *Calluna* shoots and stems showed no significant decline in respiration rate in relation to stage of decomposition but these were the litters with the slow rates of respiration and weight loss (Table 4.2) and low initial concentrations of soluble material. We would therefore have predicted a lower rate of decline in these resources in comparison with the leaf litters. In contrast with these examples of declining decomposition rate with time, Witkamp (1966) found a slight increase in the rate of respiration with time in a seasonal study of four species of tree litter but he suggested that this resulted from 'counteracting influences of chemical impoverishment and improving physical conditions', emphasising the importance of eliminating seasonal changes in physical environment from time course analysis of resource quality change.

Table 4.4. The relationship between 'decomposability' and state of decomposition. Linear regressions of rate of respiration (Y) in μl O_2 g^{-1} hr^{-1} at 10°C, and percentage dry weight loss (X) for four litter types (from Heal, Latter & Howson 1978).

Litter type	Regression	n	r	Maximum weight loss (%)	Sampling period (days)
Rubus chamaemorus leaves	$Y = 88 - 0.928X$	58	0.53***	73	1463
Eriophorum vaginatum leaves	$Y = 64 - 0.680X$	20	0.58**	47	1092
Calluna vulgaris shoots	$Y = 36 - 0.246X$	28	0.32	44	1843
Calluna vulgaris stems	$Y = 11 - 0.127X$	28	0.28	24	1851

Another feature of the progress of decay is the formation and accumulation of humic materials. This will be considered in detail in Chapter 5 but evidence already presented (Chapter 1 p. 17) shows that humus once formed has a very slow rate of decomposition. Further evidence for this can be inferred from the study of soil profiles where there is no major redistribution of organic matter between the horizons so that the vertical sequence can be taken as equivalent to a decomposition time course. In a podzol on sand under *Pinus sylvestris* the respiration rate of samples measured at 25°C in the laboratory decreased with depth (Parkinson & Coups 1963). The results (Table 4.5) are influenced by variation in moisture content, input of root material and translocation of soluble fractions through leaching, but the main trend with depth probably represents a change with age of organic matter. The organic matter in the mineral horizons was largely humus and had a respiration rate of only about 5 per cent of that of the fresh litter. Clearly the accumulation of this fraction with its slow decay rate will markedly influence the overall rate of resource decomposition, particularly in the terminal stages.

The discussion of decay-time relationships above shows that there is good theoretical and direct evidence for a decline in the rate of decomposition as organic matter decomposes. Simple linear and negative exponential models may provide reasonable descriptions for parts of the decay-time curves but are only crude approximations to the full course of decomposition. The analysis by Jenkinson (1977) of the decomposition of the tops and roots of ryegrass (*Lolium perenne*) provides a good summary of this and illustrates the use of radioisotope techniques in decomposition studies. Samples of plant material uniformly labelled with ^{14}C were allowed to decompose for

Table 4.5. Rates of oxygen uptake by different horizons in a podzol under *Pinus sylvestris* (from Parkinson & Coups 1963).

Horizon	% organic matter	Oxygen uptake $- \mu l\ O_2\ hr^{-1}$	
		Per gram dry soil	Per gram organic matter
O_0 (L)	98·5	473	481
O_1 (F_1)	98·1	280	286
(F_2)	89·3	49	55
O_2 (H)	54·6	16	30
A_1	17·2	2·7	16
A_2	1·9	0·9	48
B_1	10·6	2·0	18
B_2	5·2	0·6	11
C	1·4	0·3	19

ten years in soil under standard conditions. Analysis for labelled and un-labelled carbon allowed a distinction to be made between losses from added plant material and from the soil organic matter present before addition of litter. There was a rapid loss of 68% of the labelled carbon in the first year (Fig. 4.8) but subsequently the loss rate per year declined markedly and after ten years the total loss was about 88%. It is possible that the loss rate of the remaining organic matter from the rye-grass would decline to the same levels (2·1–4·0% loss per year) as the unlabelled carbon from the soil humus.

A striking feature of Fig. 4.8 is the possibility of dividing the decomposition curve into two or more exponential stages. Thus Jenkinson found that a double exponential function had a much better fit than a first order exponential curve (see Fig. 4.8). The fitting of two negative exponentials to a resource decay curve may be done by mathematical manipulation to minimise the errors associated with the two curves. Alternatively associating the curves with chemical components of the resources provides more explicit definition of the processes involved. For example, a curve representing an initial rapid loss of soluble compounds which comprise a known proportion of the weight, with a curve representing the slower loss of resistant compounds such as lignin, may produce a statistically poorer fit to the data but provide an hypothesis concerning the decomposition processes that might be involved, namely separating the possible effects of leaching and catabolism. In the case of Fig. 4.8 the curve fitted predicts that about 70% of the resource decomposes with a half-life of 0·25 yr whilst the remainder has a half-life of about eight years (Jenkinson 1977).

The large number of phenomena which interact to produce a decomposition curve under field conditions diminish the possibility of making meaningful kinetic analyses. The more precise kinetic studies that are possible with simple

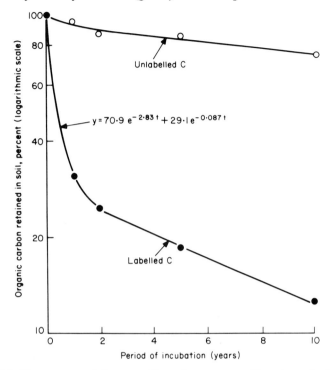

FIG. 4.8. Time-course of decomposition of rye-grass residues in agricultural soil. The fraction of ^{14}C remaining in the added residue is shown together with a similar curve for the unlabelled C of the soil organic matter. For interpretation see text (from Jenkinson 1977).

substrates under controlled conditions in the laboratory help us in our understanding and may in future be extended to field situations. In the meantime the general conclusion that we may draw from these analyses is that the relative amounts of the different components of primary resources change as decomposition proceeds and that these changes are themselves associated with changing decomposition rates.

4.3. Chemical attributes of resource quality and their influence on decomposition

We have seen that resources distinguished by taxonomic and general morphological criteria exhibit differing rates of decomposition. We have subsumed the distinguishing characteristics which determine these differences under the general term resource quality. It is now necessary to analyse the mechanistic basis of resource quality regulation of decomposition processes.

Resource quality is a composite definition of the value of a resource as food to an organism and embodies both physical and chemical criteria. The food must satisfy the physical (surface properties, texture, etc.) and chemical (phagostimulant, growth factor and nutritional) requirements for ingestion or colonisation to occur. The feeding and digestion of herbivores in relation to food quality has been widely studied (see Fiennes 1972) and many of the principles are useful in understanding decomposition processes both through vertebrate and invertebrate nutrition. Although the influence of resource quality on the rumen microflora and fauna is perhaps the best understood saprotroph system (Hungate 1966), the extrapolation of the principles to the decomposer subsystem must be treated with caution because of the anaerobic nature of the rumen. There is a large scattered literature on the nutrition of terrestrial saprotrophs, particularly micro-organisms, but comprehensive studies are generally lacking particularly for natural microbial populations and for saprotrophic animals.

In this section we analyse the chemical attributes of resource quality and in the next we turn our attention to the physical attributes. In these discussions our emphasis is on the way in which resource quality factors influence the functioning of decomposer organisms thus moderating the rate at which decomposition processes take place.

Examination of any general text dealing with the organic chemistry of plant products shows the bewildering diversity of molecular forms that any potential primary resource may contain. Similar complexity exists at the secondary resource level. We may however simplify this situation by imposing on it a classification of the molecular types and looking for categories of effect, rather than considering the effects of individual molecules in turn. The principle that the chemical constituents of plant materials are important factors controlling their rate of decay has a long history, originating largely from the work of Waksman and his colleagues in the nineteen-thirties (see Waksman 1952). Developing the arguments in these studies and those on herbivore nutrition we may recognise three main groups of compounds, classified on the basis of their significance to decomposer organisms.

1. *Carbon and energy sources.*

2. *Nutrient sources*—of elements other than carbon.

3. *Modifiers*—molecules which inhibit or stimulate decomposer activity by their chemical structure and are often (though not always) active at relatively low concentrations.

4.3.1 Carbon and energy sources

The largest quantitative demand by the heterotrophic decomposer organisms is for the energy released from organic compounds to fuel their growth and activity and for carbon for tissue synthesis. The bulk of carbon and energy in decomposer food resources is stored in a variety of polymeric compounds—polysaccharides, lipids, proteins and aromatic polymers such as lignin or humus. The composition of resources in terms of these broad categories can be determined by a relatively simple process of analysis and examples are given in Table 4.6 and in Figs. 4.5 and 4.7. The degree of resolution achieved by analysis of this kind is low. The methods commonly adopted involve successive extraction with organic or polar solvents and a summative reconstitution of the material from the fractions removed at each step (see column two of Table 4.6). The chemical definition achieved by this means is imprecise and overlap between fractions may occur. Modern developments in analytical chemistry have enabled much more detailed analyses to be carried out (e.g. see Rogers & Perkins 1968) but have yet to be applied to decomposition studies. The detailed chemistry of the polymeric substrates of decomposition is discussed in Chapter 5.

Nevertheless some patterns of difference between resource types can be distinguished from these simple analytical data. The greatest contrast is between the compositions of the primary and secondary resources. Whereas the major components of plant tissues are polysaccharides, particularly those of the cell walls, this is only a minor component of secondary resources although poly-amino-saccharides are common to the cell walls of most fungi and bacteria and the exoskeleton of arthropods. Conversely the protein and lipid contents of animals and microbes are significantly higher than those of plants. A word of caution should be expressed about the composition of microbial tissues. They exhibit a high degree of adaptation to the food source on which they are growing so that 'typical' analyses are virtually meaningless and critical arguments should always be based on specific data. Levi & Cowling (1969) have shown that the N content of the mycelium of a variety of fungi may vary as much as twentyfold in response to varying N contents in the medium. This change is probably largely due to adjustment of the cytoplasmic protein content, cell wall and nucleic acid N being more stable. Plant and animal tissues seem less flexible in their composition than those of micro-organisms, but it is clear that the composition of a particular resource is dependent on the circumstances prevailing at the time of death and entry into the decomposition subsystem. Thus the content of storage and translocatory carbohydrate in actively photosynthesising plant tissue is significantly higher than in senescent tissue—and so is the protein (compare the composition of the young and old oak leaves in Table 4.6). The storage of fat in animal tissues will also vary according to age and the physiological state of particular tissues at time of death.

Within the primary resources, woody and photosynthetic or herbaceous resources may be contrasted through the much lower content of soluble carbohydrate in the former. Storage polysaccharides such as starch may

Table 4.6. Major organic components (carbon and energy substrates) of decomposer resources.

	Deciduous leaf —young *Quercus* sp. (1)	Deciduous leaf —old *Quercus* sp. (1)	Conifer needle —old *Pinus* sp. (1)	Grass leaf *Deschampsia flexuosa* (2)	Grass stem *Zea mais* (1)	Deciduous wood Range (3)
Lipid, ether soluble	8	4	24	2	2	2–6
Storage/metabolic carbohydrate, water soluble (cold and hot)	22	15	7	13	15	1–2
Cell wall polysaccharide, hemicellulose (alkali soluble)	13	16	19	24	18	19–24
Cellulose (strong acid)	16	18	16	33	30	45–48
Lignin, residue	21	30	23	14	11	17–26
Protein, N $\times$ 6·25	9	3	2	2	1	—
Ash, incineration	6	5	2	—	8	0·3–1·1

	Conifer wood	Faeces invertebrate	Faeces vertebrate	Bacteria	Fungi	Earthworm	Arthropods	Vertebrate carcass
	Range (3)	*Glomeris marginata* (4)	Horse (1)	Range (5)	Range (7)	*Lumbricus terrestris* (7)	Various	Steer (11)
Lipid, ether soluble	3–10	—	2	10–35	1–42	2–17	13–26[8] (glycogen)	50
Storage/metabolic carbohydrate, water soluble (cold and hot)	2–8	2	5	5–30	8–60 (chitin)	11–17	14–31[8] (chitin)[9]	?
Cell wall polysaccharide, hemicellulose (alkali soluble)	13–17		24	4–32	2–15	—	5–9[8]	0
Cellulose (strong acid)	48–55	38	28	0	0	0	0	0
Lignin, residue	23–30	?	14	50–60	14–52	54–72	38–50[10]	39
Protein, N × 6.25	—	11	7	5–15	5–12	9–23	?	11
Ash, incineration	0.2–0.5	8	9					

Sources: (1) Waksman (1952); (2) Mikola (1955); (3) Browning (1963); (4) Nicholson *et al.* (1966); (5) Luria (1960)—from laboratory cultures; (6) Compiled from Cochrane (1958) and Hawker (1950)—from laboratory cultures; (7) Laverack (1963); (8) Spector (1956) *Gasterophilus intestinalis* larva; (9) Jeuniaux (1971) suggests chitin contents in adult arthropods may exceed 50%; (10) Calculated from range of N contents in final four rows of Table 4.10; (11) Giese (1962).

accumulate to a very considerable extent in woody roots—Barnes & Hava (personal communication) showed levels of up to 60% of the dry weight in some tropical trees. The degree of lignification is generally higher in woody tissues than in tree leaves, grasses or herbs but considerable overlap can occur as the senescent oak leaf data of Table 4.6 indicates. The lignins of softwoods and hardwoods have differing chemical structures and this variation may extend more widely. Cellulose accumulation is another striking feature of woody tissues. Differences in cell wall polysaccharide structure of a fairly general nature are revealed by more detailed analyses. The mannan and xylan contents of softwoods are usually of the order of 10–15% and 5–10% of the cell wall carbohydrate respectively. In hardwoods however the corresponding contents are less than 3% and 15–20%. The composition of the cell wall components also differs between tissues of different age and function within relatively massive resources such as branch, root or stem wood—see Table 5.1.

The differences between resources in terms of the relative availability of carbon and energy sources of different kinds provides a basis for some features of resource quality effect. The data of Figs. 4.5 and 4.7 support Minderman's general hypothesis of a differential rate of decomposition in which the soluble components of primary resources disappear more rapidly than the cell wall polysaccharides which are in turn more readily decomposed than the lignin. The rapid rate of disappearance of soluble components is due both to the ease with which they are assimilated and catabolised by the decomposer microflora and to leaching. The relative importance of these two effects is clearly dependent on the prevailing extent of rainfall but the two processes are often difficult to separate under field conditions. Detailed discussion of this is delayed until Chapter 6 (Section 6.6).

When leaching losses are experimentally minimised the soluble components still show the most rapid rate of fractional weight loss (Table 4.7). The absolute amounts of material mobilised may however be as great for the polysaccharide fractions as for the sugars as the second series of columns show. This imposes caution in assuming that the soluble components can markedly affect the average rate of decomposition unless they are present in very large amounts—see the third and sixth columns of Table 4.7. Thus whilst their presence undoubtedly partially accounts for the high initial catabolism rates of some resources (Table 4.4) the slope of the decomposition curve will in most cases be determined by the decomposition of the cell wall polysaccharides which account for seventy to ninety per cent of the dry weight in most primary resources. Clearly there are some exceptions in which the overall decomposition may be markedly affected by the presence of very large amounts of initial soluble materials. Thus Swift *et al.* (unpublished) found that senescent leaves of cowpea (*Vigna unguiculata*) contained up to 25% leachable material. Under conditions of heavy tropical

Table 4.7. The fraction of individual organic components lost during decomposition, and the contribution of each component to the total weight loss (from Swift 1976 after Tenney & Waksman 1929).

	Corn stalk[1]			Oak leaf[2]		
	Fraction lost	Actual weight lost (g)	Fraction of total loss	Fraction lost	Actual weight lost (g)	Fraction of total loss
Dry matter	0·36	74·0	1·00	0·22	50·0	1·00
Ether soluble	0·30	1·1	0·02	0·26	2·1	0·04
Cold water soluble	0·80	17·1	0·23	0·81	15·0	0·30
Hot water soluble	0·55	4·0	0·05	0·46	5·8	0·12
Cellulose	0·44	26·2	0·35	0·34	10·4	0·21
Hemicellulose	0·41	14·7	0·20	0·27	7·9	0·16
Lignin	0·00	0·0	0·00	0·00	0·0	0·00

[1] Loss after 27 days; original weight 203 g.
[2] Loss after 66 days; original weight 223 g.

rainfall all this material was leached out within a week. The rate of weight-loss of pre-leached leaves over a four-week period was only half that of the untreated leaves.

The cell wall polysaccharides form the bulk of the carbon and energy in primary resources. As described in Chapter 3 a wide range of decomposers are capable of attacking all or part of this component and it is unlikely that differences between resources are determined significantly by cellulose or hemicellulose decomposition. Lignin is however a far more important feature of resource quality.

Lignin and cellulose are intimately associated within the cell wall of plants and although there is probably no chemical interaction between the two, the physical proximity of the lignin may retard enzymatic attack on the cellulose. When experimentally-prepared samples with differing combinations of cellulose and lignin were decomposed by *Pseudomonas ephemerocyanea*, cellulose loss was inversely related to lignin concentration (Table 4.8). Bailey *et al.* (1968) found that delignification can markedly enhance the ability of fungi to decompose the polysaccharide components of wood cell walls. In the analysis of a number of chemical factors determining the rate of decomposition of temperate tree litters from the Eastern U.S.A. Cromack (1973) showed a close inverse relationship between the lignin content (x) and the exponential weight loss rate (Y) of five species of leaf litter. The regression equation derived was $y = 1·246 + 0·0267X$ with an r^2 value of 0·89. Whilst lignin was the best of the chemical regulators tested, significant fits were also obtained for C:N ratio ($r^2 = 0.74$) and sclorophyll index (the ratio of crude fibre to protein, $r^2 = 0.82$). This relationship of decay rate to lignin content has been seen in other types of litter but only at less significant levels. Heal, Latter & Howson (1978) obtained the value $y = 37.9 - 0.683x$ for thirteen primary resource-types from blanket bog which ranged from lignified *Calluna vulgaris* stems to relatively soft grass and herb leaves. The r^2 value in this case however, was only 0.43 indicating that although an important factor, the lignin concentration alone did not explain the differences in decomposition rate between the litters. Van Cleve (1974) confirmed the importance of lignin as a factor determining decomposition rate in primary resources at a number of Tundra sites.

We have already commented that a resource quality factor was included as a driving variable in the simulation model of microbial respiration in Tundra ecosystems (Chapter 2). Bunnell, Tait & Flanagan (1977b) have described the nature of this variable in detail. Resource quality is incorporated as a regulatory function of the rate of catabolism of C (i.e. microbial respiration) as follows:

$$R(T, M) = \sum_{i=1}^{n} r_i(T_1 M) y_i,$$

where $R(T, M)$ is the respiration rate of the total resource at temperature T and moisture content M, and $r_i(T, M)$ is the specific respiration rate of substrate i which is present in amount y_i. The equation is an extension of Minderman's hypothesis that the resource loss rate is a sum of the loss rates of the component substrates. The model was tested initially for a wide range of substrates—ethanol-soluble, cellulose, lignin, pectin, starch and volatiles— but difficulty was experienced in obtaining good fits except when only two substrate groups—ethanol soluble and ethanol insoluble—were used. Field data showed that ethanol soluble components were broken down five to six times faster than the remainder in a number of resources at Point Barrow, Alaska.

Table 4.8. The effect of lignin content on the ability of bacteria to decompose cellulose associated with it. Decomposition over a 21-day period of cellulose complexed with differing amounts of lignin prepared by extraction of lignin from jute by monoethanolamine treatment for varying times; the test organism was *Pseudomonas ephemerocyanea*. (Fuller & Norman 1943).

Cellulose content (%)	Lignin content (%)	% of cellulose decomposed
99.2	0.0	100.0
95.5	3.3	95.6
89.2	5.3	83.1
82.7	11.9	37.9
75.6	12.6	17.7

Table 4.9. Measured and simulated values of weight loss and chemical composition of *Eriophorum angustifolium*. Simulated values are based on respiration rates in relation to temperature, moisture and initial chemical composition (Bunnell *et al.* 1977b).

	Measured	Simulated
% weight loss year^{-1}		
ethanol-soluble	49	48
ethanol-insoluble	11	12
total weight	27	31
% chemical composition of resource after one year		
ethanol-soluble	7	10
ethanol-insoluble	93	90

The main expansion from Minderman's hypothesis is the incorporation of abiotic regulation of the substrate-specific rates. Field data for this were obtained by following the rate of decomposition of cellulose placed in a number of microhabitats differing in moisture and temperature relations. Significant variation in decay rate was found between canopy, litter and soil microhabitats. By combining both resource composition and micro-environmental variables in the model, good predictions of resource weight loss and change in chemical composition were obtained (Table 4.9).

The work of Minderman (1968) and Bunnell *et al.* (1977b) is probably the most detailed attempt to determine the importance of energy and C sources as resource quality factors. The latter model has good predictive value and is highly simplified in that only two substrate components are required to explain much of the variation in decay rate between litters and expressions of leaching and comminution are not considered. We may conclude from this example, and from others we have described, that the broad patterns of variation in decomposition are related to the proportions of readily metabolised materials (sugars, ethanol or water soluble components) and recalcitrant materials (lignin). At a finer level of resolution, e.g. the variation in rate of decomposition between leaves of different grass or tree species, more subtle effects of energy and C sources occur and our understanding is poor. This conclusion also applies to the effects of C and energy sources on decomposer organisms (see Section 4.5).

4.3.2. Nutrient sources

The essential features of the requirement of decomposers for macronutrient elements and the dynamic aspects of immobilisation and mineralisation have been described in Chapter 1 (Section 1.3.2). A major feature emphasised was the greater concentration of key nutrients such as N and P in decomposer tissues than in the primary resources on which they feed, and the consequent probability that nutrient limitation may be a fairly common occurrence in decomposition (Table 1.9). This obviously focusses attention on nutrient concentrations as possible sources of resource quality effects. Table 4.10 lists the macronutrient contents of a range of primary resources and decomposer organisms. Additional data have already been given in Tables 1.7 and 1.11.

Table 4.10. Nutrient element composition of primary resources and decomposers (% dry weight). All leaf material after litter fall; root and wood from living plants. Sources: (1) Daubenmire & Prusso (1963); (2) Heal, Latter & Howson (1978); (3) Allen *et al.* (1974); (4) Frankland *et al.* (1978); (5) Cromack *et al.* (1975); (6) Swift (1977b); (7) Stark (1973); (8) Luria (1960); (9) Ausmus & Witkamp (1973) quoted by (10) McBrayer *et al.* (1974).

Primary resource	Species	Source	N	P	K	Ca	Mg
Deciduous leaf	*Populus tremuloides*	(1)	0·56	0·15	0·60	2·35	
	Betula papyrifera	(1)	0·58	0·32	0·78	1·71	
Conifer leaf	*Abies casiocarpa*	(1)	0·69	0·09	0·30	1·18	
	Pinus contorta	(1)	0·51	0·04	0·15	0·55	
Sedge leaf	*Eriophorum vaginatum*	(2)	0·97	0·04	0·09	0·20	0·08
	Nardus stricta	(2)	0·53	0·03	0·10	0·08	0·08
Herb leaf	*Rubus chamaemorus*	(2)	1·31	0·07	0·09	0·85	0·53
Shrub shoot	*Calluna vulgaris*	(2)	1·38	0·07	0·09	0·34	0·06
Tree root large	*Quercus petraea*	(3)	0·5	0·06	0·2	0·4	0·08
Tree root small	*Q. petraea*	(3)	0·9	0·10	0·4	0·4	0·11
Sedge root	*E. vaginatum*	(2)	0·50	0·06	0·21	0·11	0·08
Outer bark	*Q. petraea*	(3)	0·5	0·17	0·08	0·5	0·03
Cambium	*Q. petraea*	(3)	0·9	0·08	0·4	1·3	0·15
Sapwood	*Q. petraea*	(3)	0·16	0·02	0·14	0·05	0·01
Inner heartwood	*Q. petraea*	(3)	0·10	0·01	0·06	0·06	0·01
Decomposers							
Fungus mycelium (on leaf)	*Mycena galopus*	(4)	3·60	0·24	0·57	—	—
Fungus fruit bodies (on leaf)	Mixed	(5)	—	0·68	2·90	0·07	0·07
Fungi (on leaf)	Mixed	(9)	2·80	0·24	0·12	3·30	0·19
Fungus mycelium (on wood)	*Stereum hirsutum*	(6)	1·34	0·09	0·41	0·79	0·10
Fungus fruit bodies (on wood)		(7)	1·87	0·33	0·88	0·07	0·12
Bacteria—culture	Range	(8)	8–15	2–6	1–2	1	1
Bacteria—leaves	Mixed	(9)	4·0	0·91	1·50	0·95	0·15
Oligochaeta	—	(3)	10·5	1·1	0·5	0·3	0·2
Diplopoda	—	(3)	5·8	1·9	0·5	14·0	0·2
Insecta	—	(3)	8·5	6·9	0·7	0·3	0·2
Detritivores		(10)	7·74	0·80	0·13	10·30	0·27
Fungivores		(10)	7·74	1·39	0·40	3·95	0·46

Before looking for trends a few general comments are in order. Comparability of data from different sources is again in question because of variations in analytical technique. These aspects have been reviewed in detail by Allen *et al.* (1974) who also give additional data sets. Secondly it should be remembered that elemental analysis is difficult to interpret in terms of the availability of the nutrients to the organisms. Within primary and secondary resources the nitrogen is largely in organic combination in proteins, amino acids, amino sugars, heterocyclic bases and nucleic acids and co-enzymes. In soil organic matter it is present additionally in chemically ill-defined protein-tannin complexes and humus fractions. Outside the soil solution very little nitrogen is in an inorganic state as nitrate, nitrite or ammonium. Similarly phosphorus is largely in organic form (phytin, phospholipids, phosphorylated sugars, etc.). Other elements such as Ca and K vary in their incorporation into organic fractions and may exist in ionic form in cytoplasm or vascular sap.

From the tabulated data a number of general trends can be detected. There is a generally high element concentration in herbaceous, graminaceous and deciduous tree leaf-litters, lower concentrations in coniferous and ericaceous leaf material and the lowest concentrations are found in woody tissues. However the variability between species within a taxonomic group is high and re-emphasises the danger of using taxonomic criteria for resource quality. Other features of variation may be equally important such as that due to the soil conditions in which the plant is grown and variation between the same organs growing on different parts of a plant.

An additional factor influencing nutrient concentration in the initial substrate is the time of death. The resorption of nutrients and soluble organic matter from leaves before leaf fall, has already been described (see Fig. 1.13). This may be complicated further by climatic conditions advancing or retarding abscission thereby modifying the concentration in the dead material at litter fall and in some cases even causing death before any resorption has occurred. Nutrient-rich litter may also enter the decomposer system through the activity of herbivores, particularly those such as lemmings which sever leaves at their base and consume only a small proportion. Thus whilst the usual input in temperate ecosystems occurs seasonally so that with 90% of the leaf litter falls at a time of minimum N and P but maximum Ca concentration, any intervention of the above factors can shift the position dramatically. Redistribution may also occur within or between plant tissues; for instance Merril & Cowling (1966) showed that N is removed from senescing xylem into active cambial cells during secondary growth. Localisation of nutrient may thus be a significant factor particularly in massive resources such as wood. In this case the nutrients are almost entirely associated with the living cells so that the inner bark area and the medullary rays have quite a different 'quality' to the rest of the wood. It is also necessary to mention again the considerable degree of adaptability of microbial tissues to the nutrient content in which they are growing.

The concept of the regulatory effect of C:nutrient ratios discussed in Section 1.3.2. is based on the assumption that nutrient concentrations particularly N, are commonly limiting to the activity of decomposer organisms. When nitrogen contents are low (C:N higher than about 25:1) the carbon is respired and lost from the organic matter while the nitrogen is mineralised and then converted to microbial protein. The N is thus immobilised until the death of the micro-organisms when it is recycled, thus the C:N ratio declines with time as carbon is released (Fig. 1.14). When N concentrations are high (C:N lower than about 25:1), N is not limiting microbial growth, other nutrients or environmental factors increase in importance, and N increases in availability to plants. Thus the initial N concentration (and availability) is expected to be related to the subsequent decomposition rate. Data from experimental studies tend to support this view—supplementary N added to natural materials stimulates the rate at which they decompose (Findlay 1934; Tenney & Waksman 1929; Allison & Cover 1960). In some cases nutrients other than N may be more limiting; reference to Table 1.9 shows for instance that P has a greater concentration gap between wood and wood-decaying fungus than does N. Swift (1978a) showed however that despite this N was immobilised more extensively than P. In another experiment K was indicated as the most limiting element (Swift 1973). The responses of decomposer micro-organisms to N concentrations are not always straightforward however. Park (1976) investigated the effect of varying N concentration on cellulose decomposition by forty-five different species of fungi. Whilst there was a clear relationship between N availability and the rate of cellulolysis for the majority, many of the fungi showed maximum responses at intermediate levels or were not affected by the N concentration.

In field experiments it is even more difficult to demonstrate the causal basis of nutrient concentrations as resource quality factors in decomposition processes. This is partly a practical matter—the difficulty of disentangling

Table 4.11. Regressions of weight loss against initial nitrogen concentration for a variety of litters. (1) after Voigt (1965); (2) after Heal, Latter & Howson (1978); (3) after Daubenmire & Prusso (1963).

Y	a	b	r	n	% N range	Comments
(1) Loss in 3 months	11·65	14·10	0·3691	6	1·40–2·18	Hardwood and conifer leaves in lab. at 32°C.
(2) Loss in 2 years	12·56	20·32	0·4266	14	0·44–1·38	Wood, roots and leaves in field on blanket bog
(3) Loss in 100 days	0·94	17·66	0·4782	13	0·33–0·69	Hardwood and conifer leaves in lab. at 10°C.

Table 4.12. Change in the amount of N in litter which is mineralisable by the action of pronase (after Van der Linden 1971).

Time of decomposition	% N	Release of Kjeldahl N (%)	Release of amino N and NH_3 N(%)
1 month	1·4	82	—
6 months	1·9	78	76
1 year	2·5	69	—
2 years	2·4	57	54

individual factors from multiple-factorial effects. In an investigation of the factors affecting the rate of decomposition of deciduous and conifer leaves, Daubenmire & Prusso (1963) found that weight loss correlated best with potassium concentration. Potassium was also significantly correlated with phosphorus ($r = 0·832$), lignin ($r = 0·724$), protein ($r = 0·576$) and pH ($r = 0·818$). Thus potassium may be reflecting the combination of a number of resource variables and, in itself, may be functionally of little importance. In a few cases the effect of a single nutrient such as N can be more clearly demonstrated. The regression equations of Table 4.11 show that in all cases the correlation of weight loss to initial nitrogen is positive but low. The regressions are not strictly comparable because of the differing time scales and conditions of the experiments. It is interesting to note however that Broadfoot & Pierre (1939) found a significant effect of initial N content in the early (0–2) months but not in the later (2–4 months) stages of decomposition of nineteen types of leaf litter. The absence of effect in the later stages may be due to a diminishing of difference between the species in N accessibility because of the formation of recalcitrant protein complexes. Thus Van der Linden (1971) showed that over a two-year period of decomposition of hazel leaves, whilst the total immobilised N increased there was a decrease in the amount that could be mineralised by a proteolytic enzyme from *Streptomyces griseus* (Table 4.12). This is a good illustration of the change of resource quality with time of decomposition. The increasing recalcitrance of the N fraction is probably associated with the 'tanning' of proteins by phenols released during decomposition. Handley (1954) has suggested that this process is important in humification (see Chapter 5).

The experiment of Broadfoot & Pierre (1939) mentioned above showed a further resource quality effect which may be related to the availability of nutrient elements. They showed a significant correlation between the extent of decomposition over the six-month period and the 'excess-base' content of the leaves ($r = 0·672$, $n = 19$). The 'excess-base' is the excess of cationic elements (Ca, Mg, K and Na) over the anionic (Cl, P and S). The measure is to some extent, but not strongly, correlated with pH (see Chapter 6) but is clearly a strong aspect of difference between the litters (Fig. 4.9).

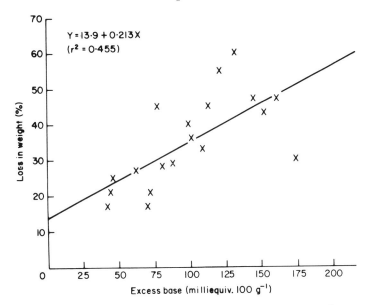

FIG. 4.9. Relationship between initial excess-base content of plant litter and the extent of decomposition after six months. Data for five conifer and fourteen deciduous tree leaf species from a laboratory experiment of Broadfoot & Pierre (1939). The correlation between the two parameters is significant ($p = 0.01$) but the linear fit is poor.

We have so far concentrated our attention very largely on the relationship of nutrient content to the immobilisation and release of nutrients by microorganisms. The role of animals on the processes of immobilisation and mineralisation may be very significant with nutrient releases being highly accelerated by their activity (see Figs. 3.14 and 3.15 and discussion in Section 3.2.8.). Nutrient concentrations may also be important selective factors for decomposer animals. One of the best documented studies of the effects of resource quality on natural populations is the work of Hughes & Walker (1970) on the Australian bushfly (*Musca vetutissima*). Freshly dropped cow faeces constitute the major food source and oviposition sites for the females as well as food for the developing larvae. Consistency, moisture content, fibre content and N were all significant features of resource quality and any one of them was suitable as a general index of quality. In South East Australia fly populations showed fluctuations which corresponded to changes in N content, with a small time lapse between cause and effect. When this situation was analysed in detail, it was found that food quality affected larval and pupal survival, and the sex ratio, fecundity and fertility of the emerging adults (Fig. 4.10). Where only low-quality dung was available as food, populations were reduced to extremely low numbers. The behaviour of the adults was

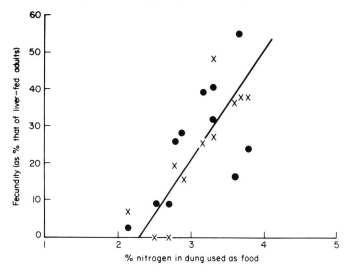

FIG. 4.10. The relationship between fecundity and the nitrogen content of dung used as adult food by the Australian Bushfly. The cultures were set up with (●) 400 or (×) 800 larvae per litre of cow dung. The emergent flies feed on the dung before oviposition and were compared with the fecundity of flies fed on an optimum quality diet of liver. The generally linear relationship of egg numbers to dung quality is apparent, though the high density series (shown by the regression line) indicates that there is some reduction in fecundity of flies reared at high density (Hughes & Walker 1970).

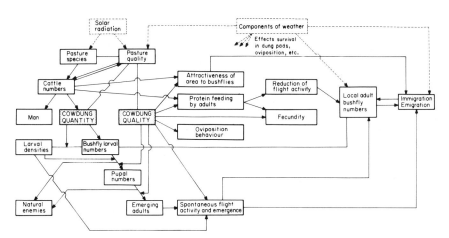

FIG. 4.11. Synopsis of the influence that food quality and quantity, acting through population density, have on the life system of the Australian Bushfly. Dung pad deposition rates are more or less constant throughout the year but resource quality shows marked seasonal variation. Food quality not only affects the survival and reproduction of the adults but also their flight activity: the flies remaining longer in areas with high-quality food and dispersing from areas where the food quality is low (Hughes & Walker 1970).

also affected by food quality; flight activity was lower in the presence of high-quality dung so that there was active dispersion from areas of poor-quality dung. Fig. 4.11 illustrates the key role of food quality and quantity in the life system of the bushfly.

This is a rare example of the detailed analysis of the relationship between resource quality and population dynamics. Unfortunately no such comprehensive studies have been carried out on soil animal populations under natural conditions. In a laboratory study Booth & Anderson (1979) were able to demonstrate the effects of variation in the quality of secondary resources on microtroph populations. The Basidiomycete fungus *Coriolus versicolor*, was cultured in liquid media containing N (as asparagine) at 2, 20, 200 and 2000 ppm and then used as the sole food supply for the collembola *Folsomia candida*. The experiment relies on the known adjustment of fungal protein in this species to the N content of the medium. The collembola showed a direct response in egg production and moulting frequency to the nitrogen content of the culture medium up to 2000 ppm nitrogen (Table 4.13). At 2000 ppm some property of the fungus, possibly metabolites, or amino acid balance appeared to be strongly inhibitory to both responses.

Table 4.13. Egg production and moulting frequency of *Folsomia candida* on a diet of *Coriolus versicolor* grown at varying N levels (Booth & Anderson 1979).

	No Food	Levels of nitrogen			
		2 ppm	20 ppm	200 ppm	2000 ppm
Egg laying rate eggs ind^{-1} wk^{-1}	1·6	9·9	14·3	28·4	6·5
Moulting rate exuviae ind^{-1} wk^{-1}	0·41	0·61	0·69	0·83	0·79

A remarkable phenomenon which has been demonstrated by Cromack *et al.* (1978) is the massive accumulation of Ca as calcium oxalate crystals on the surface of some fungal hyphae. Certain arthropod groups such as Diplopoda, Isopoda and some Cryptostigmata have a high calcium content in their exoskeleton (Table 4.10) and the Ca immobilized on fungal hyphae in acid soils could be an important dietary component. The extent to which Ca chelated by oxalic acid is available to soil animals has not been determined but the effect of this form of immobilisation might equally be to render the hyphae unpalatable and reduce Ca availability.

All the preceding discussion has been concerned with macronutrients which, particularly in the cases of N and P, can clearly be regarded as significant features of resource quality, although much more work of a comprehensive and discriminating nature is needed to establish the priority and mechanisms of effect.

Many other elements are important in the nutrition of decomposer organisms—Mo, Cu, Zn, Fe, Mn and Bo are classified as essential micro-nutrients and other non-essential elements such as Na and Al may nonetheless be of nutritional significance. These elements must therefore be considered as possible features of resource quality but little information is available concerning their significance. The data in Table 4.14 shows that in most cases they occur in higher concentrations in decomposer tissues than in the resources they decompose. The reported optimal concentrations required for active growth and sporulation in fungi are however in general very much lower than the concentrations for both plant and decomposer given in the Table (Cochrane 1958). It thus seems unlikely that these elements act as significant regulators of primary decomposition processes except in circumstances where the primary production is itself limited by their availability. A review of the significance of micro-nutrients in forest ecosystems has been given by Fortescue & Marten (1967).

Table 4.14. Micronutrient element composition of decomposers compared with the resources on which they are growing ($\mu g\ g^{-1}$). (1) Cromack *et al.* (1975); (2) McBrayer *et al.* (1974); (3) Stark (1973).

	Na	Cu	Zn	Mn	Fe
Leaf litter (1)	3·6	17·3	67	858	17·3
Basidiocarps (1)	824	44	108	157	44
Insecta (2)	0·3	50	150	30	200
Isopoda (2)	0·3	50	130	40	—
Pine needles (3)	265	4·5	19·3	88	38
Rhizomorphs (3)	219–2400	244–34·2	39·2–213·0	88–193	669–6815

A relationship between decomposer activity and micro-nutrient requirement has been shown for a limited number of invertebrate animals. For instance, the isopods (and other terrestrial crustacea) are unusual in their requirement for large quantities of copper (Wieser 1968). This is stored in the hepatopancreas in excess of the quantity needed to maintain the level of the respiratory pigment haemocyanin. The reason for this luxury uptake is not known but it may buffer the isopods against variation in the availability of copper from their food. Copper is present in rather inert compounds in fresh litter and isopods show abnormally high feeding rates on dead plant remains if they are prevented from eating their own faeces. Microbial activity in the faeces results in the mobilisation of the copper and the feeding rates decline as equilibrium levels in the hepatopancreas are re-established (Wieser & Wiest 1968).

There is also circumstantial evidence from studies of herbivores that suggests that elements such as sodium may form an important component of the diet for detritivore or microbivore animals. Arms *et al.* (1974) for example showed that the male swallowtail butterfly, *Papilo glaucus*, has a precarious sodium balance and that the aggregations of these and other butterflies seen drinking at pool margins are absorbing sodium ions from soil water. Sodium stress has also been observed in several vertebrate herbivores and may be a more widespread phenomenon than is generally recognised. Arms *et al.* suggest that the exclusion of high sodium levels from most terrestrial plants may serve to reduce grazing pressures by herbivores by forcing them to secure their requirements of sodium from sources outside their normal diet. If this hypothesis is correct then litter-feeding animals are likely to be affected to an even greater extent. Sodium is an extremely mobile ion and is rapidly leached from decaying leaves. Fungi and bacteria generally have high sodium contents and may represent an important pool of immobilised sodium in decomposer systems as far as decomposer animals are concerned.

4.3.3. Modifiers

Modifiers are chemical components of resources which because of their molecular structure influence the activity or behaviour of decomposer organisms by means other than as sources of C energy or nutrient. These chemicals may be distributed differentially between resources and thus account for some part of observed differences in decay rate. Most prominent among the compounds that have been studied are the so-called 'plant-protection' chemicals mostly of an aromatic, particularly polyphenolic nature. These afford plants some degree of protection against excessive grazing by herbivores and/or attack by plant pathogens. Other types of chemical must also be mentioned; new modifiers are produced within the decomposer subsystem, particularly the antibiotics formed by the microflora. Although these may be decomposed quite rapidly in the soil they can inhibit the activity of sensitive micro-organisms and therefore the rate at which other organic fractions are attacked. Many compounds are used by man to perform analogous functions as modifiers, e.g. various pesticides and fungicides. Others are accidental pollutants such as the heavy metals from mining activities. Any of these may enter the decomposition subsystem and exert modifying influences on soil organisms.

Not all modifications of activity are inhibitory however. Specific growth factors required by decomposer organisms such as amino acids or vitamins may be present in the primary resources or released through root exudations. Alternatively they may be synthesised by micro-organisms and become available to secondary decomposers through secretion or at death. These growth factors, whether of primary or secondary origin, can influence the

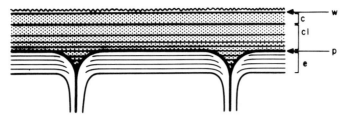

FIG. 4.12. A hypothetical plant cuticle: w, the epicuticular wax; c, the cuticularised layer; cl, the cutinised layer of the cell wall; p, the 'pectin' layer, and e, the epidermal cell wall (after Martin & Juniper 1970).

distribution of organisms and the rate at which they catabolise organic matter.

The literature on plant-protection compounds in relation to their interactions with herbivores and plant pathogens is vast (Whittaker & Feeny 1971, Harborne 1977), but there is very little which is directly relevant to decomposition processes or decomposer organisms. Some of the features of plant resistance to fungal pathogenesis are clearly of relevance, because as discussed in Chapter 3, decomposition may be initiated by necrotrophic organisms. Chemical inhibition of penetration may be exerted either as an *active* response by living cells to the presence of the parasite or may be *passive* due to the presence of pre-formed chemicals of a fungitoxic or fungistatic nature (see Strange 1972 for a review). Clearly resistance of the passive kind which may persist in senescent tissues are those that are most likely to be of significance to decomposition processes.

The first protective mechanism is encountered at the outer surface of the plant in the epidermis, or bark. The outer surfaces of plants are covered by membranes of varying thicknesses that confer both impermeability and a relative degree of resistance to decay on the plants (Fig. 4.12). The outermost layer of leaves and shoots is composed of waxes and within these are layers impregnated with cutin; suberin is an analogous substance found in the walls of bark cells (Chapter 5). These layers act as chemical inhibitors in two ways. Firstly due to the presence within them of components which have a direct fungistatic effect; for instance Martin (1964) has reported the presence of an ether-soluble acidic fraction of apple cuticle wax which is fungistatic to a number of fungi. Secondly the cuticle presents a chemical barrier because of the resistance of cutin to decomposition. Cutin and suberin are probably among the most recalcitrant of plant materials: Swift (1965), for instance, showed that the rate of suberin breakdown in root bark was much lower than that of cellulose. Little or nothing is known of the extent to which this primary barrier to parasitic penetration is maintained in the face of colonisation by decomposer fungi but this is clearly an area warranting some investigation. The thickness and chemical composition of cuticles and suberised layers is known to differ markedly between plant species and may thus well provide a basis for resource quality differences. Similar protection may be afforded by components of the outer surfaces of secondary resources. Pugh (1971) has shown that fats and oils on bird feathers inhibit colonisation and growth of several species of keratinophilic fungi, though other species such as *Arthroderma curreyi* were stimulated by the presence of feather fats. Melanisation of the cell walls of hyphae or spores of fungi renders them less susceptible to the lytic action of other microbes in the soil (Bloomfield & Alexander 1967).

Polyphenolic compounds have provided some classic examples of pre-formed resistance to necrotrophic parasites. Walker & Stahmann (1955) showed that protocatechuic acid in the dead scale leaves of onions was a factor conferring resistance against *Colletotrichum circinans*. Cole & Wood (1961) showed that a number of polyphenols, including chlorogenic acid, inactivated the polygalacturonase enzymes of a common brown-rot invader of apples. It is probably in relation to wood decay that polyphenols and other pre-formed compounds are most significant with regard to inhibiting decomposer, as opposed to parasitic micro-organisms. The resistance of wood to decomposition, compared with leaf litters, is partly a result of the low concentration of soluble carbohydrates, of nutrients and of the high lignin concentration. However the principal source of decay resistance has been attributed to toxic extractable substances particularly from the heart-wood (Scheffer & Cowling 1966). The phenolic substances are synthesised in senescent parenchyma cells from carbohydrate precursors originating in more superficial living cells, diffuse into the heartwood and are deposited within the cell walls. Analysis of extracts and tests for toxicity and decay resistance have shown that a wide range of tropolones, stilbenes, flavenoids and terpenoids all contribute to decay resistance. The wide range of toxicity and low concentrations (usually less than 0·1%) of these compounds indicates that decay resistance may derive from mixtures of several compounds. Addition of phenols to timber to reduce biodeterioration is a widespread practice, but impregnation with oil-borne preservatives distributes the toxin within the cell lumen. In this position it is more susceptible to leaching and does not protect the cell wall to the same extent as the naturally occurring toxins which are incorporated in the cell walls.

Harrison (1971) has shown that leaf-litter fungi have varying sensitivity to polyphenols extracted from leaf litter. Fourteen of nineteen species of fungi tested, including ligninolytic Basidiomycetes, showed some inhibition of their growth rate. Benoit & Starkey (1968) investigated effects of wattle tannin on the decomposition of different plant compounds. They showed that tannin had no effect on the decomposition of sodium glucuronate, moderate inhibition of attack on pectic acids and a marked inhibition of hemi-cellulose and cellulose decomposition. The strongly inhibitory effects on the high molecular weight compounds were attributed to the formation of resistant complexes between the tannin and the polysaccharides; the microbial enzymes also appeared to be directly inhibited. Other evidence of the biostatic properties of leaf-litter extracts, though phenols were not positively identified as the active agents, is provided by Bouquel *et al.* (1970). It was found that *Fagus* leaf extracts inhibited the growth of *Azotobacter* and hence N-fixation in soils under beech trees. Kapustka & Rice (1976) investigated the same phenomenon and demonstrated that phenolic acids (ferulic, synapic, vanillic, gallic, elagic, coumaric and chlorogenic acids) present in dicotyledonous species of an Old Field plant community were strongly inhibitory to N-fixing *Azotobacter* and *Clostridium*. These authors developed the interesting hypothesis that the inhibition of N-fixation by these plant compounds may delay the Old Field successionary phase by preventing the accumulation of the N 'capital' necessary to establish the climax plant community.

Polyphenols have also been strongly implicated as causes of the difference in palatability between different resource types to decomposer animals. Heath & King (1964) and King & Heath (1967) demonstrated an inverse correlation between total polyphenol content of deciduous leaf litters and the feeding activities of soil animals both interspecifically and intraspecifically with respect to time. In temperate regions litter falling in the autumn contains high polyphenol concentrations and is not attacked by soil animals. Leaching and microbial decomposition of polyphenols occurs during the autumn and winter months and soil animals usually commence feeding on the litter in spring. It has also been observed however that there is often a delay between reduction of polyphenols below the inhibitory levels and the actual onset of feeding activities. Anderson (1973a) therefore suggested that low winter temperature may be more important in limiting soil animal and microbial activity rather than the resource condition. In order to test this hypothesis *Castanea* leaves were collected at autumn leaf fall, air dried and stored at 2°C until the following June when they were placed in the field. Under the high temperature but low rainfall conditions of early summer, leaf polyphenol content dropped much more rapidly than in the autumn presumably due to enhanced microbial activity but nevertheless feeding by litter animals commenced within a week of placement of the leaves.

None of these experiments conclusively demonstrate polyphenols to be responsible for the timing of attack on litter by soil animals as microbial colonization and numerous biochemical changes will have occurred during the same period of time. This was demonstrated by Satchell & Lowe (1967) who found that while there was a strong correlation between total polyphenols and litter palatability to earthworms for unweathered leaves, many of the unpalatable leaf species were still not attacked when they had been weathered in the field and contained no extractable polyphenols (Table 4.15).

Table 4.15. Palatability of tree litters to earthworms in relation to their tannin content (Satchell & Lowe 1967).

Litter	Unweathered				Weathered			
	Leaf discs eaten as % of controls	Tannins % dry wt			Leaf discs eaten as % of controls	Tannins % dry wt		
		Condensed	Hydrolysable	Total		Condensed	Hydrolysable	Total
Elm	362	0	0	0	358	0	0	0
Alder	324	0	0	0	336	0	0	0
Sycamore	298	0	0	0	386	0	0	0
Birch	281	0·22	0·11	0·33	342	0	0	0
Spindle	275	0	0·10	0·10	338	0	0	0
Ash	260	0	1·22	1·22	279	0	0	0
Lime	247	0·40	0·05	0·45	331	0	0	0
Hazel	169	0·20	0·32	0·52	182	0	0	0
Gean	143	0·63	0	0·63	234	0·02	0	0·02
Beech	84	1·96	0·40	2·36	184	0·10	0	0·10
Pine	44	1·81	1·38	3·19	74	0	0	0
Larch	38	1·27	3·95	5·22	84	0	0	0
Oak	21	0·88	0·45	1·33	173	0·05	0	0·05

Table 4.16. Palatability to earthworms of paper discs treated with phenolic compounds (Satchell & Lowe 1967).

Polyphenol	Concentration of solution	Treated discs taken as % of control discs taken
Quercetin	suspension	107
Ellagic acid	suspension	98
Catechol	15%	84
Gallic acid	suspension	50
D Catechin	suspension	27
Phloroglucinol	10%	26
Protocatechuic acid	suspension	14
Tannic acid	5%	0
Tannic acid on birch discs	5%	0

Conversely Anderson (1973b) found that soil animals preferred to feed on *Castanea* leaf litter than *Fagus* litter even though the *Castanea* leaves contained twice the polyphenol concentration of the *Fagus* leaves. These paradoxes were resolved in part by Heath & King (1964) demonstrating that the quantities of particular plant phenols, such as gallic and protocatechuic acids, were more important in determining litter palatability than total polyphenols. Satchell & Lowe tested the selection by earthworms of blotting paper discs with eight phenolic compounds and confirmed significant differences in the behaviour of worms in relation to these compounds (Table 4.15). This experiment was not however related to the palatability of different leaf species. The essential correlation of quantitative analyses of specific phenolic acids with leaf palatability have yet to be carried out on a wider range of decomposer situations for any general conclusions to be reached. In Anderson's (1973b) study, although beech leaves were found to contain higher concentrations of gallic and protocatechuic acids than chestnut leaves, the results were not sufficiently clear cut to ascribe the differences in animal attack on these two litter types to the presence of these compounds.

In addition to their specific modifying effect on the decomposer community, polyphenols have been implicated in a differential effect of resource quality on humus formation. Handley (1954) was among the first soil ecologists to draw attention to the importance of plant phenols in decomposition processes. He showed that the accumulation of 'raw humus' beneath different tree species was reflected by the amount of water soluble tannins and polyphenols which could be extracted from the leaf litter and by their ability to precipitate proteins. Central to Handley's hypothesis of the 'mor forming' nature of different tree species, was the idea that the masking of cellulose by phenol tanned proteins was largely responsible for inhibition of leaf-litter decomposition and the development of highly organic humus forms. There was a tendency for species which are known to form mor humus to have greater precipitation ability than those giving rise to mull, but he also showed that the precipitates from mor forming species were the more resistant to microbial decomposition. The evidence suggested that basic amino groups of diamino acids were actively involved in the protein precipitation and were analogous to the action of vegetable tannins used in the leather industry. Extracts from leaf litter showed lower activity than did fresh litter and, with microscopic evidence from sections of *Calluna vulgaris*, Handley suggested that precipitation of cytoplasmic protein in the mesophyll occurs as the leaf dies, and that the precipitate masks the cellulose walls of the mesophyll from subsequent microbial attack. The evidence of Van der Linden (1971) on the accessibility of N to pronase (Table 4.2) conflicts with this and suggests that the tanning process may extend over a longer period.

Polyphenolic compounds of plant origin are thus highly implicated as modifiers of the rate of decomposition and there is plenty of evidence to suggest that differences in polyphenol spectrum and concentrations may explain many observed resource quality effects. The same practical problems remain of differentiating specific effects in multiple-factorial situations where rapid change with time is occurring and much critical work remains to be done. Nevertheless some generalisations can be drawn. (1) Polyphenols are produced in greatest diversity and largest amounts by plant species on acid, nutrient-poor soils. (2) The primary mode of action is the formation of resistant complexes with proteins. (3) The complexes produced under acid nutrient-poor conditions are more resistant than those produced where pH and nutrients are high. (4) They also act by direct inhibition of fungal and faunal activity.

The abundant evidence for the role of polyphenols is not matched for other chemicals. The significance of antibiotics to decomposition processes is unclear. This is not strictly a resource quality phenomenon as these substances are more usually associated with competition between active microorganisms than as components of resources. Resource quality factors of the types discussed above may result in the establishment of a specific decomposer microflora within the resource. If this microflora contains members with specific antibiotic capacities then the resource will gain secondary selective capacities. Despite the very wide range of soil organisms, particularly actinomycetes, that produce antibiotics in pure culture, considerable doubt as to their significance in natural habitats still exists. These doubts centre largely round difficulties in detecting significant antibiotic activity in soils. The absence of positive proof of activity in natural habitats renders discussion of roles in decomposition academic but it should be emphasised that it does not eliminate the existence of such phenomena. The very low concentrations required for activity, the instability of many of the compounds and their

Table 4.17. Some effects of pesticides on soil populations and processes. Toxic concentrations are given in ppm. Maximum non-toxic concentrations tested are shown in brackets. (1) I = Insecticide; H = Herbicide; F = Fungicide; N = Nematocide (after Hellings *et al.* 1971, Edwards 1970).

Pesticide	Type[1]	Populations			Respiration	Activities	
		Bacteria	Fungi	Earthworms		Nitrification	Nodulation
Chlordane	I	(10^2)	(10^2)	5	(10^3)	(50)	(8)
DDT	I	(10^2)	(10^2)	(200)	(13)	(50)	(10)
BHC	I	(10^3)	(10^3)	—	(10^3)	(20)	13
Simazine	H	(70)	(70)	—	(64)	(64)	—
TCA	H	(10^2)	(10^2)	—	(60)	(60)	—
Allyl alcohol	HF	9	9	—	—	25	—
EDB	IFN	—	—	—	8	33	—
PCP	IHF	2×10^3	5×10^2	—	—	5	—
Nabam	F	50	50	—	10^2	10^2	—
DD	N	2×10^3	2×10^3	—	$3 \cdot 5 \times 10^3$	50×10^3	—

tendency to absorb to soil colloids all present formidable practical problems. It is interesting to note that antibiotic activity has been detected in soils supplemented with carbohydrate, protein or wheat straw (Jeffreys *et al.* 1953, Wright 1956). The interaction with nutrients may be significant for it is clear that the general 'mycostasis' is partially at least a product of the low nutrient conditions prevailing in mineral soils and thus uncharacteristic of primary decomposition habitats (Smith 1976); perhaps specific antibiosis may be of greater significance in the latter, where nutrient limitation is absent.

The activity of man in intensive agriculture using monoculture crops has eliminated a great deal of the natural diversity in unpalatable chemicals. These natural modifiers have however been replaced by a wide range of synthetic compounds which are intended to discourage the activities of pests, parasites, or weeds. It can be readily demonstrated (see reviews by Hellings *et al.* 1971; Edwards 1970) that many of these pesticides have effects on organisms or processes of the decomposition subsystem. Some of these effects are listed in Table 4.17. Clearly the resource quality of primary resources contaminated with many of these chemicals is markedly affected. The significant question is what effect this may have on soil fertility—on the rate and pattern of nutrient mineralisation and humus formation in the soil. Data of good predictive value over the long term are currently not available; in Chapter 7 we will take up this question again with particular reference to vulnerable tropical ecosystems.

Good examples of the stimulatory effects of modifiers on decomposer organisms are few and far between. Many saprotrophic organisms are known to have specific growth factor requirements however. Among the fungi it has been noted that a great many of the Basidiomycetes, including a large number of wood decay fungi, require the addition of thiamine (vitamin B_1) to a culture medium for successful growth (i.e. they are *auxotrophic* for thiamine). Osborne & Thrower (1964) showed that extracts from Australian timbers stimulated wood decay fungi that were thiamine-auxotrophic but did not stimulate thiamine prototrophs. Although they could not produce any direct evidence for a thiamine effect they demonstrated circumstantially that the rate of decay of a range of timbers was related to the balance between stimulatory modifiers (such as thiamine) and inhibitory modifiers (such as polyphenols).

Vitamin or other growth-factor requirements are extremely common among soil bacteria but markedly less so on the rhizosphere and root surface communities. (Rouatt 1967, Table 4.18). The significance of this difference, and of the widespread occurrence of auxotrophy, is difficult to establish. Soil, litter (Schmidt & Starkey 1951) and particularly the root region are all regarded as areas in which a plentiful supply of most of the vitamins listed in Table 4.18 can be found. It is thus difficult to suppose that there is any

Table 4.18. Incidence of auxotrophy in bacteria isolated from different microhabitats in soil (from Rouatt 1967).

Growth factor	Bacteria requiring growth factor (% of isolates)		
	Soil	Rhizosphere	Root surface
Thiamine	44·9	15·2	17
Biotin	18·7	6·1	7
Pantothenic acid	3·7	3·0	3
Folic acid	1·8	3·0	4
Nicotinic acid	5·6	6·1	5
Riboflavin	1·8	2·0	4
Pyridoxine	1·8	1·0	5
Vitamin B_{12}	19·6	2·0	1
Terregens factor	1·8	<1	1
p-Aminobenzoic acid	<0·9	<1	<1
Choline	<0·9	<1	<1
Inositol	<0·9	<1	<1

important regulatory effect imposed by the resource on the decomposer. Most authors have assumed that the main significance of such auxotrophy lies in the establishment of symbioses between auxotrophic and prototrophic micro-organisms (e.g. Rouatt 1967).

4.4   Physical attributes of resource quality and their influence on decomposition

As the chemical composition of a resource influences the rate at which it decomposes, so does its physical character. As the chemical composition changes, so the physical character changes. In this section we shall examine the nature and influences of this physical character. We shall exclude from present consideration three aspects which could quite correctly be considered resource quality factors—the moisture holding capacity, the gaseous atmosphere and the pH. The prevailing state of these factors is determined by the resource—its chemical composition and physical structure—but is also subject to considerable moderation by the physico-chemical environment external to the resource. Furthermore, the principles governing the expression of these resource factors in terms of decomposer activity are the same as those for the physical environment. To avoid unnecessary fragmentation and duplication of the material we have therefore delayed discussion of these factors to Chapter 6.

The physical features of resources that we will discuss here are threefold; surface properties, toughness, and particle size. These are all fairly vague terms, less easy to define than moisture or oxygen content or pH or than chemical composition. Each of them collectively describes the combined effect of a variety of chemical or physical features, which must therefore be accounted for individually.

4.4.1 Surface properties

The waxy surface of living leaves is hydrophobic and thus restricts the development of water films. The germination of fungal spores, growth of mycelium and activity of exoenzymes may be dependent on these conditions being established. Another physical feature of the wax layer in conifer needles which probably decreases the extent of fungal attack is the formation of mats of tubules which extend over the stomata blocking them as a possible path of entry. Erosion of the waxy layer is known to occur during the life of the needle on the tree (Millar 1974) and this property is thus likely to be less significant to decomposer organisms.

Other features of the structure of the cuticle may affect the feeding behaviour of animals. There are no data for saprotrophs but some general conclusions can be drawn from observations on herbivorous insects. Bernays & Chapman (1970) showed that the selection of grasses by the grasshopper *Chorthippus parallelus* was strongly influenced by leaf thickness and pubescence. The acceptability of *Festuca* to the early grasshopper instars was reduced by the thickness of the rolled leaf blades but the adults were little affected by this factor. The immature animals also had difficulty in attacking *Holcus* because of the dense covering of trichomes on the leaf surface and even the adults were prevented from feeding on some highly pubescent for b species. Levens (1973) has shown that leaf trichomes have a wide range of defence properties against herbivores and can influence insect populations by inhibiting oviposition, reducing growth and survival of young Homoptera by restricting feeding sites and can even act as a direct source of mortality when insects become impaled on barbed trichomes. Similar features may affect soil animals after the leaves fall to the ground.

4.4.2 Toughness

Although it is easy to accept the idea that the 'toughness' or 'hardness' of a food resource is a characteristic which will affect its rate of utilisation, it is an awkward feature to establish because of the difficulty in defining the term and separating the physical effect from the chemical structures which determine it.

It is generally accepted that the penetration of plant cuticles by fungal hyphae is largely a mechanical process with little enzymatic softening of the plant surface involved. Infection pegs from germination tubes can penetrate inert substances such as gold foil or paraffin wax. Experimental determinations using penetrometers have shown a relationship between the toughness of substrate and the ability of the fungus to penetrate. This has been extended to pathogenic behaviour on leaf surfaces where cuticle thickness and toughness as measured by a penetrometer, and resistance to fungal penetration have been shown to be correlated (Dickinson 1960).

Toughness has been given considerable attention as a possible component of resource quality with regard to animal feeding activities. Both Dunger (1958) and Satchell & Lowe (1967) concluded, from feeding experiments with millipedes and earthworms respectively, that the mechanical texture of leaves was not an important determinant of its palatability. In both studies the intraspecific variation in toughness, measured mechanically with a penetrometer by Satchell and Lowe, was such that it obscured any relationship between leaf species. Satchell & Lowe (1967) also measured cuticle thickness and showed, for example, that holly leaves with a cuticle thickness of 9 μm were selected by earthworms in preference to beech with a cuticle 3 μm thick. Leaf selection experiments with *Lumbricus terrestris* are difficult to interpret, as leaves drawn into the burrow are often not immediately consumed. The selection of materials by this species is therefore not necessarily a good indication of their palatability. Heath & Arnold (1966) have shown that the thin soft leaves from shaded parts of trees, beech and oak in their study, are more readily attacked by animals and decompose more rapidly than the tough, heavily cuticularised leaves from the outer regions of the tree which are exposed to sunlight, wind and rain. King & Heath (1967) analysed the soft, medium, and hard leaves and showed that the contents of ash, total nitrogen, polyphenols, sugars, lignin, and cellulose all varied between leaf types. It is therefore meaningless to simply attribute differences in the palatability of leaves to 'toughness' unless the effects of other correlated variables can be eliminated.

4.4.3 Particle size

The relationship between surface area and volume of resource particles influences the pattern of their decomposition in a number of fundamental ways. As already discussed in Sections 3.2.3 and 3.2.8 it determines the patterns of colonisation by micro-organisms, decreasing particle size (increasing surface to volume ratio) selecting for the surface growing unicellular as against the penetrative mycelial forms. Particle size determines the ability of animals to ingest food and may also be an important feature affecting the accessibility of substrates to enzymes. Features of the physical environment such as moisture retention and gaseous diffusion are also markedly affected. It is surprising therefore how little attention has been paid to this aspect in terrestrial habitats although much more attention has been given in aquatic

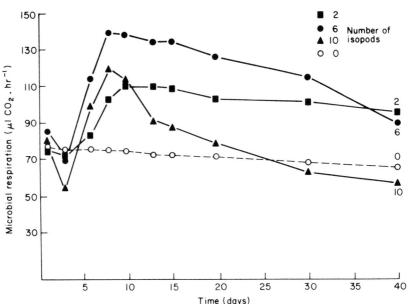

F<small>IG</small>. 4.13. Effect of soil animal activities on microbial catabolism of leaf-litter. Isopods were added to microcosms in which a mixed microbial inoculum, from isopod faeces, had been established for several days. After an initial stimulation common to all treatments, the highest isopod densities caused a suppression of microbial activity below the level of controls.

systems where particulate organic matter is so much more obviously significant (e.g. see Hargrave 1976; Berrie 1976).

van der Drift & Witkamp (1959) showed that microbial respiration in the faeces of the terrestrial caddis *Eniocyla pusilla* (Trichoptera) feeding on oak leaves was considerably higher than the rate for intact leaves. Leaves ground to a similar size as the fragments in the faecal pellets showed similar CO_2 evolution rates to faeces, suggesting that the increased surface area was the stimulus to the microbial populations. In similar experiments Hanlon & Anderson (1979a) also demonstrated that a fine balance exists between microbial stimulation and inhibition by litter-grazing animals. The addition of up to six isopods (*Oniscus*) or millipedes (*Glomeris*) to microcosms containing 1 g leaf litter caused increased microbial respiration (Fig. 4.13) but eight or ten animals resulted in decreased CO_2 evolution from the systems. Counts of bacterial cells and fungal hyphal lengths showed that the addition of the animals had stimulated bacterial growth but resulted in a decline in fungal biomass. In another microcosm experiment Hanlon (unpublished) showed that reduced particle size may produce an immediate but transitory response on fungal catabolism (Fig. 4.14). Oak litter was dried, ground and

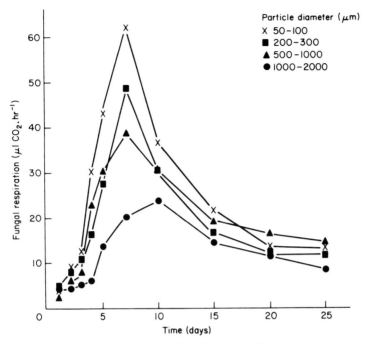

FIG. 4.14. Relationship between particle size of leaf litter and respiration of *Coriolus versicolor* (Hanlon 1978).

sifted through analytical sieves to obtain a range of particle sizes. The material was then leached to remove water soluble materials, inoculated with the fungus *Coriolus versicolor* (a cellulolytic and ligninolytic species) and the CO_2 evolution measured in respirometers. Another similar experiment was performed in which the mass of accumulated particles as well as the particle size was varied.

Equal weights (0·1 g) of different size ranges of particles were deposited on approximately 6 cm², 12 cm² and 24 cm² of glass fibre filter paper and colonised with *Coriolus*. Fungal respiration from particle sizes down to 300 μm was not significantly affected by the treatments (Fig. 4.15) but below this size range CO_2 evolution was related to the surface area of the filters on which the litter material was spread. Decomposition was strongly inhibited on the 6 cm² and 12 cm² filters. It is significant that these particle sizes represent the size ranges of particles in soil animal faeces. Earthworms and large millipede faeces contain litter fragments of about 200–300 μm, woodlice and medium-sized soil arthropods 100–200 μm and microarthropods 20–50 μm. Stereoscan electron microscopy of the litter material on the filters revealed that the fungal mycelium was restricted to the surface of compacted

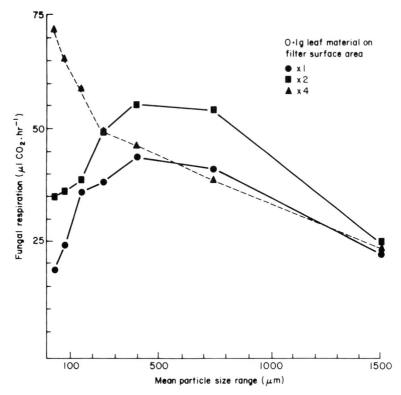

FIG. 4.15. Effect of mean particle size and size of particle aggregate (as determined by surface area of deposition) on fungal catabolism of litter (Hanlon 1978).

particles. The reason for this was not determined in these experiments but water-logged anaerobic conditions could be expected to occur in particle aggregates of greater than 3 mm diameter (Griffin 1972). Thus while van der Drift & Witkamp (1959) concluded that the increased moisture retention capacity of comminuted material would enhance microbial activity, the reverse may occur where large quantities of compact faeces are deposited by soil animals.

Another hypothesis arising from the first experiment (Fig. 4.13), that the mycelial thallus of slow growing fungi like *Coriolus*, has a low tolerance of grazing or fragmentation by soil animals, was supported by experiments where collembola (*Folsomia candida*) were added to respirometers containing ground leaf litter and *Coriolus*. Two, four or six collembola resulted in increased CO_2 evolution from the system but eight or ten collembola produced a pattern of decreased CO_2 evolution similar to that observed in the experiments with macroarthropods. The important difference between the two series of experiments is that the collembola were restricted to grazing on mycelium on the litter surface and emphasises the sensitivity of this fungus to microarthropod grazing pressures.

It is not difficult to appreciate that bacterial populations, because of the size and nature of the thallus, are more resistant to the effects of comminution by soil animals and are stimulated by higher intensities of animal feeding activities. The phenomenon is likely to be a contributory factor to the predominance of bacterial activity in soils intensively worked by earthworms and other soils animals. It is important to distinguish however between the stimulatory effect of grazing on microbial turnover and the stimulatory effect of the reduction of the particle size of the available resources. It is probable that this latter, resource quality, effect is at least as important in establishing active bacterial populations in the latter stages of decomposition.

4.5 The influence of resource quality on the structure of decomposer communities

At several places in earlier chapters we have commented on the high species diversity of decomposer communities. It is fairly clear that decomposer niches are primarily determined by patterns of variation in environment or resource availability. It is therefore appropriate to review briefly the extent to which the latter factor, resource quality, is a determinant of the composition and structure of decomposer communities. More detailed reviews have been given by Anderson (1977), Swift (1976) and Ghilarov (1978).

Some aspects of the influence of resource quality are obvious; the decomposer community of branches and leaves from the same tree are markedly different implying that some organisms are adapted to major resource-types, e.g. wood decay Basidiomycetes, xylophagous insects such as beetle larvae or wood-feeding termites. The specific mechanisms may not be apparent but encompass part or the whole range of resource qualities peculiar to wood— low nutrient content, high lignin, accompanying 'toughness', thick hydrophobic bark, presence of a specific spectrum of modifiers etc. Any one of these factors may dominate but the specialised wood decay organisms can be adapted to the whole range—e.g. the Basidiomycete *Coriolus versicolor* shows high activity at very high C:N ratios, possesses ligninases and is tolerant of a wide range of aromatic modifiers. In Chapter 3 we noted that adaptation to specific resource-types was one factor determining the distribution of decomposer organisms. It appears to be far more common among fungi than other decomposers and we have defined a resource-specific (RS) group of these organisms. The specificity may extend to resource-species level (see Table 3.3). It must be noted that environmental factors may play an important role in determining resource-specificity, e.g. the gaseous environment of a decaying log is markedly different from that of a leaf.

The role of fungi as primary saprotrophs probably explains the widespread nature of resource-specificity in this group. They colonise detritus at the stage when the chemical composition is at its most diverse. Modifier content is high and different resources show greatest differences in composition with regard both to these compounds and to C, energy and nutrient sources. This resource variability permits adaptive radiation and probably selects for specialisation. The greatest specialisation is found in parasitic organisms, particularly biotrophs; these organisms are adapted to the modifier characteristics of living as well as dead cells (see Sections 3.2.6 and 3.2.7) imposing even greater selective pressures on them.

There are however a very wide range of decomposer organisms, fungi, animals and bacteria, which do not show resource specificity. Species of Basidiomycetes, Hyphomycetes, Collembola, Acari, Diplopoda etc. may for instance be equally found in decaying wood as involved in leaf-litter decomposition. These resource-non-specific (RN) forms seem to exist by virtue of one or both of two closely linked adaptive mechanisms; specialisation to a particular stage of decomposition or to association with one or a range of more specialised decomposer organisms.

The 'succession' of fungi within a resource as decomposition progresses has possibly attracted more attention than any other aspect of fungal ecology. Similar phenomena have also been described for bacteria and animals. Until the last decade the usual explanation of fungal succession was that proposed by Garrett (1951, 1963, see p. 82) of a sequence in nutritional activity— sugar fungi, succeeded by cellulolytic fungi succeeded by ligninolytic fungi. It has become increasingly clear that this hypothesis is not adequate to explain many observed patterns of fungal succession. Neither the chemical nor the nutritional sequence can be seen in both chemical and biological data. The data of Tenney & Waksman (1929) given in Table 4.7 are a good example of the former. Whilst it is clear from the first column for each resource that water soluble components are the most rapidly depleted during the first one or two months of decay, the second and third columns show that equal or greater *amounts* of cellulose are lost during this period. Thus we must assume that cellulolytic organisms are equally active with sugar fungi during the early stages of decay. In contrast it is clear that lignin decomposition is not characteristic of the early stages of leaf decay. In woody materials however lignin may be lost right from the initiation of decomposition but this depends entirely on whether white-rot or brown-rot fungi are the colonisers (see Section 5.3.4).

Biological analysis of successions on a wide range of resources has also shown that cellulolytic and ligninolytic organisms are commonly present in the earliest stages of decomposition and sugar fungi persist throughout

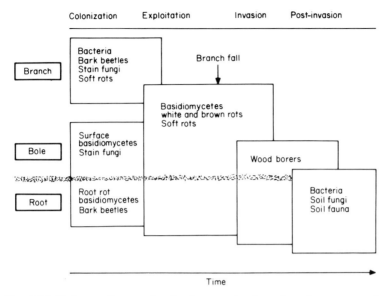

FIG. 4.16. Patterns of succession in decomposing branch and stump wood. The dominant organisms during the four major stages of the decomposition of 'typical' resources in a British woodland are indicated. The succession in any one branch or stump may differ markedly from the above pattern; some of the groups mentioned may never appear or entirely different ones dominate or the sequence may be substantially altered. Within this high degree of heterogeneity the above pattern is, however, the predominant one (from Swift 1977a).

(Hudson 1968). It is also necessary of course to consider the intervention of organisms other than fungi to understand the fungal succession. Explaining the succession of organisms in relation to decomposing resources is tantamount to explaining the decomposition process as a whole but a few generalisations can nonetheless be made. Whilst the nutritional theory can no longer be accepted as a sole explanation these and other resource quality features are probably of considerable significance.

Successions show a very high degree of variation but four main phases can generally be recognised—colonisation; exploitation; invasion and post-invasion. An example of the dominant organisms in wood-decay successions is given in Fig. 4.16. We can consider these four phases in relation to the presence of RS microflora and the adaptations of RN flora and fauna which we mentioned earlier.

The colonisation phase is dominated by fungi which are either RS necrotrophs or if RN are adapted to the specialised habitats of rhizoplane, phylloplane or branch-wounds. The most diverse fungal flora is found in the exploitation phase when the major primary decomposition of polymers occurs. Along with specialised and often RS basidiomycetes (wood) or ascomycetes (leaf and twig) which are the main agents of depolymerisation there is a wide spectrum of RN fungi. These are probably occupying the role of 'secondary sugar fungi' in association with the more specialised organisms, i.e. using sugars and other products of depolymerisation. This may be thought of as a secondary level of resource quality selection. Animal invasion in relation to time is to a considerable extent determined by resource quality—both due to the 'conditioning' of the resource (metabolism of the modifiers) but more particularly the production of high-quality microbial tissue as a food resource. Post invasion decomposition is typified by a bacterial and micro-fungal flora adapted to the particulate resources formed during comminution. The selective effects operating here are a combination of resource quality and micro-environmental factors.

We may see then that for the smaller decomposers—microflora, micro-fauna and mesofauna—the decomposer community operating at any time is probably primarily determined by resource quality factors with environment having a secondary selective effect. The activities of the macrofauna are modified by resource factors (e.g. see p. 151 to 153) but this may be secondary to features of the general soil environment. Whilst resource quality factors determine the broad character of the decomposer community considerable differences may occur between units of the same resource type. Thus the fungal community of one branch may be quite different from its neighbours (Swift 1976). This implies a considerable degree of opportunism in establishing the initial community. It is probable that the diversity of decomposer organisms reflects not only the diversity of niches created by resource quality and environmental diversity but also the sheer abundance of detrital resources.

5

DECOMPOSITION PROCESSES
AT THE MOLECULAR LEVEL

In our definitions of decomposition processes in Chapter 2 we identified catabolism as the pathway of energy flow and chemical transformation in decomposer communities. Catabolism was defined as the progressive breakdown of complex organic molecules firstly to simpler organic molecules and finally to inorganic molecules. In Chapter 4 we saw that the total mineralisation of a resource may only occur over periods of thousands of years because of the resynthesis of recalcitrant humus materials. In this chapter we shall examine the biochemical mechanisms of catabolism and resynthesis characteristic of the decomposition subsystem.

A chapter on biochemistry may seem out of place in an ecological text but we hope to show that an understanding of the chemical processes of decomposition is essential to an analysis of the functioning of the subsystem as a whole. In particular it provides some insight into a number of questions posed in earlier chapters. What is the basis of the feedback regulation of decomposition processes by resource quality factors? What is the manner of regulation by the competitive processes of immobilisation and mineralisation? In what way is humus formed, why does it decompose so slowly and what is the significance of this sink of carbon, energy and nutrient? What is the validity of the hypotheses that the niche-function of micro-organisms can be largely explained in terms of their biochemical capacities?

5.1 Community metabolism of the decomposition subsystem

The elucidation of the biochemical processes of the decomposition subsystem has been based on two different approaches. The first is the classical biochemical method—investigation of the chemistry of substrates and the mechanism of their transformation using cell-free enzyme systems. An example of this type of study is the detailed pursuit of the mechanism of cellulose catabolism described below. Enzymes have been obtained from a large number of decomposer organisms isolated from a wide range of habitats. These enzymes have been purified by chromatographic and electrophoretic separation and their activities described and measured in the laboratory during their action on purified cellulosic substrates. Most of our detailed knowledge of decomposer enzyme systems, their mode of action and their regulatory mechanisms comes from studies of this kind.

In the soil and litter however extracellular enzymes may be free in the soil solution, or adsorbed onto soil colloids or the surfaces of organisms so that the physico-chemical conditions of their action are markedly different from those of a culture flask. In the soil, enzymes do not operate in isolation but in synergistic or competitive interaction with other enzymes. As a result, an alternative approach to decomposer biochemistry has been developed which may broadly be termed 'soil biochemistry'. This approach is characterised by the attempt to relate the activity of decomposer enzyme systems to their natural environment. The approach of the soil biochemist may be analytic—dealing with individual processes and enzyme systems by the use of selective techniques—or holistic, in which the response to particular substrates of the soil as a whole is studied, for instance by the use of perfusion techniques (see Appendix p. 326). In its most extreme form this type of investigation may be based on a concept of the soil system as analogous to a tissue or organ of a living organism (for reviews see Quastel 1965; McLaren & Peterson 1967). The catabolism of organic matter by the decomposer community can be seen as balancing, in terms of energy flow and nutrient cycling, the anabolic activities of the primary producers. In the latter case considerable progress has been made in describing ecological phenomena in physico-chemical terms, based largely on the detailed analysis of photosynthesis (Gates 1962) and of the energetics of secondary levels of biosynthesis including polymer formation (e.g. see Morowitz 1968). Ultimately it may be possible to describe the decomposer system at a comparable level of detail.

We may begin our analysis by reiterating the structure of the system within which the biochemical processes must operate. We have pictured decomposition as having a 'cascade' structure in terms of its resources and organisms (e.g. see Fig. 2.4). The chemical changes during decomposition may thus be described in a similar way. Catabolic processes bring about the breakdown of the complex polymeric components of primary resources. The products of these actions have a number of possible fates; they may form the building blocks for the synthesis of the molecular components of decomposer tissues; they may act as the respiratory substrates that fuel these processes; they may be transformed to inorganic molecules. Many of the newly synthesised organic molecules in the decomposer tissues or in the soil in their turn become the substrates for succeeding cascades of the decomposition processes. In this way matter is recycled through succeeding generations of decomposer organisms, components of the output molecules of one level of decomposition becoming the inputs to the next level. Conceptually we may therefore structure the metabolism of the decomposer community in terms of processes related to primary decomposition, secondary decomposition and humus formation and breakdown. In practice this may not be possible. The levels of the cascade cannot be distinguished readily—primary resources rapidly become the site of secondary processes as microbial tissue is formed and dies. Some processes (such as proteolysis) are common to both primary and secondary decomposition. The proposed structure is therefore more useful in interpretation than as a means of analysis. From the latter point of view we can recognise a more practicable subdivision of the metabolic processes of the decomposer organisms on which we shall base the rest of the chapter.

In Chapter 4 we established that the main sources of carbon and energy for decomposer organisms are a range of polymers, mainly polysaccharides (see Table 4.6). The processes of catabolism of all the diverse forms of such polymers have some common sequential features (Fig. 5.5). The two main steps, depolymerisation and mineralisation, form the basis of two of the following sections (5.3 and 5.4), which are preceded by a consideration of the chemistry of the polymeric substrates (5.2). The processes of humus formation and decomposition are considered separately (5.4) and finally an attempt is made to interpret the preceding biochemical analyses within the ecological structure of the decomposition subsystem (5.5).

5.2 The chemistry of decomposer resources

5.2.1 Primary resources

The bulk of the energy available in plant tissues for the use of decomposers is contained within a variety of polysaccharides, sugars and lignin. The sugars and storage polysaccharides such as starch account for a minor fraction of this carbon and energy and prior to death are located within the cytoplasm and vacuoles of living cells. This is also the main site of the required mineral elements other than carbon. The main source of carbon and energy however is the polysaccharides (cellulose, hemicelluloses and pectic materials), and lignin of plant cell walls (see Table 4.6). In considering the chemistry of the plant cell wall four main features of significance to decomposition processes should be borne in mind.

(1) Many enzyme systems are highly substrate-specific, thus the more diverse the resource chemistry, the more diverse the enzyme-system required to utilise it.

(2) The cell wall is not simply a mixture of substrate molecules—the physical nature of its elaborate three-dimensional structure may determine the nature of the decomposition processes as much as the chemistry of the components.

(3) The structure of cell walls differs markedly from plant to plant and from time to time within a plant.

(4) The chemistry of plant cell walls has been elucidated for living plant

Table 5.1. Some of the major components of the walls of living sycamore cells (from Swift 1976, calculated from the data of Thornber & Northcote 1961a, b, 1962).

(a) Composition of different plant tissues (% dry weight)[1]

	Outer phloem	Inner phloem	Cambium	Sapwood	Heartwood
Pectic substances	3·1	6·6	15·0	3·8	1·3
Hemicelluloses	21·3	25·8	45·0	31·5	31·4
α-cellulose	29·0	45·0	36·7	40·0	43·0
Lignin	46·9	22·6	2·5	25·3	24·3
Mean cell weight (mg)	—	—	15·3	157	171
Mean cell length (mm)	—	—	0·52	0·72	0·68

(b) Composition of hemicellulose and cellulose polysaccharide fractions in sapwood (% dry weight)[2]

	Xylan	Glucomannan	Hemicellulose I	Hemicellulose II	α-cellulose
Glucose	18·7	20·4	8·9	30·4	95·0
Galactose	0·0	0·0	2·8	0·0	0·0
Arabinose	Trace	0·0	1·6	0·0	0·1
Xylose	69·2	33·2	73·8	15·0	1·2
Mannose	0·0	39·8	0·0	51·0	3·5
Uronic anhydride	12·6	3·9	11·4	3·2	0·0
Methylgluronic anh.	12·2	0·0	0·0	0·0	0·0
Percentage of total fraction	48	2	42	8	—

1 Note the differences in composition and the increasing density of cells during differentiation.
2 Note the complex composition of the major polysaccharide fractions of the primary cell wall.

cells (often using tissue culture). We have already pointed out that the composition of the cytoplasm may change considerably during senescence—the same may well be true of cell wall components.

Analytical data previously presented showed the cell wall as composed

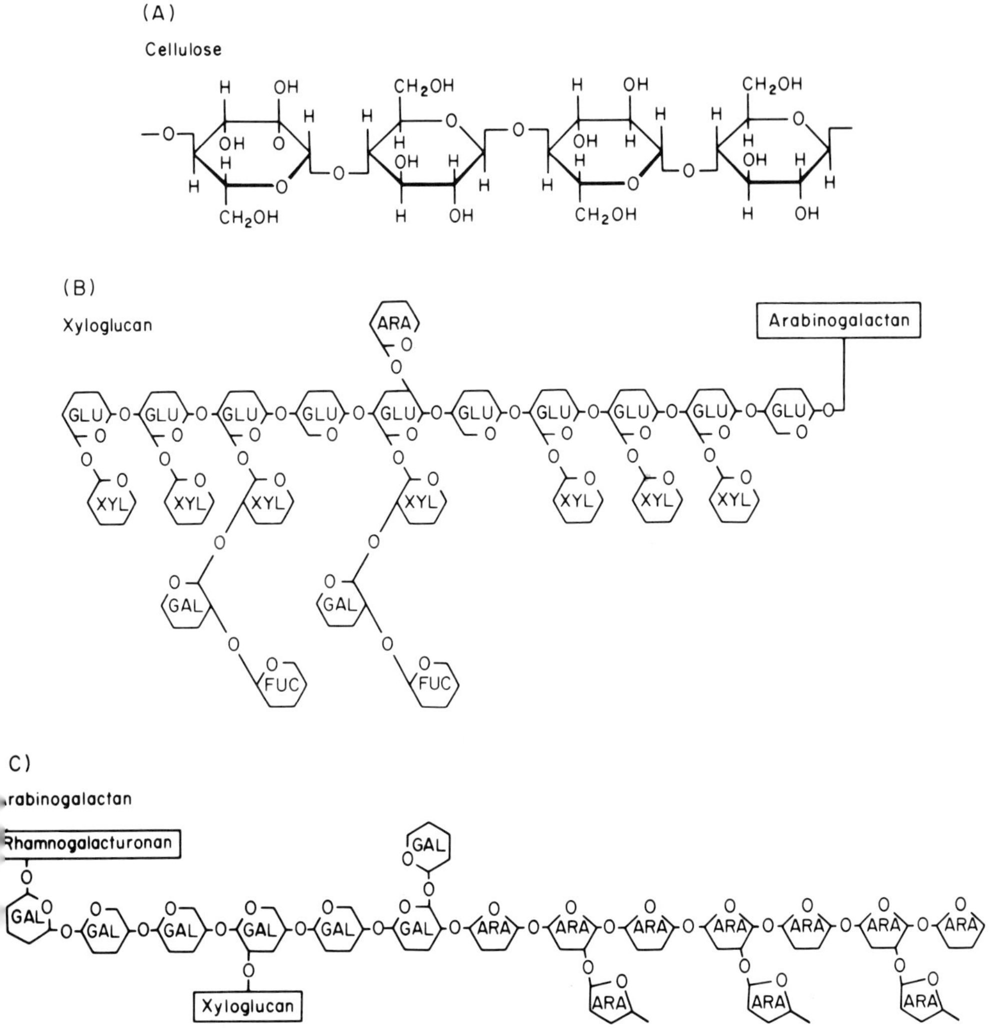

FIG. 5.1. Some polymeric constituents of plant cell walls. Polysaccharides (A-C): (A) Cellulose, unbranched chain of glucose units. This chain forms the basis of the crystalline fibrils of the cell wall (Fig. 5.2). (B) Xyloglucan—a heteropolymer of xylose (XYL) and glucose (GLU) plus a few molecules of galactose (GAL) and fucose (FUC). The polymer is highly branched. It is probable that its close association with cellulose is a result of the formation of hydrogen bonds between the glucose chain and the cellulose. It also probably bonds with

(D)

Phenyl-propane
units

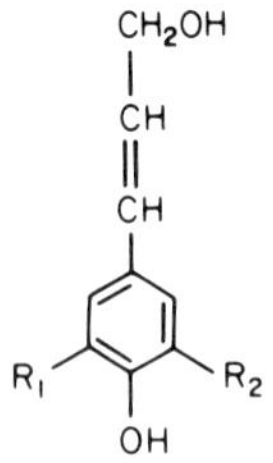

I $R_1 = R_2 = H$
II $R_1 = H, R_2 = OCH_3$
III $R_1 = R_2 = OCH_3$

(E)

Lignin

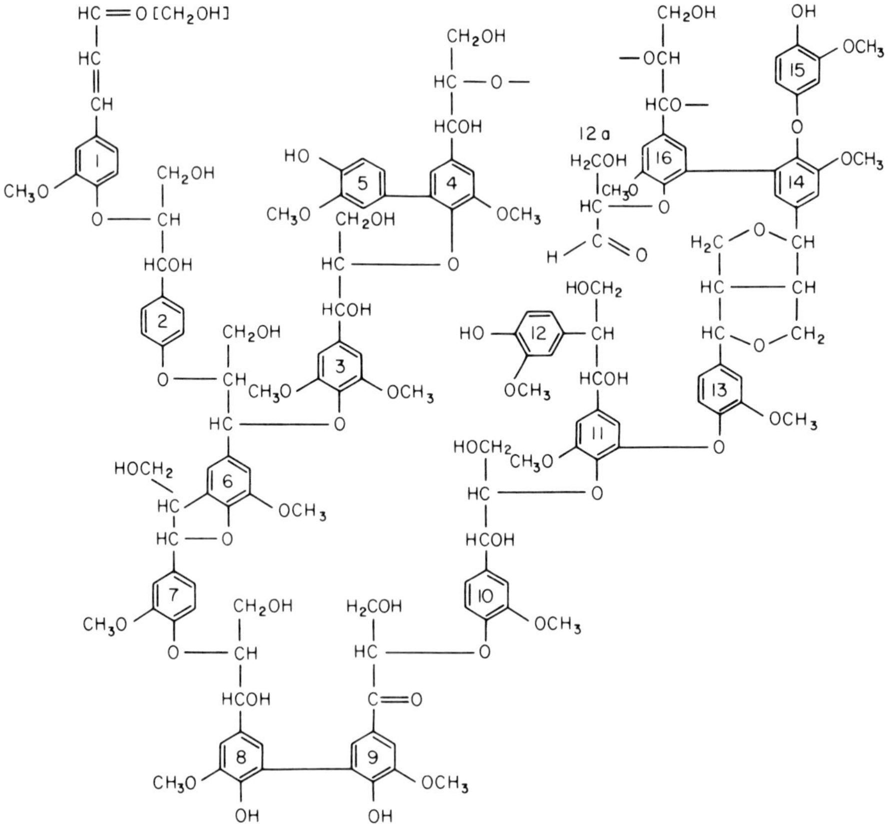

(F)
Cutin

HA   = Hexadecanoic Acid
MHHA = Monohydroxyhexadecanoic Acid
DHHA = Dihydroxyhexadecanoic Acid
DHOA = Dihydroxyoctadecanoic Acid
THOA = Trihydroxyoctadecanoic Acid

internal galactose molecules of (C) arabinogalactan which in its turn is linked with other hemicelluloses (e.g. rhamno-galacturonan containing rhamnose and galacturonic acid) and to cell wall proteins. These hemicellulose molecules form a complex three-dimensional matrix round the cellulose fibrils (Fig. 5.3).

Lignin structure (D, E): (D) Three of the commonest monomeric forms of the phenylpropane form: I *p*-coumaryl alcohol, II coniferyl alcohol, III sinapyl alcohol. (E) A molecular structure for conifer lignin as proposed by Adler (1966). Note the commonest forms of intermonomeric links; arylglycerol–arylether bonds such as that between (1) and (2)—which probably comprise about 40% of the linkages; Carbon–carbon bonds such as that between (8) and (9); and more complex coumaryl bondings such as that between (13) and (14). This intricate and heterogeneous branching molecule forms a further cage round the polysaccharides in the secondary cell wall.

A suggested structure for a cutin network in epidermal cell walls (F). The commonest molecular forms are hydroxy acids DHAA and THOA. Note the three types of linkage—ester, peroxide, and ether as exemplified in descending order between the second and third hydrocarbon chains. (After, Albersheim 1975 (A, B, C); Kirk 1973 (D, E), Van den Ende & Linskens (1974 (F).)

largely of cellulose, hemicellulose and lignin (Table 4.6). This gives a mistakenly simple impression of its structure. With the development of separation methods of greater resolution an increasing number of chemical entities have been identified from cell wall preparations. Table 5.1 shows the composition of the major polysaccharide molecules isolated from the walls of living sycamore stem tissue. All the cell wall components are arranged in a complicated three-dimensional structure the basis of which is the microfibrils of cellulose. Cellulose is an unbranched β1–4 linked polymer of glucose (Fig. 5.1) with a chain length (degree of polymerisation, DP) in the intact wall in the region of 2–14 $\times$ 10^3 glucose units. The polymer chains are in their turn arranged in a crystalline lattice maintained by hydrogen bonds between the hydroxyl groups of neighbouring chains (Fig. 5.2). The crystalline region is usually pictured as forming a core of dimension about 4 $\times$ 4 nm within the microfibril which has a total diameter of about 10 nm (Northcote 1972). The outer part of the microfibril is described as 'amorphous', that is the polymers are orientated in the same longitudinal direction as the core but without forming a crystalline structure. The microfibrils seem to be of almost indefinite length as polymer chains overlap within them. They form the fundamental framework of the cell wall within a complex matrix of the other polymers. In the primary cell the cellulose microfibrils are orientated predominantly in a network diagonal to the main (longitudinal) axis of the cells. In the outermost region (the middle lamella) longitudinally arranged fibrils are also found. In the various layers of the secondary wall the microfibrils are arranged in helical fashion (Fig. 5.2).

Northcote (1972) and Keegstra *et al.* (1973) have both recently presented models of the way in which the microfibril and the matrix may be related within the primary cell walls. The hemicellulose and pectic components of the cell walls (Table 5.1) form complex branching heteropolymers (Fig. 5.1) Some of these, such as xyloglucan are associated with cellulose in primary analysis and are believed to form hydrogen bonds with the surface of the cellulose microfibrils. By the formation of covalent bonds with the pectic side chains of the hemicelluloses, such as arabinans, xylans and rhamnans, they also provide a link between fibril and matrix (Fig. 5.3). Proteins with short length oligosaccharides (DP 8–12) linked to hydroxy-proline residues are also pictured as linking with the polysaccharides. In the secondary wall deposition of cellulose, hemicellulose and lignin takes place to the extent that the density of the wall material is increased some five- to six-fold. Pectic substances which predominate in the middle lamella (where they may act as a 'cement' between adjacent cells) decline in concentration towards the lumen of the cell. Table 5.1 shows the changes in composition and density during differentiation from cells freshly formed in the cambium to those characteristic of the xylem.

Lignin is a very complex polymer the chemistry of which is incompletely

(A)

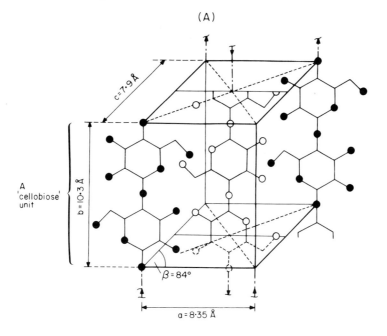

(B)

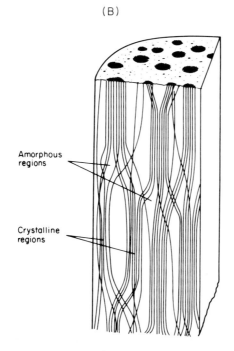

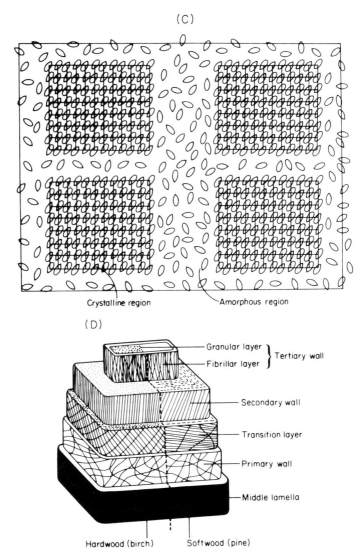

FIG. 5.2. The microfibrillar structure of cellulose: (A) Hydrogen bonds between adjacent chains maintain a regular spacing between them and give a repeating unit of two glucose molecules to the crystal. (B) and (C) The arrangement of elementary fibrils to form regions of crystalline cellulose interspersed with non-crystalline areas. (B) In longitudinal section. (C) In cross-section. The crystalline core of the microfibril is enclosed within a linearly arranged but non-crystalline cellulosic area, seen here in cross-section. (D) The cellulose microfibrils have different orientation in the different regions of a plant cell wall. The diagram also shows that different types of wall may differ in their 3-dimensional structure (A, B and C from Sihtola & Neimo 1975, D from Browning 1963).

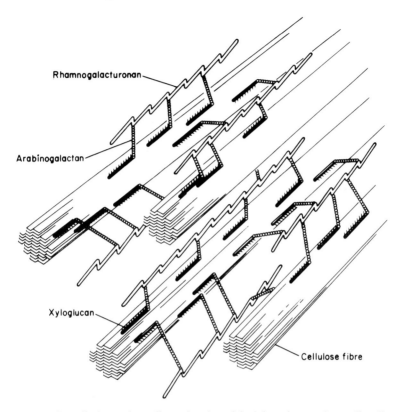

Rhamnogalacturonan

Arabinogalactan

Xyloglucan

Cellulose fibre

FIG. 5.3. A preliminary three-dimensional model of the primary plant cell wall. Showing the relationship between the aggregated cellulose microfibrils and the hemicelluloses of the matrix (from Albersheim 1975 after Keegstra *et al.* 1973). The model will undoubtedly be modified as study proceeds but it provides a current view of the three-dimensional complexity of primary resource structure.

understood. In the last stages of synthesis a polymer is formed from phenyl-propane (i.e. C6–C3) units arranged in a variety of three-dimensional forms. The units (of which three types are predominant, coniferyl alcohol, *p*-coumaryl alcohol and sinapyl alcohol, Fig. 5.2D) are linked by a variety of bonds. The most common are C–C bonds between phenyl groups and ether links between the C3 side chains and the phenyl rings, but several other types of bond have also been identified (Fig. 5.2E). The resulting polymer is heterogeneous in its monomeric constitution, its internal bonding and its physical shape and lignins from different sources often differ quite markedly in their properties. During secondary thickening lignin is laid down in the cell wall, initially in the middle lamella region and finally permeating through to the secondary walls. There is some controversy as to whether some degree of covalent bonding may be formed between the lignin and the polysaccharides (e.g. see Pew & Weyna 1962) but as a simple model the lignin may be pictured as forming a 'cage' round the carbohydrates and adding considerable rigidity to the cell wall.

An important constituent of plant cell walls is water. The amount of hydration affects the rigidity of the wall, a high water content reduces the extent of hydrogen bonding both within and at the surface of micro-fibrils. The water content changes markedly during cell wall development. Pectic polysaccharides are hydrophilic and are capable of forming gels; during early development when pectic substances predominate as fillers in the cell wall matrix the water content is high. Lignin on the other hand is hydrophobic and secondary thickening is accompanied by a decrease in water content, an increase in the degree of hydrogen bonding and a consequent strengthening of the cell wall.

Further complexity in cell wall structure may occur in the epidermal layers of plant tissues. As explained in Chapter 4, cell walls at their outer surfaces are covered and permeated by waxes and cutin or suberin. Cutin forms the framework of the cuticular layer (Fig. 4.12). The main components are long chain fatty acids and links are formed between them by esterification of an acidic group of one chain with the alcohol group of another and by peroxide and ether linkages, thus forming a complex three-dimensional structure (Fig. 5.1). In the outer layers this is permeated mainly by waxes and in the inner layers by cellulose and other polysaccharides.

5.2.2   Secondary resources and faeces

The molecular complexity of secondary resources may be quite as great as that of plant materials. An important feature is the increased quantitative importance of protein (Table 4.6) which may serve as a source of carbon and energy as well as nitrogen for organisms decomposing secondary resources. Lipids are also quantitatively more significant as they are common storage materials in both animals and microbes. Both of these categories of substrate are broken down by less-specific enzyme systems than polysaccharases and the details of their chemistry are therefore of minor significance.

Polysaccharides are found in secondary resources, both as storage compounds such as glycogen but also as structural components. Of particular importance is chitin and related polymers. Chitin is a polymer of N-acetyl glucosamine linked in $\beta 1$–4 fashion (Fig. 5.4). It can be said to have an analogous role to that of cellulose in higher plants in that it forms crystalline fibrils by hydrogen bonding between parallel chains, the repeating unit again being the dimer (chitobiose). Chitin of this kind is found in the exoskeletons of all arthropods (where it commonly comprises more than 50% of the organic dry weight), in the shells of molluscs (less than 10%) and is

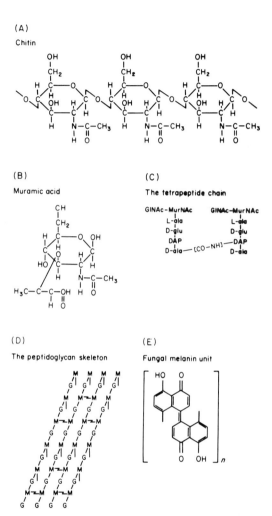

FIG. 5.4. Chemical components of microbial cell walls. (A) Chitin (β 1-4 poly N-acetyl glucosamine)—one of the main structural components of fungal cell walls. (B), (C) and (D) The structure of the peptidoglycan component of bacterial cell walls. (B) The structure of N-acetyl muramic acid, the unique amino sugar component of bacterial cell walls. (C) The arrangement of amino sugars and amino acids to form the basic tetrapeptide chains. (MurNAc = N-acetyl muramic acid; GlNAC = N-acetyl Glucosamine; L-ala = L-alanine; D-glu = D-glutamic acid; DAP = Diaminopimelic acid). (D) The peptido-glycan skeleton with glycosidic bonds between Muramic acid (M) and Gluco-samine, the tetrapeptide chains (vertical lines) and occasional cross-links between them (horizontal jointed line, see C). (E) A possible repeating unit for fungal melanin.

a minor component of many other invertebrates (Jeuniaux 1971) and the cell walls of fungi other than the Oomycetes (Rogers & Perkins 1968).

In the insect exoskeleton the chitin is commonly in close association with proteins. Covalent bonds may be formed between the two types of polymer. The stability of this complex is increased by the processes of sclerotisation which is analogous to a 'tanning' of the protein by polyphenolic compounds probably of quinonic form. In the outer layers dark melanin pigments may be deposited together with a coating of wax. The degree of sclerotisation and melanisation determines the hardness and enzymatic resistance of the cuticle. In crustaceans the protein content is lower and the chitin skeleton is permeated by calcareous materials. In mollusc shells the chitin forms a minor component in structures which are predominantly formed of a protein, cochiolin, which may again be partially tanned.

In fungal cell walls chitin is associated with a variety of other polysaccharides (glucans and mannans in particular) and protein. The microfibrillar chitin and its matrix probably forms only one of several layers in the wall (Hunsley & Burnett 1970). In bacteria a unique heteropolymer is formed with alternating units of *n*-acetyl glucosamine and muramic acid, another amino sugar (Fig. 5.4B). The muramic acid is also linked to five-member peptides and peptide links between neighbouring chains produce a complex three-dimensional peptidoglycan polymer. This polymer is the predominant molecule of the cell wall of gram-positive bacteria where it is associated with teichoic acids (a phospho-lipid in which glycerol or ribitol units are linked by phospho-ester bonds and which may form up to 50% of wall weight) and a number of polysaccharides. The peptidoglycan is less important in gram-negative bacteria where the outer layer of the cell wall may be massive and is largely composed of lipoproteins and lipopolysaccharides. The composition of the walls of different species of bacteria may vary fairly considerably and this probably confers a fairly high degree of resistance to enzymatic attack.

The cell walls of many fungi are dark in colour. This is usually attributed to melanin deposition. The exact definition of fungal melanins is in some doubt. Depolymerisation by alkali fusion of black pigments from various fungi gives catechol. A possible structure for a fungal melanin is shown in Fig. 5.4E. If this structure were general it would distinguish fungal melanin from the eumelanins of animals (e.g. arthropod cuticle) which are nitrogenous in nature being formed by polymerisation of tyrosine. The yellow and red pigments of feathers and hair are yet another form of melanin, known as phaeomelanin containing both N and S. Melanin pigments are extremely common in secondary resources and may be a significant resource quality component. Their ubiquity is of special interest in view of the central role which phenolic derivatives seem to play both in the regulation of decomposer activity and on humus biosynthesis (see Section 4.3.3). It is thus regrettable that their chemistry is so poorly understood (Harley-Mason 1965; Thomson 1976).

Faecal material is mainly a mixture of partially digested plant remains and microbial tissues. The latter may be quite significant as Mason (quoted by Webley & Jones 1971) estimated that 50% of the N in sheep faeces was microbial in origin. Thus the polysaccharides in faeces may also be a mixture of plant and microbial forms but there seems to have been no investigation of this. A feature of the faeces of many arthropods is that they are shed enclosed in a thin non-sclerotised chitinous peritrophic membrane. Enzymes, excretory products and sometimes blood are also important resource quality constituents of faeces.

5.3   Depolymerisation reactions

Reese (1968) in a comprehensive review of microbial catabolic activities has proposed a common basis to the mechanism of breakdown of polymeric substrates by decomposer organisms (see Fig. 5.5). The first action is the 'conditioning' of crystalline or other ordered structures in cell walls before enzymatic attack can occur. The first catabolic step involves the reduction in the degree of polymerisation by cleavage of intermanomeric bonds. The end-point of this action is the production of monomers or dimers (sugars, disaccharides, amino acids, dipeptides etc.) which are assimilated and mineralised by the micro-organisms. Thus whilst the initial reactions are extracellular—largely determined by the large size of the substrate molecules and their complex three-dimensional interlinking—subsequent stages may be extra- or intracellular. The existence of a prolonged extracellular phase may be a matter of some significance to the way in which the catabolic processes of decomposer organisms are regulated. The most detailed investigations of depolymerising activities have been those concerned with polysaccharides and cellulose in particular.

5.3.1   Polysaccharase enzymes

The depolymerisation of polysaccharides generally occurs by enzymatic hydrolysis of glycosidic bonds although there have been reports of non-enzymatic breakdown by a variety of agencies (Finch *et al.* 1971). The polysaccharase enzymes responsible are highly specific with regard to the constituent sugars of the polysaccharides and in most cases to the type of glycosidic bond linking them. In earlier reports this high degree of specificity was not recognised but with the development of a greater resolution in the separation methods for enzymes it has become accepted as a general feature (Reese 1968). Thus in order to bring about the solubilisation of a complex structure such as a plant cell wall a large variety of enzymes are required.

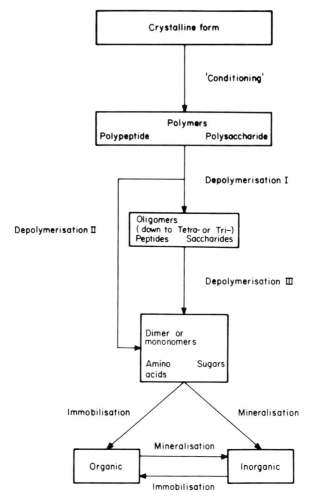

Fɪɢ. 5.5. A simple model of the main steps in the catabolism of polymeric substrates by decomposer organisms (modified after Reese 1968).

This requirement must be borne in mind in assessing the relative 'dominance' of any organism in terms of its decomposer capacity.

Two differing types of activity may also be recognised. In the *endo*-polysaccharases (I, Fig. 5.5) the cleavage of the chains is at random whilst in the *exo*-polysaccharases the cleavage proceeds in an ordered fashion from the non-reducing end of the chain with the step-wise removal of one or two units at a time (II, Fig. 5.5). The action of the endo-enzymes results in the formation of a variety of oligosaccharides of varying chain length. The polysaccharases can rarely act on saccharides with less than three constituent units and the rate of hydrolysis is generally proportional to the degree of polymerisation. Consequently the result of endo-cleavage is a mixture of short chain oligosaccharides in which the dissaccharides are dominant. The exo-polysaccharases commonly only come into action after the endo-enzymes have exposed a significant number of chain ends. The action of these enzymes may result in the direct formation of monomer units or disaccharides may be formed. The further metabolism of the di- and tri-saccharides is dependent on the action of glycosidase enzymes which can act only on substrates with DP less than three or four (III, Fig. 5.5).

This type of multi-enzyme system is characteristic of that described for cellulose. Laboratory studies using purified enzymes tested as cell free systems on purified substrates have shown the presence of at least four different types of enzyme. In addition to the exo- and endo-glucanases and a β-glycosidase a further enzyme termed C_1 has been demonstrated. In a number of experiments although a high depolymerisation activity could be readily shown when a soluble substituted cellulose (such as carboxymethylcellulose) or degraded or pre-treated cellulose were used as substrates, a much lower level of activity was found against 'natural' celluloses such as cotton fibre (Table 5.2). When investigated further it was shown that the accessibility of celluloses to attack by these enzyme systems seemed to depend more on the degree of crystallinity than on other characters such as the DP or the solubility in alkali. This led to the hypothesis that there might be a component of cellulose systems which rendered the crystalline fraction susceptible to attack by the depolymerases which were otherwise inactive. In recent years the existence of C_1 enzymes has been clearly demonstrated for a number of cellulase systems such as that of the fungus *Trichoderma viride* (Selby 1968) where this component can be clearly separated from the depolymerising enzymes. Whilst it displays little activity on its own, the synergistic action between C_1 and the other enzymes is enough to reconstitute activity equivalent to that of the whole organism. (Table 5.3). The mode of action of C_1 is still a matter for speculation. Earlier theories suggested a non-hydrolytic function

Table 5.2. The effect of crystallinity and other features of cellulosic substrates on depolymerising activity of enzymes from *Trichoderma viride* (after Reese, Segal & Tripp 1957).

Substrate[1]	Before hydrolysis			After hydrolysis		
	Crystallinity %	DP	Alkali soluble %	Weight loss %	DP	Alkali soluble %
A	90	4970	4·9	25	4200	7·6
B	70	5040	6·8	35	3040	12·3
C	40–50	4670	7·9	35	3100	15·7
D	0	3920	11·2	65	1630	35·8

[1] The substrates were all cotton subjected (except A) to various chemical treatments.

Table 5.3. The relative levels of activity of the components of the cellulase complex of *T. viride* when acting alone or jointly to depolymerise native (cotton fibre) cellulose. CMC-ase is a β-glucanase and cellobiase a β-glycosidase (from Selby 1968).

Component	Relative cellulase activity (%)
Original solution	100
C_1	1
C_x	5
$C_1 + C_x$	102
CMC-ase	4
Cellobiase	1
CMC-ase + cellobiase	2
C_1 + CMC-ase	35
C_1 + cellobiase	20
C_1 + CMC-ase + cellobiase	104

possibly involving displacement of hydrogen bonds within the cellulose micelles and the insertion of a large protein molecule between the polymer chains thus opening up access for the much smaller β-glucanases (Reese & Mandels 1971). More recent work has demonstrated that C_1 may act as an exo-glucanase (Wood & McCrae 1975) but the nature of the synergistic action and the differential activity to crystalline cellulose remains unsolved. C_1 enzymes do not appear to be universal and it has been suggested that there may be different pathways for cellulose degradation in soft-rot, white-rot, and brown-rot fungi and bacteria (Goksoyr *et al.* 1975).

Whilst the cellulose system has now been described in detail, considerably less is known of the enzyme systems involved in the breakdown of the other cell wall polysaccharides. Most evidence suggests that similar multiple-enzyme systems are involved. For instance King (1966) demonstrated the presence of a variety of enzymes in the cultural filtrate of the wood-decaying fungus *Coniophora cerebella* capable of degrading hemicellulose fractions of various kinds (Table 5.4). Reese (1968) has also listed a wide variety of such enzymes of microbial origin. These enzymes are characteristically highly specific to the substrates and at least two components are usually necessary for the breakdown of each polysaccharide; a polysaccharase acting on polymers of high DP, and a glycosidase acting on the resultant smaller saccharides. King (1966) also demonstrated that side chains in hetero-polymers might sterically interfere with activity of the main polysaccharase and necessitate the action of yet further enzymes. The requirement for a C_1 type 'conditioning' enzyme has not been demonstrated other than for cellulose but it is possible that such enzymes are essential for the initiation of attack on intact cell walls.

Table 5.4. Depolymerizing enzymes detected in cultural filtrates of *Coniophora cerebella*. G = Glucose; X = Xylose; M = Mannose; Gal = Galactose; Gal U = Galacturonic acid (King 1966, King & Fuller 1968)

Polysaccharide	Bond attacked	Polysaccharase	Glycosidase
Cellulose	G 1–4G	+	+
Laminarin	G 1–3G	+	
Xylan	X 1–4X	+	+
Glucomannan	G 1–4G	+	
	G 1–4M	+	
	M 1–4G	+	
Arabionogalactan	Gal 1–3Gal	+	
	Gal 1–6Gal	+	
Pectin	GalU 1–4GalU	+	

Multiple enzyme systems are also probably involved in the breakdown of microbial cell walls and insect exoskeletons. Chitinases are polysaccharase enzymes similar to those described above. The action of the polysaccharase enzymes may however be preceded by other enzymes which disrupt the complex cell wall. For instance an amidase which hydrolyses the amide bond between muramic acid and alanine in bacterial peptidoglycan has been demonstrated in *Bacillus subtilis*. This enzyme is autolytic and it is interesting to speculate whether the 'conditioning' of microbial cell walls is a largely autolytic process, the component polysaccharides then becoming more readily accessible to extracellular polysaccharases of other decomposer organisms. A further possible process of 'conditioning' may occur during the passage through animal guts. As well as the disruptive effect of mastication, enzymic lysis such as by lysozyme may play an important role not only in rendering microbial polysaccharides available to the animal but also in producing readily catabolised components of faeces. This is an area requiring a great deal more investigation.

The mechanism of regulation of polysaccharase enzymes presents an interesting problem of some relevance to their functioning in decomposer systems. The substrates are polymeric and insoluble; it would therefore seem likely that these enzymes would be constitutive in nature. However there have been a number of reports that enzymes of this type are induced. For instance, Mandels & Reese produced evidence in 1960 that the cellulase system of a number of fungi was adaptive in nature and induced by a variety of substances but most effectively by β(1–4) glucans of varying size including cellulose and cellobiose. Hulme & Stranks (1971) however have demonstrated the production of Cx enzyme in *Trichoderma viride* in the absence of any inducer. Their evidence points to regulation of a constitutive enzyme purely through end-product repression by, in the case of cellulases, glucose and cellobiose. The activity of the polysaccharase enzymes may also be depressed by the presence of a wide variety of inhibitory substances (Mandels & Reese 1965). Among the most active of the naturally occurring groups of compounds are the phenols, particularly in oxidised form.

Environmental factors also regulate depolymerisation for instance decomposition of cellulose may take place under anaerobic conditions. Although there is no reason to suppose that the extracellular cellulolytic enzymes of fungi will not retain their activity in the absence of oxygen the production of cellulase systems under anaerobic conditions is confined to bacteria. The most frequent free-living cellulolytic organisms are probably *Clostridium* species but the most extensively studied flora is that of the herbivore rumen which is dominated by such genera as *Ruminococcus* and *Butyrivibrio* (Hungate 1966). The intestinal flora of invertebrate decomposers has not been studied to a comparable extent. The mechanism of depolymerisation probably resembles that in fungi but the metabolism of the liberated sugars differs markedly from the aerobic pathways (see Sections 5.4 and 7.2).

5.3.2 Depolymerisation of lignin and cutin

A very wide range of micro-organisms are known to be able to metabolise the types of aromatic molecule which comprise lignin polymers (see Fig. 5.1 and Section 5.4), but only a restricted range of organisms, mainly basidiomycete fungi, have the demonstrated capacity to metabolise the intact molecule. Indeed there is at present no clear view of the enzymatic basis of the initial depolymerising steps. The simplest hypothesis would be some mechanism of extracellular cleavage of the inter-unit bonds. In support of this hypothesis there have been a number of demonstrations of the presence of phenylpropane units or their derivatives in materials decayed by fungi. Henderson (1955) isolated vanillic acid and syringic acid from birch sawdust rotted by *Polystictus versicolor* and the same two substances plus ferulic acid and dimeric molecules such as dehydrovanillin were found in prepared lignin cultures degraded by white-rot fungi by Ishikawa and his co-workers (1963). The latter workers showed in a series of experiments the ability of these fungi to metabolise these molecules including cleavage of the glycerol β-aryl ether bonds (Fig. 5.1) in 'model' dimeric compounds such as guacyl-glycerol-β-guacyl ether.

There is however very little other evidence for the cleavage of inter-monomeric bonds in lignin in the manner postulated above. Kirk (1971) has pointed out that despite the quantitative importance of the ether-type linkages in the lignin molecule cleavage of these bonds by the postulated 'β-etherase' enzyme would liberate only a relatively few monomeric units. An alternative theory is that the primary attack comes not on the inter-monomer bonds but on the common underlying feature, the aromatic ring. This idea has attractions for it suggests a single mechanism for the initial opening up of the polymer rather than the multiple enzyme system necessary for initial depolymerisation based on cleavage of the intermonomer bonds. There is no experimental evidence for this mechanism in relation to intact polymers although the capacity for ring cleavage is widespread among decomposer micro-organisms (see Section 5.4).

A group of enzymes that has often been implicated in lignin breakdown is the phenolases. This stems from the use of the 'Bavendamm Reaction' for the detection of 'white-rot', i.e. lignin-degrading ability, in wood destroying fungi. An almost total correlation has been found between the ability to oxidise gallic acid in agar to its brown-coloured quinonic form and the possession of ligninolytic ability by Basidiomycetes (e.g. Lindeberg 1948). This oxidation is catalysed by phenolase and these enzymes have thus been implicated in lignin degradation. The correlation does not however necessarily imply that the phenolases are involved in the primary degradation of the lignin polymer and their involvement was challenged by a number of workers (e.g. Henderson 1960) on the grounds that their main mode of action, the oxidation of phenolic rings to quinonic form, does not seem readily to fit them for a degradative role, and also because of evidence that phenolase activity does not necessarily correlate with ligninolytic ability in fungi other than Basidiomycetes. Henderson (1968) later demonstrated the ability of a range of moulds from soil to utilise lignin-related monomers in pure culture without any detectable phenolase activity. Lignin degradation during decay of wood has also been demonstrated in non-Basidiomycete fungi (Kirk 1973). The latter author has however produced compelling evidence that phenolase enzymes from Basidiomycetes can degrade 'model lignins' by oxidation of alkyl-phenyl carbon to carbon bonds within the phenyl propane monomers (Kirk *et al.* 1968). No 'β-etherase' activity was detected in these experiments. Kirk *et al.* (1968) further showed that this form of cleavage would be at least as efficient as 'β-etherase' in disrupting the lignin molecule.

The mechanism of lignin depolymerisation thus remains unsolved; it is possible that several different enzyme systems are involved and that these may be differentially distributed among ligninolytic micro-organisms. The recent production of ^{14}C labelled model lignins offers the possibility that this problem may be rapidly elucidated in the near future (Kirk *et al.* 1975).

Cutin can be used as a sole carbon source by a number of soil micro-organisms, both fungi and bacteria, although most of the work on the elucidation of the enzyme system has been performed using a single species of mould, *Penicillium spinulosum* (Heinin & de Vries 1966). Cleavage of the cutin esterase and reduction of the peroxidase links by a second enzyme, carboxycutin peroxidase. The liberated fatty acids are oxidised by more familiar enzymes such as oleic, linoleic, and stearic oxidases acting on liberated chains of different lengths. Comparable systems may exist for suberin degradation but little attention has been paid to this polymer.

5.3.3. Protease enzymes

In comparison with the enormous amount of work which has been invested in the later stages of N mineralisation (e.g. nitrification Section 5.4) very little consideration has been given to the first stage—the depolymerisation of the protein macromolecule. Thus whilst the general principles of the bio-chemistry are clear, virtually nothing is known of the proteolytic enzyme systems of decomposers and the ways in which they may be regulated.

Primary attack is catalysed by the group of enzymes known as endo-peptidases. These promote the hydrolysis of the peptide linkages between the amino-acids. The attack may be specific—confined to only one, or a limited range of amino-acid sequences—or of a more general kind. For instance an enzyme produced by *Clostridium histolyticum* is specific to collagen. This appears to be associated with a limiting of the attack to peptide bonds between amino acid pairs which are flanked by the aromatic amino proline. In contrast an endo-peptidase from *Aspergillus oryzae* was able to attack all proteins tested (Cochrane 1958). A major difficulty in the improve-ment of our understanding of the specificity, mode of action, distribution and regulation of endo-peptidases is lack of study of the chemistry of proteins of importance in decomposer resources. Thus most enzyme testing has been carried out on easily prepared 'standards' such as gelatin, casein or egg albumen. One exception to this is keratin. This is a protein of widespread occurrence as a component of mammalian hair, hoof and horn, and the feathers of birds. The protein is regarded as being of a resistant nature and is only decomposed by a specialised 'keratinophilic' microflora largely composed of those fungi classified as dermatophytes (Mathison 1964). The resistant nature is probably due to the highly crystalline structure of keratin and to the formation of disulphide bridges between polypeptide chains. It has been shown that cystine may form up to 15% of the keratin molecule. In the case of the actinomycete *Streptomyces fradiae* and the clothes moth *Tineola biselliella* primary attack involves the production of highly alkaline (pH > 10) and highly reducing (redox potential below -250 mv) conditions under which the disulphide bridges were reduced to produce free thiol groups, possibly under the influence of an enzyme resembling glutathione reductase, see p. 189.

After this reduction the protein is far more susceptible to proteolysis. Mathison (1964) found evidence for some kind of conditioning of keratin prior to proteolysis by dermatophyte fungi but could not confirm the presence of a reducing system of the type described above. He suggested however that the main difference between keratinophilic and non-keratinophilic organisms lay in these preliminary reactions rather than in the possession of any specific proteases. Pugh (1971) also found that a fatty acid component of feathers inhibited the action of keratinolytic organisms. Although specific information is very sparse it is probably safe to assume that most decomposers have an

$$
\begin{array}{ccc}
\begin{array}{c}
| \\
NH \\
| \\
CO \\
| \\
CH-CH_2-S-S-CH_2-CH \\
| \\
NH \\
| \\
CO \\
|
\end{array}
& \xrightarrow{+\,2H} &
\begin{array}{c}
| \\
NH \\
| \\
CO \\
| \\
CH-CH_2-SH \quad + \quad HS-CH_2-CH \\
| \\
NH \\
| \\
CO \\
|
\end{array}
\end{array}
$$

array of endo-peptidases but that the distribution of specific forms may be more limited.

Further depolymerisation of oligopeptides is carried out by *exo-peptidases* which will hydrolyse only those peptide bonds which are adjacent to free amino or free carboxyl groups and is therefore limited to attack at chain ends. These enzymes are responsible for the liberation of individual amino acids which are then metabolised by decomposer organisms as described in Section 5.4.

5.3.4 The ecological significance of depolymerisation reactions

The first question we may consider is the extent to which a knowledge of the chemistry and enzymology of resource polymers helps in interpreting the observed differences in their rates of decomposition. In Chapter 4 evidence was presented to show that not only was lignin the most slowly decomposed of plant cell wall components but that the lignin content of a resource affected its overall rate of decomposition. The chemistry of lignin provides a basis for its recalcitrance to enzymatic attack. The presence of a variety of different bonds, the steric interference provided by aromatic monomers with a variety of side chains, its highly branched and folded structure and its hydrophobic nature, all constitute hindrances to enzymatic depolymerisation. Very little is yet understood of the mechanism of break-down but the involvement of enzymes such as phenolases suggest that it may yet prove to be chemically fairly straightforward. One of the most striking features of lignin breakdown seems to be the relative rarity of the capacity among decomposer organisms. It is perhaps significant that the organisms possessing the ability—the white-rot Basidiomycetes—are also among the best equipped for polysaccharide breakdown. It is possible that the energetic cost of lignin catabolism is too great for it to be of value except as a means of improving the accessibility of the cell wall polysaccharides. In Chapter 4 evidence was presented to show that lignin may mask attack on cellulose or hemicelluloses. Clearly the white-rot Basidiomycetes are at an advantage if they can disperse the lignin cage. Lignin may be of significance in secondarily thickened cell walls not only because of its own chemical recalcitrance but because it masks the other more readily decomposed components, such as the polysaccharides (Section 4.3.1). The protective cage formed by complex branching molecules in the outer surfaces of plants, fungi, and arthropods—like cutin or melanin—may also delay the depolymerisation of cellulose or chitin. It is interesting to note that the distribution of cutin- or melanin-degrading enzymes also seem much less wide than that of the polysaccharides they cover.

With regard to the carbohydrate constituents the difference between the soluble sugars and the polysaccharides is clearly one of accessibility. Differences between the polysaccharide components are unclear. Minderman's hypothesis suggests that hemicelluloses are decomposed more rapidly than cellulose (Fig. 4.4) but the data sets of Figs. 4.5 and 4.7 and Table 4.7 indicate little difference between them. On the basis of the chemistry of the cell wall and the known specificity of polysaccharase enzymes for sugar monomers and band configurations many of the hemicellulose components must be regarded as more complex substrates than cellulose. On the other hand the crystalline nature of cellulose represents an additional problem of accessibility of the glycosidic bonds as is evidenced by the requirement for a specific enzyme, C_1. In the absence of any data concerning the relative rate of degradation of chemically-defined hemicellulose components no firm conclusions can be drawn. Similar consideration must apply to the bacterial and fungal cell walls and the insect exoskeleton, where even less is known of the enzymology. It is worth re-emphasising however, that chemical considerations apart, the extent to which a resource component is used may simply relate to its relative abundance.

What is clear is that the heterogeneous chemistry and complex three-dimensional conformation of these structures necessitate the production of a multi-enzyme system for their breakdown. Organisms with only a limited range of such enzymes can obtain only relatively small gains in energy from such complex structures. Unfortunately very little is known of the distribution of enzyme systems at present but it is clear that a great many decomposers are deficient in this respect—including most animals and probably the majority of bacteria. Among the fungi, diversity of polysaccharase enzymes has only been demonstrated in wood-decaying Basidiomycetes but little attempt has been made to search elsewhere. Among these, considerable variation in mode of attack has been demonstrated. The mode of attack on cellulose differs markedly between different organisms (Fig. 5.6). The white- and brown-rot Basidiomycetes also differ in that whereas the cellulose enzymes appear to be bound to the fungal cell wall in the former, in the latter they diffuse some distance from the hyphae. This produces quite

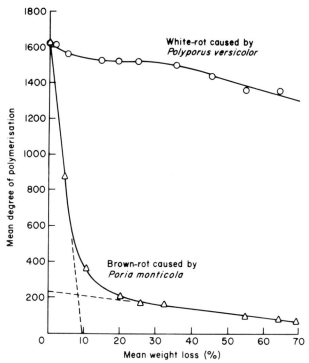

Fɪɢ. 5.6. Variation in the mode of cellulose depolymerisation between different Basidiomycete fungi. The degree of polymerisation of sapwood cellulose shows a dramatic decline as a result of attack by a brown-rot fungus *Poria monticola* but only a gradual erosion by the cellulase system of *Coriolus* (*Polyporus*) *versicolor*. The author suggested that whereas the *C. versicolor* enzyme attacked randomly at points along the cellulose fibril the attack of *P. monticola* was confined to amorphous regions initially (see Fig. 5.2B). Extrapolation of the lower part of the brown-rot curve to the *Y* axis suggests rapid release of fragments of average DP about 250 which is in good agreement with the lengths of crystalline regions (after Cowling 1961).

marked differences in the microscopic appearance of cell wall decay (Wilcox 1970). Cellulolytic bacteria resemble the brown-rots in having surface-bound enzymes. Lacking the penetrative ability of fungi this restricts their depolymerisation to an erosion of the surface in the immediate vicinity of the cell which possibly accounts for the much slower rate of cellulolysis associated with bacteria. White-rot Basidiomycetes, enzymatically the most totally equipped, of all decomposers, also show marked variations in the pattern of attack on cell wall components (Campbell 1952). Some species show a simultaneous attack on all components, lignin, hemicellulose and cellulose (e.g. *Coriolus versicolor*) whereas others degrade the lignin and hemicelluloses before the cellulose (e.g. *Fomes annosus*) and in others the polysaccharide degradation precedes lignin breakdown (e.g. *Armillaria mellea*). The sequence in which enzymes are produced may differ in different organisms and under differing circumstances and this may be a matter of some significance with regard to the pattern of depolymerisation observed in a given resource. There is very little evidence in this respect but Albersheim *et al.* (1969) have shown that an ordered sequence of production may occur in some necro-trophic fungi.

The possibility that only a minority of organisms possesses the enzymatic capacity to extensively degrade plant cell walls suggests that a large number of decomposers will have a secondary role dependent on these primary decomposers. This is an agreement with Garrett's (1963) group of 'secondary sugar fungi' (see Section 3.2.3). There are a few isolated pieces of evidence to support this suggestion at the enzymatic level. Selby (1968), among others, has shown that the C_1 enzymes from one species of fungus can act syn-ergistically with the Cx enzymes from a second species to produce effective breakdown of crystalline cellulose (Fig. 5.7). This suggests that the inability of one component of a multi-enzyme system may not necessarily debar an

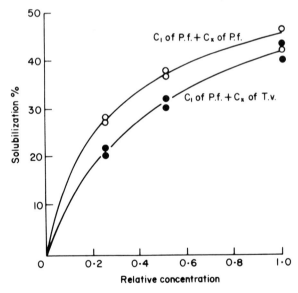

FIG. 5.7. Co-operation between enzyme components in cellulose decomposi-tion. The upper line shows the effect of the enzymatic complement of *Penicillium funiculosum* on the breakdown of native cellulose; the lower line shows that a mixture of C_1 enzyme from *P. funiculosum* and Cx from *Trichoderma viride* is just as efficient as when both components come from the same organism (Selby 1968).

organism from participating in cell wall depolymerisation. Another example of co-operation in cellulose breakdown is given by Enebo (1963). He showed that the rate of anaerobic cellulose breakdown by a species of *Clostridium* was greatly enhanced by the presence of other bacteria capable of utilising the organic acid end products of the clostridial attack. Synergistic participation of Basidiomycetes in lignin breakdown has also been demonstrated (Sundman & Nase 1972) although the enzymatic basis is unclear.

The possibility that the regulation of enzyme production occurs mainly through catabolite repression offers intriguing possibilities of co-operation between organisms. Enzyme production in one organism may be subject to control by the activities of other organisms. Both the activity of the depolymerising enzymes of other organisms (resulting in the production of repressors) and the uptake activity of organisms (in removing repressors) would act as regulators on primary depolymerising activity. If there are quantitative differences in response of the enzyme systems of different organisms or qualitative differences, with some organisms possessing inductive enzymes and others constitutive, then the potential permutations of interaction are manifold.

The present state of knowledge of the enzymatic basis of depolymerisation processes during decomposition has done little more than reveal the potential complexity of the system and can support only the few superficial generalisations offered above. The very complexity does however lend weight to the hypothesis advanced at the end of Chapter 4 that one possible basis for the species diversity of the decomposition subsystem is the diversity of possible nutritional strategies open to decomposer organisms. It is also clear that many problems of interpretation of decomposition processes, particularly those of the differences in rates of decomposition between resources and between resource components, and those of the roles played by different decomposer species or groups of species, will only be settled by an increased understanding of the biochemistry of the depolymerisation processes.

5.4 Immobilisation and mineralisation

The end product of the depolymerisation reactions described in the previous section is an extracellular pool of simple organic compounds such as sugars and amino acids. These molecules are now small enough to be absorbed into decomposer cells through the cytoplasmic membrane where they enter the pool of metabolic intermediates. Molecules entering the cells may have been transported to the cell surface by leaching, by the product of the decomposer's own depolymerising enzymes or the product of the activity of other decomposers. This emphasises a general point of some significance—that up to the point at which the soluble products enter the oxidative pathways within the decomposer cells the catabolic processes of decomposition impose an energetic deficit on the decomposer organisms. There is an energetic cost to the organism in the production and secretion of the depolymerising enzymes but no energy gain from the depolymerising activities themselves. There is a further expenditure of energy during the uptake of materials by active transport processes. These costs can be placed under the heading of maintenance energy which we shall consider later. An interesting product of this cost is the advantage gained by organisms that absorb sugars and other compounds that are produced by the depolymerising activities of other organisms. This energetic benefit may explain one of the selective benefits of the absence of depolymerising enzymes.

In the discussion of immobilisation and mineralisation in Chapter 2 we emphasised the inseparable nature of the two processes. The biochemical basis of this can be illustrated by reference to the metabolism of amino-acids. Inorganic nitrogen may be immobilised by micro-organisms by the *oxidative amination* of keto-acids with ammonia. An example of this is the reaction catalysed by Glutanic Dehydrogenase.

$$NADH + NH_4 + \alpha\text{-ketoglutarate} \rightleftharpoons NAD + glutamate + H_2O.$$

N immobilised in this fashion enters the amino acid pool of the cell and may be utilised directly in the biosynthesis of protein. Alternatively other amino acids may be formed by a series of *transamination* reactions. Immobilisation of N may also occur by uptake of amino acids. In this case the amino acid may be incorporated directly into protein or the amino group subjected to transamination, e.g.

$$Amino\ acid + \alpha\text{-ketoglutarate} \rightleftharpoons glutamate + keto\ acid.$$

In this case the keto acid, i.e. the carbon skeleton of the amino acid, may be mineralised via the tricarboxylic acid cycle (TCA). If the cellular demand for N is low then the amino acid may be *deaminated*, for instance by a reversal of the glutamic dehydrogenase reaction described above. The product of this reaction is NH_4^+ which may be liberated from the cell and α-ketoglutamate which can be oxidised via the TCA or utilised biosynthetically. The examples given represent only a few of the possible transformations but the significance of them all is the same. The carbon and nitrogen of an amino acid may be mineralised and appear as CO_2 or NH_4^+; or be immobilised and form microbial (or animal) protein, polysaccharide etc. A considerable number of experiments have been performed which demonstrate the occurrence of these actions in decomposer habitats. We may use the example of Greenwood & Lees (1956 and 1960). They used the technique of percolation—the repetitive washing of soil with a solution of a known substrate (in this case an amino acid) in such a manner as to saturate all possible sites of its utilisation and to enable sampling of soluble products formed by microbial action. Care was taken to keep the soil aerobic during the percolation and

respiratory analysis of oxygen consumed and carbon dioxide formed was
also carried out. This enabled them to estimate the extent to which the
complete mineralisation of an amino acid, as given below in the example,
was quantitatively achieved:

$$CH_2(OH)CH(NH_2)COOH + 2\tfrac{1}{2}O_2 \longrightarrow 3CO_2 + 2H_2O + NH_3$$
(Serine)

Fig. 5.8 shows the time course of a typical experiment. Amino-N dis-
appeared within twenty-four hours and there was a corresponding rapid
increase in mineral-N. The maximum mineralisation of N always fell short
of the total amino-N input however; no other form of extracellular N could
be detected and the authors concluded that the difference was due to
immobilisation in microbial tissue. When different amino acids were compared it was seen that the amount of mineralisation was related to the C:N
ratio of the molecule ranging from 77% mineralisation for glycine (C:N =
1·7) to only 34% for phenyl alanine (C:N = 7·7). Similarly only about 40%
of the amino acid was recovered as CO_2 and it was assumed that the remainder
was immobilised as microbial-C. This experiment emphasises the regulatory
role of C:nutrient ratios in determining the balance between immobilisation
and mineralisation at the metabolic level. The data of van Driel (1961)
emphasise this point again. In a similar experiment he incubated twelve

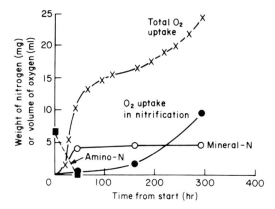

Fig. 5.8. Aerobic mineralisation of an amino acid. The amino acid DL-
serine was percolated through 10 g of soil as a solution containing 45 mg in
50 ml of water. The rapid disappearance of the NH_2-nitrogen is shown
accompanied by the establishment of a pool of mineral-nitrogen. The difference
between the two may be attributed to nitrogen immobilised in microbial tissue.
Oxygen uptake due to nitrification indicates the further oxidation of the
mineral nitrogen. The difference between this curve and that for total oxygen
uptake shows that oxidation of carbon skeletons continued for little longer
than N-mineralisation (after Greenwood & Lees 1956).

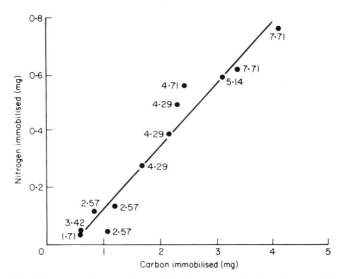

FIG. 5.9. Nitrogen and carbon immobilized upon incubation of twelve different amino acids with soil in quantities to supply 1·4 mg of nitrogen per 20 g of soil. The number beside each point is the number of milligrams of carbon present per milligram of nitrogen in the original amino acid (van Driel 1961).

different amino acids with soil and showed the varying degrees of immobilisation of C and N in relation to the initial C:N ratio (Fig. 5.9)—note the similarity to Fig. 1.15). It should be noted that these experiments are relatively short term and the dominant feature of immobilisation is the synthesis of microbial tissue. In the next section we shall consider the mechanism and the efficiency of the immobilisation reactions; in the last part of the chapter we shall consider another aspect of immobilisation—the formation of humus.

5.4.1   Immobilisation reactions and energy transfer

We can thus picture the cellular metabolism of decomposers in a simple modular way with the balance of their mineralisation and immobilisation reactions controlled by the composition of the available C and nutrient sources outside the cell (Fig. 5.10). The equilibrium pictured here is between extracellular resources and the intracellular pool of intermediate metabolites. These metabolites may be utilised for cellular growth or be oxidised to supply the energy for biosynthesis.

In Table 5.5 the fate within cells of decomposition intermediates such as sugars, amino acids or fatty acids is illustrated. The details of the intermediary metabolism need not concern us in detail, it is generally the end products of these processes which are of ecological significance. The Table

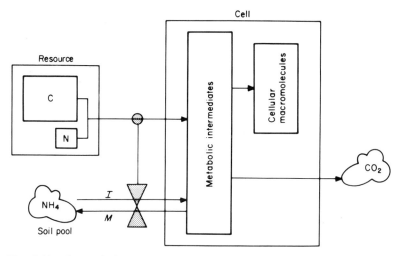

FIG. 5.10. The regulation of mineralisation and immobilisation of nitrogen by the C:N ratio of the extracellular organic resources available to a decomposer cell. Also shown are the fates of the assimilated C and N in the cell.

emphasises the basic unity of the oxidative pathways. Thus in the first stage of oxidation the majority of substrate molecules, whether sugars, fatty acids, amino acids, or aromatic molecules, are converted to a small number of common intermediates. Under aerobic conditions these are oxidised via the tricarboxylic acid cycle to CO_2 and water with oxygen acting as the terminal electron acceptor. Under anaerobic conditions oxidation is less complete and among the organic compounds formed some must act as electron sinks. Coupled with all these processes are phosphorylation reactions which provide the energy transfer from these catabolic systems to those of anabolism in the form of ATP. Whilst emphasising this basic unity however we should also remember that the details of intermediary metabolism may vary greatly, particularly with regard to the fermentative metabolism of bacteria. Details of pathways concerned may be found in biochemistry text books. The essential basis of all these actions is that the oxidation of the organic molecules taken up by decomposer cells provides energy for the biosynthesis of essential cellular macromolecules, i.e. for the production of decomposer tissue. It is mainly to the efficiency of this process—which determines the amount of energy and nutrient passed on to the next trophic level in the form of secondary resources—that we shall address our discussion.

In our earlier discussion of the quantitative ecology of decomposer organisms (Section 3.4) we drew attention to the lack of reliable data on the biomass or production of decomposer groups in natural environments. A clear understanding of the functioning of the decomposition subsystem requires some quantitative estimate of the flow of energy and nutrient

Table 5.5. Stages in the intracellular metabolism of depolymerisation products. Only some of the major steps are shown to indicate that the fate of C from different types of molecule is fundamentally the same. The intermediates formed as the result of preliminary oxidation may be diverted to biosynthesis. Amino acid groups: (1) = alanine, serine, cysteine; (2) leucine and allies, tyrosine, phenylalanine; (3) aspartic acid; (4) glutamic acid, histidine, proline, arginine. Phenolic groups: (1) following ortho-ring fission; (2) following meta-ring fission (based on Morowitz (1968); see Stanier, Adelberg & Ingraham (1976) for an account of respiratory and fermentative metabolism of sugars, acids and amino acids; see Dagley (1967) for the metabolism of phenolics).

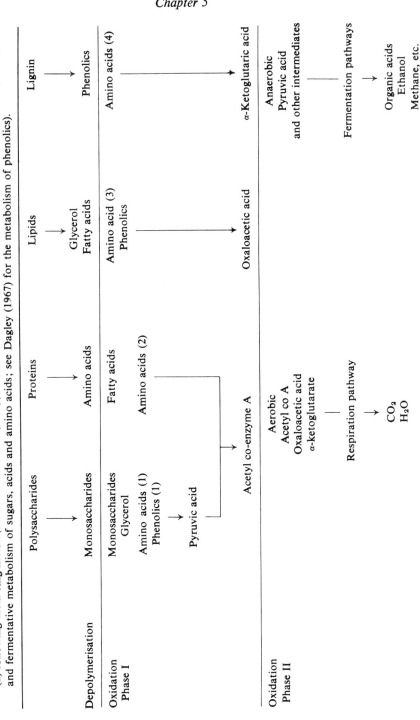

	Polysaccharides	Proteins	Lipids	Lignin
Depolymerisation	Monosaccharides	Amino acids	Glycerol Fatty acids	Phenolics
Oxidation Phase I	Monosaccharides Glycerol Amino acids (1) Phenolics (1) Pyruvic acid	Fatty acids Amino acids (2)	Amino acid (3) Phenolics	Amino acids (4)
	Acetyl co-enzyme A		Oxaloacetic acid	α-Ketoglutaric acid
Oxidation Phase II	Aerobic Acetyl co A Oxaloacetic acid α-ketoglutarate		Anaerobic Pyruvic acid and other intermediates	
	Respiration pathway		Fermentation pathways	
	CO_2 H_2O		Organic acids Ethanol Methane, etc.	

through the major compartments of the food web. In a number of simple models (Figs. 1.4, 2.5) we have suggested that the major energy transfer in primary decomposition is between the plant detritus and the microbial (particularly fungal) saprotrophs. Information from laboratory studies on micro-organisms is of some relevance in this respect. The efficiency of conversion of substrate to cell material has been studied fairly extensively in the last two decades, but almost exclusively under laboratory conditions utilising soluble substrates. If the yield (Y) is expressed by a simple equation $Y = M/S$ where M is the weight of cellular material formed and S is the weight of substrate (energy source) consumed then a very constant efficiency between 0·55 and 0·65 has been found for unicells (Payne 1970). This refers to aerobic conditions of low energy concentration and optimal conditions for growth. Several authors have also predicted on thermodynamic grounds that 0·6 approximates to the expected maximum yield (e.g. see Wiegert 1976, Mayberry *et al.* 1967). In practice however the yield often drops below this value as substrate may be assimilated but re-secreted as extracellular product. Thus Whittaker (1951) in a study of species of wood-decay fungi found values for Y ranging from 0·09 to 0·53 g g^{-1}. Assimilated energy may also be utilised not for growth but for the maintenance of the existing population biomass. In actively growing micro-organisms this is an almost negligible fraction of energy utilised but the amount of energy required to maintain the viable biomass of a decomposer community can be considerable (see later). For decomposer animals the maintenance cost is very much higher because of their mobility and the high energy expenditure on feeding behaviour.

The laboratory study of growth yield in micro-organisms is beginning to pay dividends in the fields of cellular energetics and the molecular control of biosynthesis (Stouthamer 1976). It remains to be seen whether microbial ecologists can exploit these approaches to unravel some of the quantitative aspects of the functioning of decomposer communities. There are very few estimates of microbial growth yields on complex natural substrates or in decomposer environments. Visser & Parkinson (1975), Frankland *et al.* (1978) and Swift (1973, 1978) have made estimates of fungal growth on leaf and wood which suggest efficiencies ranging from about 0·10 to 0·50 g g^{-1}, that is within the range found for simple substrates under laboratory conditions. By observing the changes in bacterial populations and mycelial lengths following amendment Shields *et al.* (1973) concluded that about 60% of the ^{14}C-glucose added to soil was initially immobilised in microbial tissue. Evidence from other isotopic amendment experiments (e.g. those of Jenkinson 1968, Mayaudon & Simonart 1959a, b) supports the conclusion that initially some 40–60% of substrate may be converted to microbial tissue. The subsequent fate is presumably further mineralisation and the formation of humus but a single component ($< 10\%$) of added carbon remains in microbial biomass even after incubation for two or three months (Jenkinson 1977).

These investigations on microbial growth efficiency raise some interesting questions concerning the activity of microbial populations in the soil. Babiuk & Paul (1970) and Gray & Williams (1971) both consider the question of the amount of energy required merely to maintain an observed microbial population level in soil in relation to the estimated energy input (as primary resources). In both cases very little energy remained for production and the authors concluded that for most of the year the major part of the population must remain in an energy limited dormant state. This seems a perfectly reasonable conclusion for mineral soil where the population is dependent on 'pulses' of organic nutrient leached from above, producing a zymogenous type of growth, such as is simulated in the amendent type of experiment. In regions of the soil where major inputs of primary resources occur, such as the litter layer or rhizosphere, it is difficult to support the hypothesis that the population is energy limited for more than a small period in any year. The fact that the sums of energy input and energy utilisation do not tally is indicative of the very superficial understanding we currently have of the energetics of microbial growth in the decomposition subsystem.

5.4.2   Mineralisation reactions and elemental cycles

In this book when considering nutrient cycles we have confined our discussion almost entirely to the transformations between organic and inorganic forms of the essential elements. Earlier in this chapter we described the main mechanism whereby N is transformed from organic to inorganic form by deamination reactions resulting in the formation of ammonium. In a similar way phosphorus is liberated from organic molecules such as nucleotides or phospholipids by the action of phosphatase enzymes. In the case of P the only types of reaction which are biologically mediated are those of immobilisation and mineralisation—the reversible transformation between organic and inorganic forms. Elements such as N and S may, in contrast, occupy a variety of inorganic forms, the transformation between which are due to the activity of micro-organisms. Not all these organisms are strictly decomposers; indeed, some of them are autotrophs. Many of the reactions do have significance for the decomposition subsystem however, either because they affect the availability of a particular element, or because of their effect on the physical-chemical environment of decomposer organisms.

These transformations result in changes in the oxidation–reduction state of the element. The existence of a variety of such forms and of specific pathways linking them has enabled cycles of transformation of each element to be constructed (e.g. see N and S cycles, Fig. 5.11). Different forms of the elements are characteristic of biological pools or the various atmospheric or lithospheric abiotic pools. The concept of the biogeochemical cycling of the elements based on such considerations has proved of considerable value in

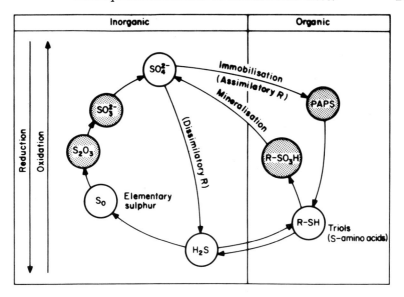

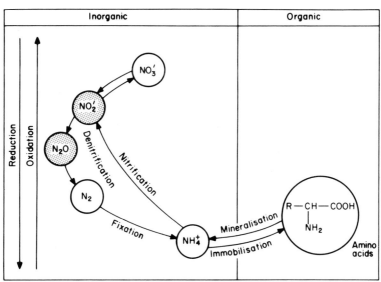

FIG. 5.11. Elemental states of S and N and some of the microbial transformations between them. Molecules or ions in shared compartments are rarely present in more than negligible amounts (S diagram after Doetsch & Cooke 1973).

assessing the magnitude of man's effect on the balance of chemical forms on a global scale (e.g. see Svensson & Söderlund 1976). On the smaller scale of ecological investigation the biogeochemical cycle is not a particularly illuminating concept and the reactions within the cycle are best considered independently. This is probably a realistic viewpoint; whilst equilibria undoubtedly exist between many of the linked reactions, at a given site over a given period of time one or more of the reactions may be dominant and its consequences relatively independent of the others. Thus for instance, whilst it may be of global significance to determine the balance of gains and losses of N from the biosphere due to N-fixation and denitrification respectively, within a given biological system there is no necessity to consider any link between the two processes.

We may consider three types of reaction—reductions, oxidations and fixations (i.e. the conversion of a gaseous form of an element to a soluble or insoluble form). We shall concentrate on N and S transformations which have the most dramatic biological effects and mention other significant elemental transformations in passing.

In aerobic soils of neutral to high pH, N commonly exists in the form of NO_3'. Assimilation of NO_3' by micro-organisms is accompanied by reduction to NH_4^+ (via NO_2') under the influence of a nitrate reductase enzyme. Under anaerobic conditions the same initial reaction may proceed but as an alternative to the formation and assimilation of ammonium a number of gaseous products may be formed including nitrous oxide and nitrogen gas. The alternative pathways can be designated thus:

A detailed account of the biochemistry is given by Campbell and Lees (1967). In the anaerobic reaction (dissimilatory nitrate reduction) the nitrate is acting as a terminal electron acceptor in the place of oxygen. For this reason this is sometimes known as nitrate respiration. Dissimilatory ammonium formation occurs in a number of *Clostridium* and *Bacillus* species and in *Escherichia coli* and is probably not of any great ecological significance. The activity of the *denitrifying bacteria* (the organisms forming a gaseous product) is however of considerable significance for it may result in a considerable loss of N from the area in question.

Similar patterns of assimilatory or dissimilatory sulphate reduction are also brought about by microbial activity. Assimilation of SO_4^{2-} under aerobic conditions results in the formation of an organic intermediate, PAPS (3-phosphoadenosine-5'-phosphosulphate) formed by action between

ATP and the sulphate. This intermediate acts as the source of the formation of sulphur-containing esters or amino acids in the cell. This is of course an immobilisation reaction as is assimilatory nitrate reduction. Dissimilatory sulphate reduction is largely brought about by species of the bacterial genus *Desulphovibrio*, the end product of the process being H_2S. Sulphate is used as the electron acceptor for the oxidation of organic compounds such as pyruvic, lactic or malic acid and ethanol, all formed by the fermentative activity of other organisms. These bacteria are thus decomposers like the nitrate reducers such as *Pseudomonas aeruginosa* and *Bacillus licheniformis*. It should be noted however that some denitrifiers—such as *Thiobacillus denitrificans* or *Micrococcus denitrificans*—are chemoautotrophs.

Desulphovibrio species are most characteristic of marine or freshwater sediments and are largely responsible for the formation of the black deposits of ferrous sulphide. *D. desulphuricans* is probably the most important of these organisms in waterlogged soils but H_2S formation more commonly occurs by another pathway in these habitats. During anaerobic protein decomposition species of *Clostridium* and other bacteria can break down the amino acid cysteine in the following manner:

$$HSCH_2—CHNH_2—COOH + H_2O \xrightarrow[\text{desulphydrase}]{\text{Cysteine}}$$

$$H_2S + NH_3 + CH_3—CO—COOH$$

liberating H_2S and NH_3 and forming pyruvic acid. Sulphate reduction may be significant in terms of the removal of the available form of sulphur from the decomposition subsystem and H_2S formation may be more important in other environmental respects. Firstly, a number of metallic ions may be taken out of solution and deposited as insoluble sulphides. Secondly H_2S may be quite toxic to other organisms and thus further reduce the level of decomposer activity in anaerobic soils. In contrast chemoheterotrophs such as *Desulphovibrio* capable of utilising the products of fermentations under highly reduced conditions may play a key role in maintaining the mineralisation of N, P and other elements.

The formation of H_2S is not an irreversible process moreover even under anaerobic conditions. The photoautotrophic sulphur bacteria such as *Chlorobium* and *Chromatium* use H_2S as a source of reducing power in the fixation of CO_2 and deposit elemental sulphur in or outside their cells. These organisms are relatively infrequent in terrestrial habitats except at the surface of bare waterlogged soils. Sulphur oxidation in soils containing sulphide or H_2S is largely dependent on the resumption of aerobic conditions. The most active group of sulphur oxidising organisms are the members of the chemoautotrophic family *Thiobacillaceae*. The energy released in the oxidation is used for the fixation of CO_2 (Table 5.6). Different members of

Table 5.6. Oxidations of inorganic substrates employed by chemoautotrophic bacteria as a source of energy (from Doetsch & Cooke 1973) $\Delta G°$ is the change in free energy occurring during the oxidation which gives a measure of the energy made available for work in the cell.

Inorganic energy substrate	Formal equation for reaction		Bacterium
NH_3	$NH_4^+ + \frac{3}{2}O_2$	$NO_2^- + H_2O + 2H^+$ $\Delta G° = -66$ kcal	*Nitrosomonas*
NO_2^-	$NO_2^- + \frac{1}{2}O_2$	NO_3^- $\Delta G° = -17\cdot5$ kcal	*Nitrobacter*
H_2	$H_2 + \frac{1}{2}O_2$	H_2O $\Delta G° = -57$ kcal	*Hydrogenomonas*
$S°$	$S° + \frac{3}{2}O_2 + H_2O$	H_2SO_4 $\Delta G° = -118$ kcal	*Thiobacillus*
Fe^{2+}	Fe^{2+}	$Fe^{3+} + e^-$ $\Delta G° = -11$ kcal	*Ferrobacillus*
CO	$CO + \frac{1}{2}O_2$	CO_2 $\Delta G° = -66$ kcal	*Carboxydomonas*

the genus *Thiobacillus* utilise sulphur at varying degrees of oxidation ranging from sulphide and elemental sulphur to thiosulphate (see Doetsch & Cooke 1973 for details), but the end result of all the activity is sulphate formation. Of particular interest is *Thiobacillus denitrificans*, the only member of the group able to operate under anaerobic conditions where it can use NO_3^{2-} as a terminal electron acceptor whilst oxidising sulphur as a source of energy.

$$5S + 6KNO_3 + 2H_2O \rightarrow K_2SO_4 + 4KHSO_4 + 4N_2$$

The Thiobacilli are only one of a group of chemoautotrophs which bring about analogous oxidations (Table 5.6). Among these transformations those which have the greatest significance to decomposition processes are undoubtedly the stages in the oxidation of ammonium known as *nitrification*. The two stages of the process are brought about by two genera of bacteria *Nitrosomonas* and *Nitrobacter*. Nitrification has been regarded as a key process in N cycling because of the probable increased availability of NO_3' to plants in comparison with NH_4^+. A correlation has been assumed between the nitrifying capacity of soil and its fertility and there has been much investigation into the conditions promoting or inhibiting nitrification (e.g. see Russell 1973, Allison 1973). Under high rainfall conditions however nitrification may result in loss of N from soils because whereas the NH_4^+ ion is adsorbed to soil colloids NO_3' is found in free solution and readily leached (see also Section 7.6).

There are some further environmental consequences of these oxidation processes. The production of large amounts of H_2SO_4 or HNO_3 can result in a marked decrease in pH. This not only has direct biological effects but also affects the solubility of other elements. For instance the insoluble ferric states of iron tend to predominate above pH 5 but increasing acidity favours the ferrous state and the formation of soluble salts. High acidity may also result in the direct mineralisation of P, K, Mg, Ca, Al, Fe, Mn and Si by action on primary minerals or clays. Changes in the solubility of iron are also brought about by oxidation/reduction reactions such that the availability of iron (as the soluble Fe^{2+} ion) depends on a number of complex interactions (see Alexander 1971a). Oxidation to the ferric state may be brought about by the chemoautotrophic *Ferrobacillus* and in most well drained soils, particularly those of neutral or alkaline pH, iron occurs in oxidised and insoluble forms. If a soil becomes waterlogged soluble ferrous hydroxides are rapidly formed. This is entirely due to biological agencies but is probably only rarely due to a direct microbial transformation (e.g. the use by *Bacillus circulans* of Fe^{3+} as an electron acceptor in anaerobic respiration). The indirect effects are mediated in a number of ways; by the production of acid conditions due to the acidic products of anaerobic decomposition and by the lowering of the redox potential by microbial activity. As sulphur reduction occurs however ferrous sulphides may be produced in the manner described above. This process is probably largely responsible for the formation of gleys (patches of sticky grey clay) in soils. Solubilisation of iron is of course an important part of the process of podsolisation with ferrous salts being washed down and precipitated as organo-metallic complexes lower in the soil.

The above account of some of the oxidation/reduction reactions in soil emphasises first of all the close integration of the heterotrophic soil microflora (the decomposers) with the activities of the autotrophs. Secondly it reveals the crucial importance of the oxygenation state and the pH of soil to the elemental states and mobilities within it. Both these features of the edaphic physical environment will be taken up in the next chapter.

A small number of free-living prokaryotic organisms are able to 'fix' atmospheric N—that is convert molecular nitrogen N_2 to NH_4^+ which is then rapidly incorporated into organic form. Nitrogen fixation is carried out aerobically by members of the Azotobacteriaceae and by many Cyanobacteria. Anaerobic N-fixation has been shown in a few species of the Enterobactericeae, in a large number of *Clostridium* species and two of *Bacillus* (both genera of the Bacillaceae) and also by most species of *Desulphovibrio*. The biochemistry of N-fixation has attracted a great deal of attention in recent years but the details need not concern us here (see Postgate 1971a and b for reviews). The main feature of significance is that the process is an energy-demanding one and the organisms carrying it out are true decomposers utilising organic residues in the soil as the source of the required energy. Among the N-fixing *Clostridium* distinct saccharolytic and proteolytic groups have been recognised but understandably less attention has been paid to the decomposer function than the fixation process.

In the mid-nineteen-sixties the significant discovery was made that the nitrogenase enzyme of N-fixation also reduces acetylene to ethylene—a reaction very easily followed by gas chromatography. The application of this assay for nitrogenase activity had been of enormous benefit to both biochemical and ecological investigations of N-fixation. The technique is not without its criticisms, particularly if used to obtain quantitative estimates of the extent of N fixed within a given ecological system. It is however the only method available for making such estimates and has greatly facilitated the development of N-budgeting for ecosystem analysis (e.g. see Paul 1975, Svensson & Söderlund 1976).

The significance of N-fixation for the decomposition subsystem is obvious —it is a source of N additional to that in organic form in primary resources or humus. The quantitative importance of N-fixation by free-living organisms is however still a matter of considerable doubt despite the fact that the acetylene-reduction test has enabled the detection of nitrogenase activity in a wide range of sites. There is considerable danger in extrapolating from the small soil samples used in acetylene-reduction assay to inputs at the ecosystem scale. Nonetheless the general conclusion is that the amount of N fixed is generally very low in relation to the total N already present within biological pools, e.g. $0 \cdot 1$ to $300 \, mg \, N \, m^{-2} \, yr^{-1}$ for grassland (including symbiotic fixation by legumes) and about $2500 \, mg \, N \, m^{-2} \, yr^{-1}$ in forests (Paul 1975). The conversion of data to the ecosystem level is probably misleading however for N-fixation is most likely to be at the microsite level in supplementing the N available to decomposer populations operating in N-deficient environments. In this respect it is interesting to note recent reports of N-fixation in decaying wood (Sharpe 1975, Ausmus 1977) and in the hindgut of the wood-feeding termites (Breznak *et al.* 1973, French 1975).

The high energy cost of N-fixation is undoubtedly one of the most significant features determining its occurrence in natural environments. Mulder (1975) has calculated that between 50 and 100 mg of organic-C substrate is required for every mg N fixed by the aerobic *Azotobacter*; one hundred to 200 mg C are required by the anaerobic *Clostridium*. If we add to this the information that the NH_4^+ ion inhibits the nitrogenase enzyme even at very low concentrations then certain restrictions on N-fixation became apparent. It is only likely to take place and be of any significance when the N is mainly in organic form and the C:N ratio above 100:1. At C:N levels below about 25:1 N is non-limiting and ammonification is likely to occur as soon as any C mineralisation takes place. Between that level and about 100:1 N may be limiting but the amount of energy required to fix N would bring the C:N ratio down to non-limiting levels anyway. N-fixation is thus only likely to be of significance in the decomposition subsystem at times and in sites of energy and carbon excess.

5.5 Humus chemistry

In other parts of this book we have stressed that the formation of humus during the biological decomposition of plant matter is a feature of central importance to ecosystem structure and function. Humus is a highly stable component of soils with a very slow rate of decay, as revealed both by ^{14}C dating (Section 1.2.2) and respiratory measurements (Table 4.5 and Fig. 4.8). Furthermore its colloidal structure is an important determinant of ionic equilibria in soils. In this section we shall discuss current theories of the chemistry of humus, its possible mode of formation during decomposition processes and finally consider the possible reasons why it is so stable and resistant to decomposition.

5.5.1 The composition and structure of humus

In Chapter 2 we defined humus as an amorphous organic fraction of soil, physically and chemically distinct from that 'cellular' fraction of SOM which is composed of the residue of plant, microbial, and animal remains. The chemistry of humus remains elusive; although there are features common to all humus extracts it is probable that it is chemically heterogeneous and should be regarded as a category of molecular form. Whilst it is clearly polymeric in nature, it shows no evidence of crystallinity or of any repeating pattern of bonds or monomeric constituents. Schnitzer & Khan (1972) have reviewed in detail the methods used in the chemical and physical analysis of humus. Most approaches to its chemical determination originate from the recognition of a number of distinct molecular categories which are separated by the use of a series of extractions. A small *bitumen* fraction, which consists largely of fats, waxes, and other lipid- or hydrocarbon-like compounds (possibly largely of plant origin) may be extracted with benzene and methanol. The humus proper forms a blackish solution in dilute (0·5 M) NaOH plus an insoluble black precipitate termed *humin* (HU). Acidification of the solution with HCl (to about pH 1) results in the precipitation of the main fraction of the humus which is called *humic acid* (HA). A second, minor, component of the precipitate is *hymatomelanic acid* (HYA) which unlike HA is soluble in ethanol. Concentration of the acid solution by dialysis yields the second major component, *fulvic acid* (FA).

Humic acids are polymeric structures of a highly heterogeneous nature. Most information regarding their structures has come from degradative studies which reveal the nature of possible components but very little of the way in which they are arranged. Humic acids have N contents which range from 0·4 to 5%, but most commonly at the upper end of this range. Of this the major part (20–50%) is in bound amino acid form, about 3–10% in amino sugars and a small amount ($< 1\%$) in heterocyclic, purine or pyrimidine form. The remaining 50% or so remains unidentified as does the exact way in which the above monomers are combined within the HA complex. Other humus forms contain similar nitrogenous components (Table 5.7). Indeed the N content of humus is relatively constant at 5–5·5% with C:N ratios in the region of ten to twelve (Allison 1973). Whilst there is evidence of peptide linkages, attempts to isolate intact proteins from humus have been largely unsuccessful. The existence of significant amounts of amino sugars plus such 'non-protein' amino acids as diaminopimelic acid suggest that a significant component of the HA may be residues of microbial cell walls. Polymerised carbohydrates are also found in HA preparations and again it has been suggested that these polysaccharides may be of microbial origin. It is probable however that there are also significant fractions that are residues of higher plant polymers. There is considerable evidence that much of this carbohydrate fraction is resistant to further degradation either because of 'recrystallisation' of the polymer into an unaccessible form or by 'tanning' of the surface by phenolic-protein complexes originating in plant cytoplasm (Handley 1954 see Section 4.3.3). About 20–50% of the phosphorus in soil is commonly in organic form in association with humus, but the phosphorus content is more variable than the N and the C:P ratio ranges from 60 to above 200. The P component is even less well characterised than N but the major fraction (up to 35%) appear to be a variety of forms of inositol phosphates—several of which are unknown in plants; a further 2% may be in nucleic acid form and about 1% in phospholipids and phosphates. Over 50% of the S in soils may be in organic form—most of it probably as constituents of the peptide component of the humus.

There is little doubt that the major component of humic acid is aromatic in nature. Wright & Schnitzer (1961) estimated the aromatic component to be as much as 49% of a HA preparation. Haworth (1971) has pictured the structure to have a 'core' that consists of heterogeneous aromatic polymer with carbohydrate, peptides, phenolic acids and metals attached 'peripherally' (Fig. 5.12). The attachment of the amino acid, peptide, or protein component may be stable against biological attack, due to hydrogen bonds formed by reactions analogous to those of tannins with proteins. This 'tanning' action

Table 5.7. Distribution of nitrogen in various fractions of humus (Khan & Sowden 1971, 1972).

	Humic acid	Fulvic acid	Humin
Total N (%)	4.0	3·9	4·6
Amino-acid N (% total)	28·3	26·4	36·1
Amino-sugar N (% total)	1·3	3·6	1·6
Ammonia N (% total)	19·8	15·1	22·1
N accounted for (% total)	49·4	45·1	59·8

(A)

Humic acid molecule

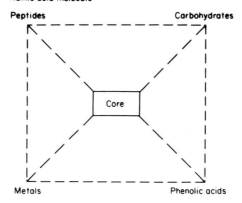

(B)

Fulvic acid core

FIG. 5.12. Some postulated structures for humus components. (A) Diagrammatic representation of the humic acid molecule proposed by Haworth (1971). The core is supposed to be largely aromatic with the other components bonding to its reactive surface. (B) A structure proposed by Schnitzer and Khan (1972) for the aromatic core of fulvic acid.

has been mentioned before as possibly important in the sequestering of N into inaccessible form during humus formation. Tannins are polyphenolic polymers based on either gallic acid or catechin monomers. During tanning, hydrogen bonds are formed between the phenolic group of the tannin and the peptide bonds of proteins. Such complexes are far more resistant to enzymatic proteolysis than the original protein (Table 4.12).

The nature of the aromatic core of HA is uncertain. Hurst & Burges (1967) suggest that there are units of two basic types, those derived from lignin breakdown and those derived from flavenoids. However, there is certainly a polycyclic fraction which is often suggested as quinonic in origin. Other authors favour a large component derived from microbially synthesised aromatics (see Section 5.5.2). The doubts as to the exact nature of the aromatic component derive partly from the harsh nature of the degradative treatments that are used to demonstrate their presence. Schnitzer & Khan (1972) have proposed a complex aromatic core for FA (Fig. 5.12), based on hydrogen bonding between phenolic and benzenecarboxylic acids. This structure is highly 'porous' and enables the acquisition of other organic molecules by internal adsorption as well as by peripheral bonding as in the HA model. The relationships between the main humic components are still far from clear, but some consistent differences have been noted between FA and HA. Both forms show a range of molecular weights but FA has a lower range (1000–10,000) than HA (5000 to greater than 1,000,000). FA is less condensed, has a higher carbon content (44–48% v. 35–37%) and greater frequency of functional groups (particularly carboxyls) than HA. Humin seems to closely resemble HA and may differ from it only in a closer association with inorganic materials leading to more ready precipitation.

The humic components as a whole however appear to be characterised by high molecular weight spherical molecules with a high proliferation of cross-linkages and a reactive surface capable of bonding a wide variety of other compounds. This type of structure accommodates both the overall similarity and the variable nature of humic materials. The elucidation of the details of the chemistry is hindered by this heterogeneity and as Hurst & Burges (1967) remarked 'the concept of humic acid (remains) almost a personal one, varying with the vantage point and prime interests of the observer'.

5.5.2 Formation of humus

In view of the uncertainty surrounding the chemical structure of humic materials it is not surprising that the mechanisms of synthesis of these macromolecules remain shrouded in mystery. Allison (1973) in a review of the formation of humus quotes the words of pioneering workers in the field Schreiner & Dawson (1927), '(humic) substances are rarely homogeneous products of any one type of reaction or any one group of parent substances, but...represent accumulations of the more resistant end-products of a variety of reactions taking place under natural conditions, either directly or indirectly through biological processes'. Regrettably little in terms of detailed understanding of the sequential process of humus formation can be added to this general statement. This may be due to the fact that research into humus formation has been largely treated as a chemical, rather than a *bio*chemical problem. Placing the problem within the context of the biological processes of decomposition may aid in elucidating the essential features of the process. The main problem concerns the origin of the aromatic core of humic materials such as HA. Earlier theories, such as that of the ligno-protein origin of humus proposed by Waksman in 1938, supposed that humus originated largely by relatively minor modifications of the resistant residues from the decomposition of plant matter. This 'plant alteration' hypothesis has been replaced more recently by the generally accepted view that the aromatic content of humus has little direct similarity to the lignin molecule. There are however considerable differences of opinion as to the origin of the aromatic constituents.

Haider, Martin & Filip (1975) have summarised the accumulating evidence for the similarities between many components of humus and a variety of aromatic pigmented compounds synthesised in pure culture by fungi and bacteria. Many of these compounds are formed from aliphatic carbon skeletons which are the products of carbohydrate metabolism. This seems particularly characteristic of Fungi Imperfecti. These aromatisation pathways seem to be less characteristic of Basidiomycetes which in contrast, are the main agents for the release of aromatic compounds by lignin decomposition.

Evidence for the incorporation of precursors of both plant and microbial origin comes from ^{14}C studies. Table 5.8 shows some typical results taken from data of Mayaudon & Simonart (1959a and b). This shows that a large component of ^{14}C-labelled plant materials may be incorporated into all three major humic components within thirty days. Similar studies using ^{14}C-labelled microbial tissues confirm the incorporation into humus of substances of microbial origin (Mayaudon & Simonart 1963, 1965). Table

Table 5.8. Incorporation of carbon into humus from decomposing primary resources. [14]C labelled materials mixed with soil and residues estimated after 30 days (from Mayaudon & Simonart 1959a, b).

	% of added ^{14}C found in			
	Total residue	FA	HA	HU
Glucose[1]	11	2	4	6
Water solubles[2]	22	8	6	8
Hemicellulose[2]	20	4	8	8
Cellulose[3]	20	5	6	9
Lignin[3]	75	7	34	29
Protein[2]	25	4	11	10

[1] added as pure substrate.
[2] extracted from ^{14}C labelled rye-grass.
[3] extracted from ^{14}C labelled rice.

5.8 confirms that lignin products may be major contributors to humus formation but the modern view is that the lignin-type phenols are considerably modified—presumably by microbial action—prior to incorporation in HA or FA. The same data also show that, whilst most of the glucose is rapidly mineralised, there is a fraction of the carbon which is incorporated into humus. Carbon transfer from glucose and cellulose to humus-bound amino acids has also been demonstrated by ^{14}C techniques (Sorenson 1972).

From all these studies a picture is emerging of humus formation as a continuous process during decomposition with carbon from all the major plant components contributing to humus carbon either by direct transformation or, more commonly, after extensive 're-synthesis' activity by microorganisms. The suggestions of Schreiner & Dawson (1927) that humus might be formed by a variety of pathways from a variety of precursors are thus largely borne out by modern research. Fig. 5.13 summarises some of

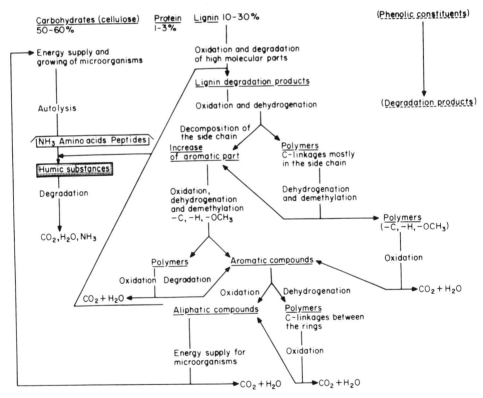

FIG. 5.13. Possible pathways of humus biosynthesis. Note the importance of microbial transformation of lignin degradation products (from Flaig 1966).

the probable pathways of formation of humus precursors based on the extensive studies of Flaig (1966) and his co-workers. What is not yet clear is the way in which the aromatic building blocks, whether of plant or microbial origin, are assembled into the humus polymers. Views on this vary from suggestions that the polymers are synthesised enzymatically within microbial cells to the idea that they form by purely chemical oxidative condensations following cellular autolysis (Felbeck 1971). The less condensed and more reactive nature of FA compared with HA and HU have led to suggestions that they may be formed in sequence but there is little evidence to support this. ^{14}C is rapidly incorporated into all three fractions (Table 5.8) which is more suggestive of simultaneous formation. In any case it is probable that the variation within each of these fractions is at least as great as between them.

5.5.3 Decomposition of humus

In an earlier chapter we described the method of ageing soil organic matter by the ratio of carbon isotopes (Section 1.2.2). This method has shown that the mean residence times for humus in the soil are to be reckoned in hundreds or even thousands of years, implying that these materials are almost inert. This type of analysis may be extended to date the individual humus fractions. The most extensive study of this kind is that of Campbell *et al.* (1967) on a number of Canadian soils. Their main findings are illustrated by the data in Table 5.9. In both soil types the different fractions had markedly different residence ages implying that FA was decomposed more readily than HA which was in turn younger than the humin. Other analyses (see review by Schnitzer & Khan 1972) show similar differences. Most revealing perhaps

Table 5.9. Mean residence times (t_r) for fractions of the humus in two Canadian soils (after Campbell *et al.* 1967).

	Grey-forest podzol		Chernozem	
	Content (% C)	t_r (yr)	Content (% C)	t_r (yr)
Total humus	—	870	—	250
Fulvic acid	15	495	29	0
Humic acid	(40)*	—	37	195
(residual)	33	1400	—	—
(hydrolysable)	7	25 ± 50	—	—
Humin	(31)*	—	32	485
(residual)	24	1230	—	—
(hydrolysable)	7	465	—	—

* Totalled from the two fractions.

was the fractionation of the podzol HA by the use of hot 6N HCl. The hydrolysable fraction had a very short residence time, ranging round twenty-five years (and in fact too low for the sensitivity of the method). This readily decomposed fraction probably contained much of the 'less-humified' component of the HA such as peptides and carbohydrates. The second feature to emerge from Campbell's data, which is also confirmed by other studies, is that both the total and the individual humus fractions have different residence times in the two soils. Whether this reflects differences in their composition (i.e. a resource quality factor) or the effect of differing physical environments, we shall consider in the final section of this chapter.

Little information is available concerning the organisms capable of degrading humic materials or of the biochemical pathways involved. This is hardly surprising in view of the poorly understood chemistry, the high degree of resistance and the experimental difficulty involved in separating the degradative effect from possible re-synthesis processes. A number of authors (e.g. Hurst, Burges & Latter 1962; Mathur & Paul 1967) have used enrichment cultures with humic materials as sole carbon or nitrogen sources to isolate potential humus-degraders from soil. A number of organisms, including Basidiomycetes, Fungi Imperfecti and bacteria have been shown to decolourise humic and fulvic acids. The attack appears to be predominantly on the aromatic core of the molecule rather than on the nitrogen fraction but the mechanism of action remains obscure. Clearly little can be concluded from these largely qualitative studies, using extracted substrates, concerning the pattern or rate of decomposition in the field. Nor is it clear why humic materials are chemically so stable as both the probable bonds and the aromatic rings are known to be susceptible to microbial enzymes in a manner analogous to lignin catabolism. It is possible that the resistance is most dependent on the physical nature and the heterogeneity of the structure. The complex three-dimensional structure formed by the cross-linking and the non-repetitive pattern of the polymer may reduce to virtually zero the probability of matching steric patterns between enzymes and substrate.

Some evidence for the importance of this accessibility factor may come from consideration of circumstances under which humus degradation is markedly enhanced. Among the conditions in which this occurs is that in which physical disruption of soil structure takes place. In the final pages of Chapter 1 we described how the content of soil carbon generally declines when a soil is converted to agricultural use and regularly ploughed. This is further exemplified in Fig. 5.14 showing the general decline in soil N in a variety of soil types, originally under prairie grassland, when they were ploughed for the growth of arable crops. The cause of such decline is undoubtedly complex and many features contribute to it. One suggestion for which there is some experimental evidence is the physical destruction of the soil crumb structure permitting greater accessibility of microbes and their

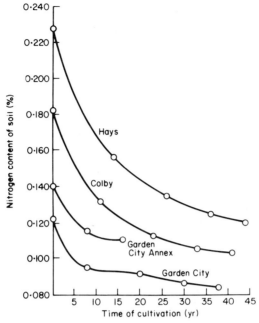

FIG. 5.14. The effect of cultivation on the N content of soils. The four different soils were all under prairie grassland in Kansas prior to ploughing and cultivation (from Hobbs & Brown 1957).

enzymes to the organic substrates—an effect comparable with the stimulation of microbial catabolism by the comminutive action of animals (Fig. 4.14). The same effect can be produced by subjecting a soil to repeated drying and wetting cycles (Birch 1958), grinding (Waring & Bremner 1964) or autoclaving (Stanford, Legg & Chichester 1970).

Both in the case of cultivation and in the experimental procedures it is probable that the portion decomposed after the disruption corresponds to a more labile fraction than the aromatic core. This may correspond to the 'hydrolysable' fraction of the HA and HU which has been shown to have a short residence time (Table 5.9). The point is, however, that whilst this fraction is chemically accessible to microbial enzymes it may be physically protected from attack by the complex stereochemistry of the soil colloids.

5.6 Decomposition at the molecular level

The preceding survey of the biochemistry of decomposition processes goes some way towards the provision of explanations for some of the resource quality effects cited in Chapter 4. We must also ask whether it provides anything like a unified description of decomposition at the molecular level

which will help in understanding the ecological phenomena derived from
larger scales of resolution. In Fig. 5.15 we present a highly simplified unitary
model of the biochemical processes of decomposition. The cascade structure
of decomposition is reduced to three levels—that of primary decomposition
feeding a secondary level of decomposer growth, and decay of both of these
supporting the third level of humus formation. These three levels have a
sequential relationship but in reality the processes are coincident, as we have
emphasised in earlier models (e.g. Fig. 2.4). The outputs from each level are
of two kinds—those of mineralisation and those of immobilisation—the
latter being pictured as transfers to the succeeding level. The most important
biochemical distinction however, is between three types of metabolic pathway
—that of carbon, that of nutrients such as N, and that of aromatic compounds
such as polyphenols or lignin components. The three are intimately linked
in practice—protein or lignin may be sources of C and energy for decomposers and carbohydrates may be precursors of aromatics but the dominance

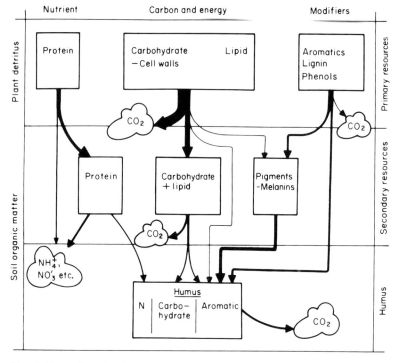

FIG. 5.15. Major chemical pathways of decomposition processes. The model
illustrates in a simple way the changes in composition of resources through
two 'cascades' of decomposition. The increasing dominance of aromatic
components is emphasised and indicates the regulatory influence of these
components over the dissipation of energy and the flow of nutrients.

of function is most probably as shown. The most significant feature is the regulatory effect of one pathway on another. The carbon path gives the energy dynamics to the system and in particular, by the regulatory effect of C : nutrient ratios, determines the nutrient dynamics. The aromatic pathway acts as the major resource quality factor at all levels—both by forming enzymatically recalcitrant molecules and also by the direct toxic effects of the phenolic or quinonic monomers. It is interesting to note the increasing dominance of this path as decomposition proceeds. It is possible that this provides an essential stability to the decomposer subsystem. The sequestering of a component of C, N and other nutrients into stable forms from which they are only released at very slow rates or under conditions of extreme change in the physical and biological environment may be regarded as a type of chemical homeostasis. The feedback nature of resource quality effects can be seen here as having a general role rather than just a series of specific roles.

Regulatory mechanisms occur within every part of the system (catabolite regulation of polysaccharase activity for instance)—between different pathways within each level (e.g. the effect of lignin on polysaccharase enzymes) and between levels (e.g. the tanning of protein by humic acids). Taken as a whole however, the formation of humus can be seen not only as a product of decomposition processes but as the major pathway of internal regulation of the whole subsystem.

How then does this model fit to our earlier descriptions of decomposition processes? Firstly it provides an extension to the discussion of decomposition rates given in Chapter 4. The suggestion of Fig. 4.8 that there may be two or more distinct phases of decomposition characterised by differing rates of breakdown is supported by the biochemical analysis. The initial rapid rate coincides with the utilisation of accessible simple molecules such as sugars and the depolymerisation of cell wall and storage polysaccharides. This is also predicted by other data of Chapter 4 (Figs. 4.5 and 4.7, Table 4.7). During this period microbial resynthesis occurs (Fig. 4.6) probably converting some 40–50% of the carbon of primary resources into microbial tissue. But, both as a result of the synthesis of microbial aromatics, and by direct conversion of primary resource components, this phase is also characterised by the synthesis of humic materials (Table 5.8). Thus the second, slower phase of decomposition in Fig. 4.8 is due not only to the accumulation of relatively resistant residues such as lignin but also to the increasing proportion of humified material. Jenkinson's data (Fig. 4.8) show however that this second rate of carbon mineralisation is still much faster than that of the native SOM. Even this half-life estimate of twenty-five years is much below the times estimated by isotopic C dating (Table 5.9). This range of different decay rates implies the presence of distinct organic components within the soil organic matter and probably within the humus itself. This is good evidence for the cascade nature of decomposition and confirms that the three cascade structure of the model in Fig. 5.15 is far too simple.

There is regrettably little analytical data from soils against which to test this model. This is largely due to the major uncertainties still attendant upon the quantitative analysis both of humus itself and also of other components such as protein or polysaccharide in soil (e.g. see Finch *et al.* 1971). Most data that is available comes from Russian studies. Table 5.10 shows the

Table 5.10. Composition of the organic layers of a soil under spruce, showing the increase in humus components and the decrease in primary resource components with horizon age (from Remezov & Pogrebnyak 1969 after Gradusov 1958).

	Percentage ash-free dry weight					
	Hemicellulose	Cellulose	Lignin	Bitumen	HA	FA
L (O_1)	13	26	27	12	3	4
F (O_2)	7	18	22	10	6	14
H (O_3—A_1)	7	16	19	11	7	15

vertical distribution of HA and FA in a forest soil compared with that of plant polysaccharides and lignin. This illustrates neatly the inverse relationship of change with time between primary resource components and humus. It is interesting however to note the extent of persistence of the polysaccharides in the 'humus' horizon. In Chapter 1 we suggested that a major component of the accumulation of SOM in mor humus was probably of partially decomposed plant origin. Handley (1954) was the first to propose this and suggested that it was due to the masking of the cellulose from attack by the deposition of tanned proteins in the microfibrils. Handley was able to find only a few data to support his contention of a higher cellulose content for mor than mull humus forms and there is little more today.

Neither are there many data which enable us to answer the obvious but interesting question as to whether different types of soil differ significantly in the quantity and composition of their humus. The most detailed discussion of this has been by the distinguished Russian soil scientist Kononova (1966, 1968). As her data shows (Table 5.11) zonal soils may differ not only on the amount of humus that is accumulated but also in its composition. Kononova distinguished clearly between the type of humus formed in forest areas with acid soils such as the podsol and the tropical latosol where the HA/FA ratio is less than one, and that characteristic of the drier steppe soils such as chernozems where the ratio is above one. The grey deciduous forest soil lay in an intermediate position. Kononova suggested that the larger HA content of the chernozem soil suggested that it would be much more stable than that

Table 5.11. Humus composition of a range of different soil types (after Kononova 1966 and 1968).

Soil	Ecosystem type	Humus content %	Component C as % of total		
			HA	FA	HA/FA
Tundra	Tundra	1·0	10	30	0·3
Podzol (well developed)	Boreal forest (taiga)	2·5–3·0	12–15	25–28	0·6
Grey forest soil	Deciduous (oak) forest	4·0–6·0	25–30	25–27	1·0
Chernozem (deep)	Grass-herb (steppe)	9·0–10·0	35	20	1·7
Chernozem (shallow)	Grass-herb (steppe)	7·0–8·0	40	20	2·0–2·5
Lateritic soil	Sub-tropical forest	4·0–6·0	15–20	30	0·6

of the forest soils. This agrees with the estimates of residence time obtained by Campbell *et al.* (1967), (Table 5.9), although the HA/FA ratios obtained by them (their grey-forest podsol has a ratio of 1:3) do not fully accord with those of Kononova.

The biochemical study of decomposition processes is still in its infancy. Many problems of method need to be solved before major progress can be made. Nevertheless there are areas of investigation for which the methods are available but which await exploration. It should be of enormous interest for example to extend the types of analysis illustrated in Tables 5.9, 5.10 and 5.11 to an intensive study of different soil and vegetation types, but including an investigation of the resource quality factors of the litter input (e.g. the polyphenol spectrum—see Coulson *et al.* 1960) and the biological nature of the decomposition processes.

6

THE INFLUENCE OF THE PHYSICO-CHEMICAL ENVIRONMENT ON DECOMPOSITION PROCESS

6.1 The environments of decomposer organisms

The decomposition module of Fig. 2.6 shows the processes of decomposition regulated by the combined effects of the resource quality and the physico-chemical environment on the community of decomposer organisms. The macro-climatic variables—rainfall and temperature—are, clearly, important regulators of decomposition processes. In Chapter 1 we saw that differences in the rates of turnover of organic detritus in different ecosystems correlated with a latitudinal gradient in climate. The dominant influence of climate was also seen in relation to its role in soil-forming processes (Fig. 1.8). Major differences in the physico-chemical make-up of soil may also determine the distribution of decomposer organisms and the rate of decomposition processes. The impact of the physico-chemical environment on decomposer organisms is, felt however, at scales much smaller than those at which major differences in climate or soil-type are described.

Reference to Figs. 3.2 and 3.3 shows that the size of decomposer organisms ranges from a few cubic microns to several cubic centimetres. The environmental features relevant to these organisms must be set at the same scales. The habitats of decomposer organisms are largely on or in the soil, although some are located in the canopy. The first four sections of this chapter are concerned with an analysis of the regulatory influence of the abiotic micro-environments of litter and soil on the decomposer organisms. The emphasis is on the regulatory mechanisms of individual features of the environmental complex—moisture, aeration, pH, temperature, etc. In the latter part of the chapter we consider the influence of the environmental complex as a whole with a particular emphasis on the dominating effect of climate and its interaction with edaphic features.

Both litter and soil may be regarded as porous structures—a solid phase comprised of particles, with a ramifying system of pores between them. In the litter layer the solid phase is largely organic and the particles are of large size varying from substantial branches to minute seeds. Pores occur both within the plant tissues—the microscopic lumina of the vascular tissues—and between the different litter components. In the later stages of decay, cavities are excavated within the larger resources by decomposer animals. In the soil layers the organic component is progressively more amorphous and is increasingly associated with mineral components such as clays or sands to form soil crumbs. In each case the pores are filled by the gaseous and liquid phases of the environment which interpenetrate not only the major pores between crumbs or litter components, but also the micro-pores within them. Some of the varying microhabitats created by these combinations of particle and pore are illustrated in Fig. 6.1.

The dimensions of the pores, and thence the moisture-holding capacity and the extent of aeration of the soil, is determined by the particle size. Thus a predominance of small particles, such as clay, greatly increases the surface area of the solid phase and the porosity of the soil. Chapman (1965) calculated that the internal surface area of a 1 cm^3 of a good loam soil might be as much as 67,000 cm^2. The proportion of this area due to minerals of small size such as clay was 98% compared with the larger sand particles which contributed only 0·2% of the internal area, although comprising 31% by weight. The drainage and aeration characteristics of a soil are as much affected by the sizes of the soil pores as by the total pore volume so that a soil of very high pore volume but consistently small particle (and thence pore) size may be subject to waterlogging (e.g. heavy clay soil). The SOM content is again important here the formation of soil crumbs by complexing between clay and organic colloids results in a varied particle size and improved drainage and aeration.

As emphasised earlier (Chapter 3) the internal surface area of the litter and soils is of particular significance to the bacteria which are largely adsorbed to particle surfaces. The habitat of micro and meso fauna is within pores of varying size whereas that of the fungi varies from the penetration of particles to ramification across their surfaces and the interparticle pores (see Figs. 3.5 and 6.1). These differences in organism size and microhabitat dimension determine to some extent the manner in which the individual environmental factors affect these different groups of decomposer organisms.

6.2 Moisture

The direct effect of rainfall as the agent of leaching is considered later (Section 6.6), here we consider the influence of water within the micro-habitats of litter and soil. The moisture environment is important directly in supplying water for tissue growth and a medium for the activity of 'aquatic' organisms such as nematodes and protozoa, and indirectly in the way it influences other factors such as aeration and pH. These secondary aspects will be dealt with under the appropriate headings.

The moisture environment of litter and soil (i.e. the amount and distribution of water within it) depend on the supply of water to it by precipitation,

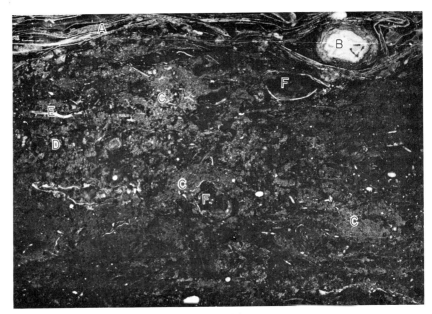

FIG. 6.1(A.)

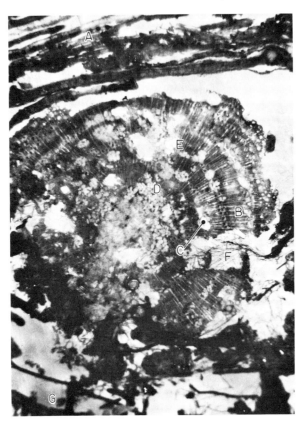

FIG. 6.1(B.)

Details of gelatine embedded sections of woodland soil profiles which emphasise the range of microhabitats present in the organic layers at different scales of resolution.

(A) Top 5 cm (A horizon) or a mor-like moder showing the different combinations of primary resource mass and voids which locally affect gaseous diffusion, water tension, pH and the activities of soil organisms. Compact leaf litter A with larger cavities around twig B, aggregates of fine leaf fragments produced by feeding activities of mesofauna C, open structured regions D, and mycorrhiza E, formed by large leaf fragments and faeces of macrofauna, large

FIG. 6.1(C.)

voids within twigs which have been extensively decomposed to leave tubes of bark F. (B) Cross-section of a rotten twig in the fermentation F sub-horizon of the same soil shown in (A). Compact masses of sweet chestnut leaves (*Castanea sativa*) with dark fungal stroma A, vessels and medullary rays B, Cryptostigmatid mite in cavity excavated by feeding along the vessels C, faeces of Cryptostigmata D, Collembola eggs E, fungal hyphae innervating twig from surrounding leaf litter F, mycorrhiza G. (C) Cavities in compact soil organic matter. Juvenile Cryptostigmatid mite A, dematiacious fungal hyphae spanning the void B, mite faeces C, mycorrhiza D, insect exoskeleton E.

the interception by vegetation and the losses from it by evapotranspiration and drainage. The relative importance of these factors is influenced by the temperature environment, the physical nature of the soil and the character of the vegetational cover. Losses also occur by lateral run-off which is determined by the topography of the area. In dry periods the only loss from soil is by evapotranspiration. In a bare soil the change in moisture content is usually small. Once a dry layer is created at the soil surface the diffusion path of the water molecules is lengthened and the resistance so greatly increased that evaporative loss may rapidly drop to negligible proportions so that under these conditions the moisture content of the soil remains fairly constant. The presence of vegetation however ensures the continuance of transpiration loss and reduces water storage (i.e. creates a water deficit) until the uptake of water by plants is no longer possible and wilting may occur.

During periods of heavy rainfall, water may be lost from a soil by drainage under the influence of gravity. When precipitation ceases, the soil moisture content rapidly stabilises at the *field capacity*, a value which is fairly consistent for any given soil. The ability to hold water against gravity is largely due to capillary retention in the soil pores. Adsorption of water to the surface of clays and other colloids is a contributory factor but it is neither theoretically necessary nor practicably possible to distinguish between the two. These forces may be represented in terms of negative pressure (= *suction*) or potential energy. The tendency to retain water is commonly termed the *matric potential*, but the concept of suction is probably easier to use in describing the system and its significance for soil organisms. The suction pressure can be regarded as the force needed to remove the water from the soil matrix. This can be measured in terms of the equivalent column of water in centimetres, which can be conveniently represented on a logarithmic scale where $pFx = 10^x$ cm water. An alternative unit more closely related to the energetic concept of soil water is the bar (1 bar $= 100$ joules kg$^{-1} = 1022$ cm water, Griffin 1972).

The nature of the soil matrix thus markedly affects its capacity to retain moisture and the relationship between the moisture content and the matric potential of a soil reflects this (Fig. 6.2). As Griffin (1963, 1972) has emphasised, the availability of soil moisture to inhabitant micro-organisms is related to its matric potential and this is not predictable either on the basis

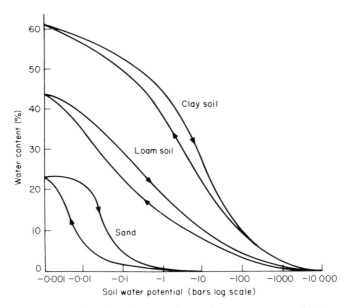

FIG. 6.2. The relationship between the moisture content and the matric potential of three soils. Note that a different relationship is found during wetting (lower curve) and drying (upper curve) (From Griffin 1972 after Yong & Warkentin 1966).

of a simple gravimetric moisture content determination ($gH_2O \, g^{-1}$ dry weight soil) or by estimating the moisture content in relation to the saturation or field capacity (e.g. % moisture holding capacity) unless the moisture characteristics as shown in Fig. 6.2 are known. Unfortunately the moisture relationships of soils are still commonly expressed in these simpler forms, thus creating problems of interpretation.

In summary, we can picture the soil water in a well-drained soil as penetrating the crumbs and forming as films and continua within the solid matrix, gaseous pockets only being found in the larger pores. As the moisture content falls below the field capacity the water phase shrinks back into progressively finer pores, the matric potential decreases exponentially and the amount of free water is markedly reduced.

The patterns of absorption and retention of water in the litter layer, or in dead standing resources in the canopy, have been little studied. Information

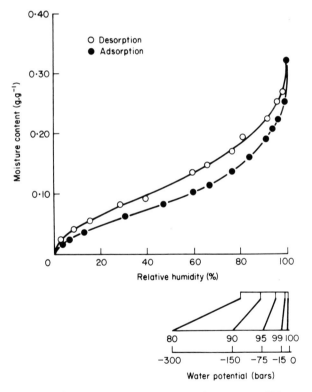

FIG. 6.3. The equilibrium moisture content of wood at different relative humidities (after Browning 1967). The approximate relationship between water potential (pF) and the RH is shown by the lower scale taken from Griffin (1963). Note that as with soil there is a difference between wetting and drying curves.

on the activity of water within plant materials is available however based on studies carried out in the timber industry. Freshly cut timber may have moisture contents up to 200% of the dry weight but if left to dry under normal conditions of temperature in a temperate climate a stable moisture content of about 30–35% will be reached. If the drying is carried out in atmospheres of differing RH this equilibrium moisture content will differ (Fig. 6.3). The equilibrium between the vapour pressure of the atmosphere and the moisture content of the material provides a means of determining the potential of the water in the material for it will equal the osmotic potential of the solution used to control the RH (Griffin 1972). From Fig. 6.3 it is clear that below about 30–35% moisture content, the content and the potential are related but above this value potential is very low and independent of content. This equilibration point at 100% RH is known as the *fibre-saturation-point* (fsp). The water in timber below this content is all held within the cell walls. Cellulose is a hygroscopic material which holds water by capillary retention within its fibrils. Energy is required to remove this water—hence its potential. Above the fsp however water is not retained by capillary action but is located within the cellular lumina and intercellular cavities as 'free-water'. Whether this water exerts any potential against uptake by micro-organisms is uncertain.

6.2.1   The influence of moisture on decomposer organisms

Griffin (1963, 1972) has been responsible for directing the attention of soil microbiologists to the relationship between the matric potential of water in the microenvironment and the activity of micro-organisms. The potential determines the force required to take up the water. Micro-organisms absorb moisture directly through their permeable outer covering and must thus directly overcome the increasing tendency of the soil or resource to retain it as the potential decreases.

Both bacteria and fungi are able to adjust their internal osmotic environment to maintain it hypertonic to the surrounding medium over a very wide range of external potential and it is probably by this means that they are able to maintain their moisture content at about 80% (Christian & Ingram 1959, Burnett 1968). The importance of this potential gradient can be illustrated with reference to fungi; the growth of fungi by apical extension of hyphae is dependent on a hydrostatic pressure being maintained within the hypha, 'stretching' the freshly synthesised and 'soft' apical wall. Thus any drop in the hydrostatic pressure will result in immediate cessation of growth. The ability to adjust to a varying water potential during growth through litter and soil is clearly a characteristic of considerable significance to the maintenance of the activity of fungi.

Laboratory experiments with micro-organisms show that the optimum potential for the apical growth of most fungi lies between − 50 and − 100 bar, the growth of most being severely reduced at the lower end of the range. The allowable range for spore germination is usually slightly wider (Griffin 1972). The usual method of assessing response by linear growth may seem less ecologically valuable than would the effect on production but the intimate relationship between hyphal extension and the water environment makes it a very good activity parameter as well as being easier to measure. Although testing of bacteria seems to have been less comprehensive the majority show detectable growth to potentials of about − 75 bar (Walter 1956, Lanigan 1963).

In general the laboratory predictions for fungal activity are borne out by studies in soil. An active mycoflora is maintained down to about − 150 bar potential (Chen & Griffin 1966). This however contrasts strongly with the bacteria which seem to become inactive at potentials below − 10 to − 15 bar. A good illustration of this is given by Kouyeas (1964) who followed the growth of soil microbes on microscope slides buried in soils at different moisture potentials. Bacteria proliferated on the slides only at potentials higher than − 1 bar, whilst fungi and actinomycetes persisted beyond − 20 bar. Specific bacterial activities such as nitrification and sulphur oxidation have also been shown to be significantly reduced at potentials of − 5 to − 10 bar. Griffin (1972) has put forward a theory to account for the apparent inability of bacteria to survive soil potentials well above their limits as defined in laboratory experiments. He has shown theoretically that the mobility of soil bacteria, either self-propelled or by passive dispersal by soil water movements, is dependent on the presence of continuous water channels of at least 1.0μm thickness. Channels of this diameter will be drained in soils that have potentials below − 1 bar. Thus below these potentials bacterial activity will rapidly decline as the substrates providing the food resources at bacterial microsites within the soil are used up without there being an opportunity for re-location on fresh substrate. Fungi and actinomycetes are not limited in this way because of their ability to grow across the soil matrix and indeed across air-filled pores to colonise fresh substrates. It is interesting to speculate how important this factor may be in the litter layer where substrate distribution is less localised.

Although fungi are active through a wide range of moisture potential the structure of the community may change markedly in relation to relatively small changes. Thus zoosporic forms are largely evident only above − 1 bar (Kouyeas 1964) and the majority of common moulds decline in frequency beyond − 50 bar. However *Aspergillus* and *Penicillium* species may maintain activity down to − 400 bar or beyond (Chen & Griffin 1966). Within this *xerophilic* flora marked distribution patterns also occur and show a clear interrelationship with temperature (Fig. 6.4). These studies show that some

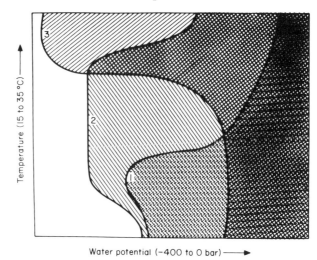

Water potential (−400 to 0 bar) ⟶

FIG. 6.4. Diagram of the interaction between temperature and water potential in determining the pattern of activity of fungi in colonising pieces of hair lying on soil. Below about −145 bar the population consisted almost entirely of species of *Penicillium* and *Aspergillus* which could be grouped according to their reaction to potential and temperature. Group 1 includes *Penicillium chrysogenum*, *P. frequentans*; Group 2 *P. citrinum*, *Aspergillus versicolor*; Group 3 *A. flavus*, *A niger*, *A. terreus* (from Griffin 1972, after Chen & Griffin 1966).

elements of the microflora are capable of remaining active at very low moisture potentials. Reference to Fig. 6.2 indicates that in soils other than those with very high clay content, fungal activity is only seriously affected below 10% moisture content. However it should be remembered that this is probably a common occurrence in the litter and surface layers of soils even in temperate climates and that this is the site of most significant decomposer activity. Many of the xerophilic species of *Aspergillus* and *Penicillium* have a cosmopolitan distribution in soils and may contribute importantly to decomposition during the final phases of drying, particularly at high temperature. The influence of moisture relations on the nature of the mycroflora may also be seen very strikingly in standing dead material. Decomposing branches, twigs, or more particularly the stems and leaves of herbs and grasses are extremely subject to desiccation. Webster (1957) and Webster & Dix (1960) have shown that the gradients of moisture content within grass tussocks may be an important determinant of the composition of the fungal flora of decomposing stems and leaves.

The microfauna, particularly protozoa and nematodes, also live within water films and are directly affected by moisture potential. Limitations on the activity of nematodes occurs at a moisture potential of about −1 to

— 10 bars, probably coinciding with the onset of dehydration. Many micro-fauna have adaptations which enable them to survive desiccation however, such as encystment, anabiosis, or the production of desiccation resistant eggs. The resistant spores of fungi and bacteria perform a similar function and enable all these micro-organisms to respond rapidly to even extreme change in the moisture environment. The microfauna are motile and may migrate to areas of more favourable moisture conditions but the theoretical limits to activity of this kind, as previously detailed for the bacteria, will be even more stringent because of the greater size of the organisms.

The meso- and macro-fauna differ from the organisms so far considered in that they occupy air-filled pores rather than water films. Water intake is prob-ably largely through ingestion with the food, though there may in some cases be direct absorption through the outer integuments. These organisms are thus not so much affected by water potential as by the water content and relative humidity of the soil pores. The arthropods have in general a well developed waterproof epicuticle which confers a measure of control over evaporative water loss. The soft-bodied animals such as the Oligochaetes are more susceptible to moisture stress and are usually restricted to wetter conditions. A number of physiological and behavioural adaptations assist their survival under varying moisture stress. Reabsorption of moisture by the nephridia minimises direct water loss and enables the moisture content of the worms to be maintained at about 80—85%. Losses of up to 50–80% of body water can be sustained however before irreversible desiccation occurs. The ability of the larger soil animals to migrate over large distances both vertically and horizontally confers a further advantage in avoiding moisture stress. *Lumbricus terrestris* normally lives near the surface of soils but may migrate to depths of two metres in dry summer months. *Allolobophora* species show a further adaptation by undergoing periods of diapause within a mucus lined cavity during dry periods.

The Collembola provide an interesting example of the variation in moisture relationships that may occur within arthropod groups. The sub-order Symphypleona respires by the use of tracheae and the heavy cuticle provides good protection against desiccation. Members of this sub-order are able to occupy relatively dry habitats and are dominant in grassland. The second sub-order, Arthropleona, are atracheate and the body surface is used for respiratory exchange. They are thus more subject to desiccation and require microhabitats with humidities maintained above 90% for survival. They are more characteristic of the damper temperate soils found in woodlands. Similar ranges of adaptation are found in other arthropod groups (see Wallwork 1970 for details). Perhaps termites show the most remarkable adaptations to avoid desiccation. Termites foraging above the soil surface in arid climates construct runways of soil which retain a moist environment round the colony as it gathers detritus and brings it back to the nest.

6.2.2   Influence of moisture on decomposition rate

Evidence from studies of timber-decay by Basidiomycete fungi suggests that below the fibre-saturation-point very little decay activity is found. Below about 25% moisture content even the survival of wood decay fungi is very limited. Lopez-Real & Swift (1975) found that when decayed wood dried below about 35% the growth of fungi within the cells was inhibited. Thus the permissible range of moisture potential for wood decay seems to be lower than would be predicted from a knowledge of the organism's response to moisture potential (e.g. Bavendamm & Reichert 1938, showed that the linear growth of many wood decay fungi was still high at -100 bar). This discrepancy may relate to the activity of the extracellular enzymes essential for the breakdown of the cell wall components. It has been suggested that the hygroscopic water is not in a physical state to permit the diffusion of these enzymes and that some free water is necessary for active depolymerisation of the cell wall to occur. This inhibiting factor may be expected to be disrupted during decomposition however as the cell wall becomes more porous and the potential of its bound water is raised. However Scheffer (1936) could find little change in the RH equilibration relationships of timber during extensive decay although the ability to absorb free water and thus maintain a favourable moisture content for decay was progressively increased. Once decay has been initiated moisture limitation becomes less likely, even under relatively dry atmospheric conditions, because of the generation of large amounts of 'metabolic' water by hydrolysis of the polysaccharides, (Peterson & Cowling 1973).

In contrast to the general observation that the moisture content of wood increases as decomposition proceeds, Witkamp (1963) found a negative correlation between the extent of decay of mixed deciduous leaf litter and its moisture content. Floate (1970b) studied the effect of three moisture contents (25, 50 and 100% mhc) on the mineralisation of carbon, nitrogen and phosphorus from the leaves of two grasses, *Festuca* and *Nardus*. Only minimal reductions were found for all three processes at the lowest moisture content over a twelve-week period. It is difficult to assess the moisture potential of the grass but it is highly probable that free water was present in the tissues even at this level. In contrast sheep faeces produced from the same grasses showed very significant reductions in the amount of mineral-N released at both 25 and 50% mhc. As Floate pointed out however this is largely a reduction in the amount of NH_4^+-N released and may be indicative of increased immobilisation rather than any decline in microbial activity in the drier materials. The necessity for assessing the moisture relations of the materials is emphasised by this example. It would be of interest to know, for instance, what differences there were between the leaves and the faeces in the relationship of mhc to potential.

Field studies of the relation between moisture content and decomposition are complicated by the interactive effects of other factors. Van Cleve & Sprague (1971) overcame this by removing woodland litter from Boreal forest sites at regular intervals during the year and measuring the uptake of O_2 in the laboratory at constant temperature but field moisture. The respiratory levels were consistent through the year except for a period of lowered activity in June–July which corresponds with moisture contents dropping below 50%. An interesting feature of these results is the differing amplitude of response to moisture change at different temperatures of incubation, a much greater response being found at lower temperatures. In contrast Douglas and Tedrow (1959) found that the oxygen uptake of four Arctic soils showed a diminished response to moisture change when incubated at 14°C and 7°C as compared with 19·5°C. In a more detailed study of the effect of moisture potential on decomposition processes in soil Miller and Johnson (1964) found a maximum rate of CO_2 evolution from four soils incubated at 30°C to lie in a range from $-0·5$ to $-0·15$ bar, the rate at -50 bar was less than 10% of the maximum. The rates of nitrification and total-N mineralisation showed similar maxima and were negligible below -15 bar. These data are in good agreement with the predictions from bacterial activity discussed in the previous section.

Present evidence suggests that moisture availability may limit the catabolic capacity of micro-organisms in soil at potentials below about -10 to -50 bar. There are few data on which to predict the effects of moisture content or potential on the catabolism of litter or on the comminutive activities of decomposer animals.

6.3 Aeration

The composition of the atmosphere in decomposer environments is a product of both biological and physical phenomena. Under aerobic conditions oxygen is used up and carbon dioxide released, resulting in a progressive reversal of the normal atmospheric ratio of the two gases. Under anaerobic conditions the gaseous component declines and gases other than O_2 and CO_2 begin to predominate, e.g. CH_4 and H_2S. The biological effects within litter and soil are corrected by diffusion between the pores and the external atmosphere. The composition of the gaseous environment within litter and soil is probably largely determined by the factors which control this diffusion rate. The intervention of a liquid phase is of particular importance in this respect. Oxygen and carbon dioxide differ markedly in their properties, and in their effects on decomposer organisms, and so are most conveniently considered separately. It is possible however that interactive effects between the two may be more significant than those of either molecule alone.

6.3.1 Oxygen

The highest levels of decomposer activity are to be found in aerobic environments; anaerobiosis usually results in incomplete degradation of substrate and the accumulation of organic matter. A critical feature of the relationship of oxygen to decomposition processes is therefore reached at the point at which concentration falls below that necessary for the maintenance of aerobic activity. A simple classification of decomposer organisms into aerobic and anaerobic forms is misleading as a continuous range of relationships exists from obligate aerobes through facultative forms to obligate anaerobes. Although fungi and most meso- and macro-fauna are strict aerobes they may have the capacity to survive anaerobic conditions. In contrast many bacteria and some nematodes and protozoa are active in fully anaerobic conditions. Many facultative anaerobes utilise fermentative pathways even in the presence of relatively high levels of oxygen. Greenwood (1968) showed that the switch from aerobic to anaerobic respiration in the soil occurred at an O_2 concentration of about 3 μM. This is about one hundred times higher than the Michaelis constant for saturation of the terminal oxidases of aerobic respiration. This indicates the operation of limiting factors other than just the concentration of oxygen in the soil atmosphere, e.g. the rate of diffusion of gases to cellular sites. Strictly anaerobic processes such as sulphate reduction will occur only in the absence of molecular oxygen. Thus changing oxidative conditions will produce a varying spectrum of response through the decomposer community.

Anaerobic organisms are known to exist even in the upper horizons of apparently well-aerated soils and it has been postulated that local pockets of anaerobiosis may occur. The basis for this lies in the poor solubility (Table 6.1) and slow rate of diffusion of O_2 in water. As most micro-organisms, particularly bacteria, exist within the water phase in soil the continued oxygenation of their sites depends on the diffusion of O_2 through the water film. If the pore space of a soil is substantially filled by water the extent of oxygenation at the surface of soil crumbs may be negligible. Even in much drier soils the oxygenation may be limited by moisture content. Greenwood (1968) calculated that on the basis of normal microbial respiration rates, the rate of diffusion of O_2 in water would be insufficient to replace O_2 lost over distances greater than 3 mm. Thus the centre of crumbs of greater radius than this would be expected to be anaerobic even in well-aerated soils. The significance of this suggestion is however difficult to assess as the extent of microbial activity within crumbs has been brought into question by some authors. Allison (1968) for instance has suggested that most bacteria within crumbs originate from the time of their formation and that no subsequent multiplication takes place. Evidence for this is shown in the combined presence of undecomposed protein and polysaccharide within soil aggregates.

Table 6.1. Characteristics of O_2 and CO_2 in relation to other physical factors of the environment.

(A) Solubility of gases in water in relation to temperature (from Brock 1966 after Lange 1956).

Temperature	Solubility (mg 1^{-1} at 760 mm Hg)	
(°C)	O_2	CO_2
0	69·4	3346
10	53·7	2318
20	43·0	1688
30	35·9	1257

(B) Ionic forms of CO_2 in water in relation to pH (Brock 1966 after Hutchinson 1957).

pH	Percentage of total CO_2		
	CO_2	HCO_3^-	CO_3^{2-}
4	99·6	0·4	—[1]
5	96·2	3·8	—
6	72·5	27·5	—
7	20·8	79·2	0·03
8	2·5	97·2	0·3
9	0·3	96·6	3·1
10	0·0	75·7	24·3

[1] = negligible.

Substantial suppression of aerobic decomposition in soil may occur at times of waterlogging. In many soils this will be only a temporary state but the extensive accumulation of organic matter in areas of high rainfall and poor drainage can be largely attributed to this mechanism. At the Moor House wet mire for instance decomposition rates for cellulose were reduced five-fold below the water table which fluctuates within 0 and 30 cm of the surface (see Section 7.4).

Within soils the O_2 content varies with depth, season, and the type of soil (Fig. 6.5). Very little information is available for litter layers but Brierley (1955) showed that even in beech litter as deep as 15 cm the oxygen content did not fall below 19·5%. There is no practical or theoretical basis on which to predict the moisture content of plant materials at which significant reduction of oxygenation may occur, but periodic soaking of leaf litter may result in the development of anaerobic conditions. Oxygen depletion is probably more common in relatively massive resources such as the branches, stems or roots of trees as Table 6.2 shows. In general these show more extreme deviations from atmospheric ratios than soil, reflecting both higher

Chapter 6

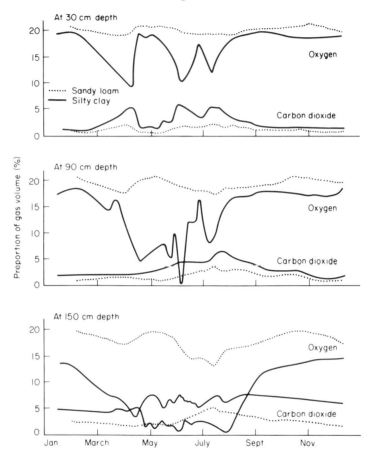

Fig. 6.5. The oxygen and carbon dioxide content of the soil air at three depths in a sandy loam and a silty clay apple orchard (from Boynton 1941).

levels of respiratory activity and slower rates of diffusion. Griffin (1972) reviewed the response of soil fungi to varying concentrations of oxygen and concluded that, measured in terms of their germination or linear extension growth rates, most species were unaffected by reduction of the oxygen partial pressure in the surrounding gaseous atmosphere to below 4%. It should be emphasised that in all the studies reviewed the change in oxygen concentration was not accompanied by any significant change in the carbon dioxide concentration. Below the 4% oxygen level the growth rate of fungi is in general severely limited. Soil animals appear to have similar demands though little specific information is available. Most insects show lowered oxygen consumption at atmospheric levels of less than 5% O_2 (Wigglesworth 1965) although some can operate at lower levels, e.g. *Zootermopsis nevadensis*

Table 6.2. Atmospheric composition of decaying wood. Data from 43 branches of *Fagus grandifolia* sampled on nine occasions between May and October (from Paim & Beckel 1963).

	Concentrations (% of total gas)	
	CO_2	O_2
Minimum range	0·25–2·25	3·0–9·0
Mean range	3·0–9·0	9·0–19·0
Maximum range	5·25–14·0	17·0–19·0

(Cook 1932). As with fungi many invertebrates can survive periods of anaerobiosis in an apparently dormant metabolic state.

The data from Table 6.2 suggest that it is likely that inhibitory or even anaerobic O_2 levels may occasionally prevail in decaying wood, particularly if the water content is high. Most available evidence suggests little decrease in the activity of wood decay organisms above O_2 levels of about 2%. Below this level Paim & Beckel (1963) could find no evidence of invasion by cerambycid beetle larvae. Wood decay fungi may be more tolerant of oxygen depletion; Lopez-Real & Swift (1977) showed that hyphal growth of fungi within wood blocks could occur when they were in equilibrium with an external atmosphere containing only 0·5% O_2. When the blocks were placed in an atmosphere of N_2, fungal activity ceased after a short interval of time indicating that inhibition occurs at the time of total consumption of O_2, i.e. at the onset of anaerobic conditions.

6.3.2 Carbon dioxide

The dynamics of carbon dioxide within decomposer habitats are markedly different to that of oxygen. CO_2 is about thirty-five to fifty times more soluble than O_2 in water over the range of temperatures common in soils (Table 6.1). This means that high moisture contents do not affect the rate of diffusion in litter and soil as much as with O_2. Another important characteristic of CO_2 is its tendency to dissociate in water into a number of different ionic forms:

$$CO_2 \text{ (atm)} \rightleftharpoons CO_2 \text{ (dissolved)} + H_2O \rightleftharpoons H_2CO_3$$
$$H_2CO_3 \rightleftharpoons H^+ + HCO_3' \rightleftharpoons H^+ + CO_3^{2-}$$

The relative frequency of these forms is greatly affected by pH (Table 6.1). Thus for water in equilibrium with a given atmospheric partial pressure of CO_2, only dissolved CO_2 is significant below pH 5 but the bicarbonate ion is the predominant form between pH 7 and 10. The solubility and the ionic equilibrium are affected by the presence of other ions and by temperature. It thus becomes clear that for organisms such as the bacteria and fungi that are usually bathed in a water film, measurement of the gaseous CO_2 concentration of the atmosphere in equilibrium is not a sufficient measurement for interpretation of the environment to which they are being subjected. Simultaneous pH measurement can, however, largely correct this. The critical biological question in this respect is whether these organisms show any differential response to the different forms of CO_2. Griffin (1972) concluded that bicarbonate is the most significant form with regard to fungi. If this is generally true then it would be predicted that the response of micro-organisms to changes in the partial pressure (pp) of CO_2 would be greater in alkaline environments (where the bicarbonate ion would be more likely to be present at biologically significant concentrations) than in acid ones. Most data from cultural studies on micro-organisms are with respect to gaseous levels of CO_2 without reference to pH. Varied types of highly differential response have been recorded but it is not uncommon for levels of CO_2 above atmospheric to stimulate fungal growth. However, levels in excess of 0·05–0·1 atmospheres may be inhibitory when O_2 is held at normal atmospheric levels (Hintikka & Korhonen 1970).

For soil animals inhabiting air filled pores only the pp of gaseous CO_2 is of environmental significance. High levels of CO_2 are known to have a narcotic effect on most insects but there is evidence that many soil invertebrates are adapted to withstand higher than average levels.

Greatly varying concentrations of atmospheric CO_2 have been recorded in soil environments and may reach levels approaching 0·1 atm at depth or under conditions of high moisture content but in the upper regions of most soil it rarely exceeds 0·02 atm (Fig. 6.5). It is therefore possible to conclude that only rarely will CO_2 levels have a marked effect although in view of the demonstrated complexity of interaction between CO_2, soil water and pH, this conclusion may be incorrect with regard to litter. Here again, in bulky materials such as decaying wood, the high levels of decomposer activity result in extreme levels of CO_2 accumulation being occasionally recorded (Table 6.2). Wood decay fungi may be stimulated by CO_2 levels as high as 0·2 atm and most are little affected up to this level (Thacker & Good 1952). These data only relate to gaseous CO_2 and the effects of ionic forms have not been considered. The pH of most litters is acidic however and so the infrequency of the bicarbonate ion may be one reason for the low sensitivity to CO_2 fluctuations. In a similar way, wood-boring cerambycid beetle larvae showed no quantitative change in their respiratory activity at 0·14 atm CO_2 although the pattern of spiracular movement was affected (Paim & Beckel 1963).

Further evidence for the effect of high levels of CO_2 on microbial decomposer activity comes from the work of Clark (1968) who drew attention to the well-known phenomenon of the depression of decomposition rate caused by increasing the volume of soil or material being tested. He showed that increasing the bulk of soil amended with 0·5% alfalfa meal from 25 g to 100 g resulted in a decline in CO_2 evolution from 6·5 mg g^{-1} to 4·5 mg g^{-1} over a three-week period. Conversely, similar increases in bulk or decreases in surface/volume ratio resulted in increased rates of nitrification. He was able to demonstrate that the latter effect was due to the stimulatory effect of increased CO_2 accumulation within the bulkier samples but it was not clear whether the depression of carbon mineralisation should be attributed to the same cause.

6.3.3   Combined effects of O_2 and CO_2

The combined effect of reciprocal changes in O_2 and CO_2 content has been less commonly investigated than the effect of either gas alone and yet the joint change is the norm for decomposer habitats. Because of such physical considerations as their differing solubilities, the changes are rarely equal (Fig. 6.6, Table 6.2). When such combined changes have been investigated the results have not always been predictable in terms of the observed effects of either gas alone. Lopez-Real & Swift (1977) showed that whereas fungi growing in wood were not particularly sensitive to changes in CO_2 or O_2 alone, reciprocal changes in the concentrations of the two gases had more dramatic effects. This type of synergistic effect has been reported for other fungi (see Tabak & Cooke 1968) and warrants further investigation in relation to decomposer activity.

The account we have given of the effects of O_2 and CO_2 concentrations on the activity of decomposers gives a general impression of a mosaic of microsites in the soil differing in relation to these factors and within which the conditions may vary rapidly with time. The conditions in the litter layer are probably more homogeneous and stable but little evidence is available to confirm this. The main factors regulating this pattern of aeration are the moisture environment and the particulate structure of the soil. It is thus difficult to separate, on a scale larger than the micro-habitat, the effects of climate and soil type from those of aeration.

6.4   pH

The role of pH is among the most difficult of environmental factors to understand but is of central importance. The difficulty lies partly in the complex interaction of other factors (such as the nature and size distribution of the particulate phase, the concentration of cations and the precipitation) which determine environmental pH; partly in the highly varied distribution of pH;

and partly in the complexity of biological and chemical effects that pH can have. We consider first the patterns of pH found in decomposer habitats and the factors which determine them. We shall see that, once again, far more information is available for the mineral layers of soil, than for litter but that the activity of decomposers in the organic layers is an important determinant of the pH regimes in the underlying soil.

The pH of a soil, or soil horizon, is usually given as a single value, or range of values, based on standard measurements made with a glass electrode. The soil samples used vary in weight and are suspended in variable volumes of distilled water or calcium chloride solutions of various strengths. The variations in technique used produce some small variations in pH value obtained. Far more important ecologically however are the marked variations in pH between microsites that may occur within such a soil sample and which are not measured by these methods. The merit of this approach to pH determination is that it produces a '*Bulk pH*' (pH$_B$) measurement which is an average of all the *Local pH* values (pH$_L$) and may have a quantitative relationship to them which is useful at least for the purposes of comparing different soils or horizons.

In order to understand how this relationship may exist let us first consider the factors which determine the bulk pH. It is determined primarily by the balance between hydrogen ions (and to some extent Al^{3+1}—see Fig. 6.6) and basic cations such as Ca^{2+}, K^+, and Mg^{2+}. This balance is modified by other environmental features such as the nature of the solid matrix of soil and the precipitation. The significance of the solid matrix lies in the proportion of negatively charged particles which it contains. Among the mineral fractions clays have charged surfaces because of unsatisfied valencies in the clay minerals (Fig. 6.6). Many organic colloids of the humus fractions are also negatively charged. The presence of the negative charge at the surface of these materials leads them to attract a 'cloud' of cations to the surface region. The capacity of a soil to attract cations in this way has already been described (Chapter 1) as the Cation Exchange Capacity (CEC) and is usually measured as milli-equivalents of H^+ bound per 100 g of soil.

In acid soils H^+ tends to predominate at the colloid surfaces but in base-rich alkaline or neutral situations the main components may be basic cations such as Ca^{2+}. These states are by no means permanent. The hydrogen ion is adsorbed more strongly than other cations so that when an excess of H^+ is produced (e.g. during the formation of organic or mineral acids such as H_2CO_3 by the decomposition of organic matter) they will *exchange* with the Ca^{2+} or other basic cations on the surface:

$$Ca \text{ (adsorbed)} + 2H_2CO_3 \rightleftharpoons 2H^+ \text{ (adsorbed)} + Ca(HCO_3)_2.$$

This equilibrium may shift to the left if excess of Ca^{2+} occurs in the soil solution due to release from $CaCO_3$. If the soil is subject to heavy leaching

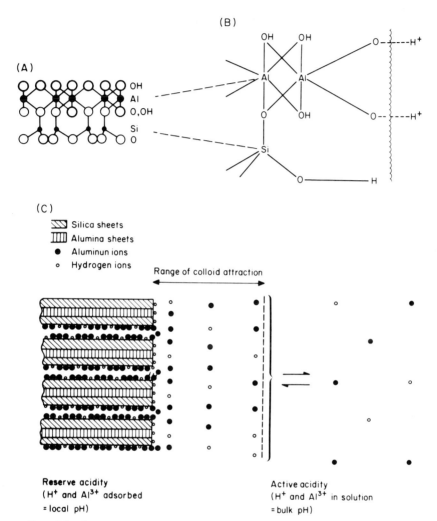

FIG. 6.6. The structure of clay and the relationship between its charge and the pH of the soil solution. (A) The structure of kaolinite clay showing a 1:1 (silica: alumina layers) structure. Note the layers of hydroxyl ions or oxygen atoms at the edge of the crystal. (B) These atoms can be sources of negative charges, particularly at broken edges, which will bind H^+. At high pH this binding is loose and exchange with basic cations occurs readily. (C) The density of H^+ (or of Al^{3+} which can create acidity when hydrolysed to give $Al(OH)^{2+}$ and H^+) is always greatest at the site of the crystal creating a difference between the local pH and the bulk pH and also giving a strong buffering capacity to the clay minerals. In this case the clay has a 2:1 structure (e.g. montmorillonite). The greater surface exposure gives these clays a higher cation exchange capacity than 1:1 clays.

then the removal of the soluble calcium bicarbonate throws the equilibrium strongly to the right and leads to build up of hydrogen ions on the clay surfaces. The hydrogen ions held at the surfaces determine what is known as the *reserve acidity* in contrast to the *active acidity* of the solution, which is in equilibrium with it (Fig. 6.6c). Changes in the active hydrogen ion concentration are adjusted by adsorption or loss of H^+ from the colloid surfaces. The amount of H^+ in the reserve acidity of a soil is usually greatly in excess of the active acidity and so acts as a very considerable buffering agent for the soil—it needs very large additions of H^+ or Ca^{2+} to the soil solution to give a significant change in pH_B—and the greater the CEC the greater the buffering effect whether the soil is alkaline or acid. The pH_B is more closely related to another quantity—the *base saturation*—that is the proportion of the CEC occupied by the neutralising bases, such as Ca^{2+}, Na^+, or Mg^{2+}. Clearly the lower the base saturation the greater the reserve acidity and thus the lower the equilibrium pH_B of the soil. In most temperate mineral soils with a reasonably high CEC, base saturation of 50% gives a pH_B of about 5·5 and changes by about 0·10 units for every 5% increase in base saturation.

The excess of hydrogen ions associated with clay or humus colloids produces a local pH (i.e. in the immediate region of the surface) which is significantly lower than that of the surrounding solution. This can be seen diagrammatically in Fig. 6.6C. It can also be appreciated that this difference is likely to remain more-or-less permanent, except perhaps in extreme alkaline soils where reserve acidity is negligible. This then is one of the major sources of difference between pH_L and pH_B.

The bulk pH of litter or other detritus is an even less solid concept than that of soil. It is usually determined by measuring the pH of a liquid in equilibrium with either intact or macerated resource material. The pH of plant materials is usually acid; the leaves of temperate deciduous trees lie generally within the range of pH 5·0–6·5 whilst those of conifer needles are more acid (pH 3·5–4·2, Broadfoot & Pierre 1939; Plice 1934). The pH of woods is largely 0·5–1·0 lower than that of leaves with conifers again being lower than hardwoods (Campbell and Bryant 1941).

The pH of leaf litter is highly correlated with that of the soil on which the vegetation is growing (e.g. see Fig. 6.7) although for alkaline soils it is generally below that of the soil. The pH of plant material is determined by the balance between the hydrogen ion and basic cations. The H^+ concentration is probably largely a product of the organic acids found in the cell vacuoles which may vary markedly in concentration. In contrast the content of basic cations in plant tissues very much reflects their availability to the plant from the soil. Thus a soil with low cation content will produce vegetation with low base content and high acidity. Thence, largely speaking, the differences between the pH of hardwood and coniferous litters may largely reflect the soil types on which they grow. Additionally, plants growing on

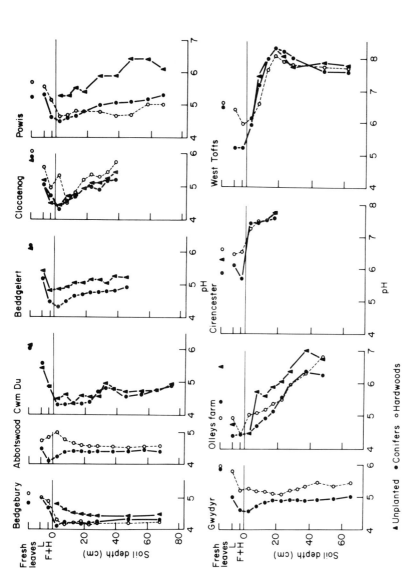

▲ Unplanted   ● Conifers   ○ Hardwoods

FIG. 6.7. pH profiles of a variety of temperate forest soils in the United Kingdom. The sites all have a mixture of tree species, largely conifers planted by the U.K. Forestry Commission: significant correlation coefficients were found between the pH of fresh leaves and that of the soil ($r = 0.80$, conifers; 0.89 hardwoods). For leaf litter the correlation coefficients were lower but significant ($r = 0.53$, conifers; 0.85, hardwoods). The acid shift in decomposing leaves between fresh litter and that in the F layer can be clearly seen in the profiles (from Ovington & Madgwick 1957).

base deficient soils allocate a greater proportion of NPP to plant protection chemicals, and thus increase the acidity by production of phenolic acids.

The cell wall components of plant materials may exhibit a considerable degree of cation-exchange behaviour in a manner analogous to that of soil colloids. Impure cellulosic products possess a net negative charge which is attributed mainly to the pectic acids and lignin components but partially modified celluloses may also have this property due to the presence of carboxyl groups. The negative charges on these polymers will attract cations which in acid conditions exchange with hydrogen ions in obedience to the mass action law (Sookne & Harris 1954). Thus, local differences in pH may occur within decomposing litter.

The products of decomposition are predominantly acid but unless the local pH change is extreme the plant materials will have a tendency to retain a high proportion of basic cations. The rates of loss of cations from decomposing litter shows an inverse relationship with their affinities for charged fibres (i.e. Ca^{2+} binds more strongly than Mg^{2+}, K^+, or Na^+). Plant litters thus remain alkaline in comparison to the highly fragmented materials of the fermentation and humus layers (Fig. 6.7). Indeed the generalisation has been made (Williams & Gray 1974) that the initial impact of decomposition is to cause an increase in the bulk pH of plant materials. Sjors (1959) for instance found an average rise of 0·7 pH units from a starting mean of pH 5·7 for twelve hardwood leaf litters after one month's decomposition. The decrease in H^+ concentration implied by this is usually attributed to the leaching out of acidic materials particularly the components of the vacuolar sap. Nykvist (1963) found the highest acidity and content of organic acids in leachates during the earliest stages of decay and Frankland *et al.* (1963) have shown a strong correlation between rainfall and pH change in decomposing litter (Fig. 6.8). The main increase in alkalinity during the first year of decay came during early summer—some six to eight months after the main litter fall—following a period of increasing acidity during the relatively dry period of February to May. The biggest losses of basic cations corresponded more closely with the periods of increasing acidity than with rainfall. This study gives a crude but illuminating impression of the shifting equilibria between the generation of acidity by decomposition, the exchange of H^+ with cations on the plant materials, and the leaching effects of rainfall.

Most soils show the greatest acidity in the horizons (such as the O_2 or A_1) which contain the products of primary decomposition—the fragmented cellular fraction of the soil organic matter (see Figs. 1.9 and 6.8). At this stage of decay the formation of acidic products is high and mineralisation is proceeding at such a rate that the buffering effect of bound cations is destroyed. Below the immediate 'decomposer horizons' however the buffering capacity of most soils is such that the leaching of acids from the litter has no noticeable

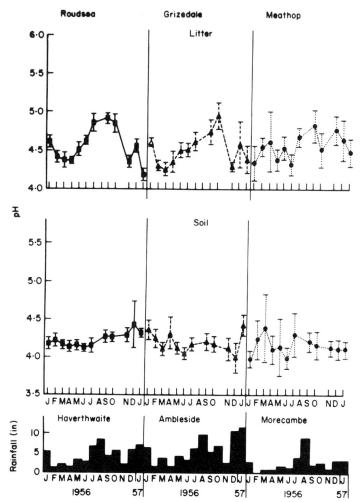

FIG. 6.8. Seasonal change in the pH of litter and mineral soil in three wood-lands of the English Lake District shown with the rainfall at the nearest meteorological sites. Note the decreasing acidity of the litter coincident with the heaviest period of rainfall. As there was no significant correlation between the content of basic cations of the litter and pH the changes are probably attributable to the leaching of acid materials formed by decomposition (Frankland *et al.* 1963).

effect on the bulk pH, as is shown by the mineral soils of Fig. 6.8. Neverthe-less, it should be re-emphasised that marked local changes in pH occur. Pockets of acid pH may be found in the soil associated with the site of decomposition of roots or stumps.

Having considered the ways in which decomposition processes help to determine the pH environment we can now analyse the ways in which this environment regulates the activity of decomposition processes.

6.4.1   The effects of pH on the activity of decomposer organisms

The microflora and microfauna, living within water films at the surface of charged particles in the soil, are clearly the most susceptible of decomposer organisms to the effects of variation in local pH. These organisms are not totally at the mercy of the pH_L however, for many of them have the capacity to modify the pH in their immediate environment. In micro-organisms grown in pure culture it is commonplace to observe a pH shift in the medium, usually resulting in increased acidity. This may be due to selective uptake of cations (e.g. of NH_4^+ in preference to NO_3') or production of organic acids and CO_2. An extreme example is the action of the sulphur-oxidising bacterium *Thiobacillus thiooxidans* which generates sulphuric acid in its immediate environment and can readily flourish at hydrogen ion concentrations as high as pH 1·0. Conversely, the production of ammonia during proteolytic activity can create an environment with decreased acidity. The extent to which this type of pH change is significant in natural habitats is unclear. In pure culture many fungi form large amounts of organic acids (see Whittaker 1951 for an example of this in wood-decay fungi). Such activity is known to lower efficiency of growth by a factor of 50% in some cases however, and it is doubtful whether it occurs in more exacting natural conditions. On the other hand such behaviour may possibly be a necessary adaptation to regulate extracellular pH.

The local pH round microbial cells can also be influenced by the negatively charged nature of the cell surface (Marshall 1971), which may attract a local concentration of H^+ in a manner analogous to that of clay minerals. This concept of a local, self-determined, pH in the vicinity of micro-organisms is one that must be allowed for in interpreting microbial activity in relation to environmental pH. There is a large body of information detailing the effect of pH on various aspects of the physiology of bacteria, fungi and invertebrate animals (e.g. see Doetsch & Cooke 1973, Cochrane 1958, Dickinson & Pugh 1974). Examples of this are usually tabulated to show the minimal, optimal and maximal pH values for the activity in question. Some broad generalisations can be drawn; bacteria tolerate relatively narrow ranges of pH at the alkaline end of the spectrum in contrast to fungi which have generally broad optima but are most active at acid pH. The physiological basis of such effects are probably complex and a variety of different mechanisms may operate. In the bacteria direct effects may be quite important as it is probable that the pH of the cytoplasm closely reflects that of the external medium. Optimum pH will therefore relate closely to an 'average' pH for the maximising of enzyme function within the cell. Fungi seem able to regulate their internal pH and maintain it between pH 5 and 6. The same is true of most invertebrates where the digestive system is in most cases maintained at a weakly alkaline pH.

Effects on enzymes may be indirect in the sense that extracellular enzymes are involved. Proteins, being amphoteric in nature, usually have very specific pH optima although some can operate over a broad spectrum. There is little information on the pH spectra of the extracellular depolymerising enzymes of decomposers. It is also necessary to exercise caution in extrapolating from information obtained from liquid systems in the laboratory. McLaren (1960 and 1962) and others have pointed out the considerable differences in behaviour of enzymes that occur when particles or other solid surfaces are included in the systems. Enzyme molecules, with their amphoteric qualities have a high probability of adsorption onto charged surfaces, which means that the pH and ionic environment in which they act is different from that of the soil solution. McLaren and other workers have demonstrated for a number of different enzymes (proteases, phosphatase, urease) that the optimum pH for the enzyme in question was higher when adsorbed on clay minerals than in solution, presumably because of a decreased pH at the adsorbent surface (e.g. see Skujins 1967). Adsorbance has also been shown to affect the persistence of enzymes in the soil (e.g. their resistance to proteolysis). The local pH may play a role in this, acting to stabilise the ionic environment (Skujins 1967). Substrate molecules may also adsorb to charged particles and although local pH is probably only one of a number of factors involved, this can result in markedly different rates of catabolism when compared with controls lacking particles. The responses are complex however and both stimulation and inhibition of the rate of mineralisation has been observed (see reviews by Stotzky 1974 and Marshall 1971).

There are also many reports of differences in the activity of microorganisms between sorbed and non-sorbed sites and some of these have been attributed to local pH effects. A much quoted example is that of nitrification. McLaren & Skujins (1963) showed that *Nitrobacter agilis* had a pH optimum for production of NO_3' from NO_2' of about 6·0 in liquid suspension but that this was increased to 6·5 or 7·0 in the presence of soil colloids or ion-exchange resins. They interpreted this in terms of the NO_2' ion being repelled by the negatively charged surfaces which at the same time accumulated H^+ thus creating a decreased pH_L. Decreasing the H^+ concentration in the solution raised the pH_L at the surface region and also permitted readier access of the NO_2'. In this case then pH change is not the whole story but the pH equilibrium is a key factor in determining the rate of reaction.

The current information relating the effects of local pH in the vicinity of charged surfaces to the activity of decomposers or their enzymes has been gathered from studies on soil particles such as clays. There is reason to

Chapter 6

suppose however that local pH effects of the same kind also operate within plant detritus at the surface of charged fibres.

The pH of the soil solution may affect the availability of essential elements to decomposers because of variation in solubility and ionic form (e.g. the forms of dissolved CO_2—see Table 6.1). Perhaps the most dramatic effect is that on iron which will precipitate out as insoluble hydroxide from ferric salts at pH 3 and from ferrous salts at pH 5. In general more acid conditions increase the solubility of a wide range of necessary elements including P, Ca and K. In soils where these elements exist largely in insoluble form the influence of pH in making them accessible may be of great significance.

The effects of pH on the microflora and fauna are thus often subtle and expressed at a local level. The meso- and macro-fauna are largely able to avoid such effects both because of the scale of their habitats and by possession of an external integument. Nonetheless pH_B can often be distinguished as a factor determining the distribution and activity of these organisms. The earthworms have probably been studied more intensively in this respect than any other group. The soft-bodied nature of these organisms makes them relatively sensitive to contact with solutions of varying acidity and as a result they show marked distributional discontinuity in relation to pH (Chapter 3). Bornebusch (1930) found that forest soils in Denmark of pH 4·3 or less contained few earthworms while neutral or basic soils supported the largest worm populations. Bornebusch's results are supported by Satchell

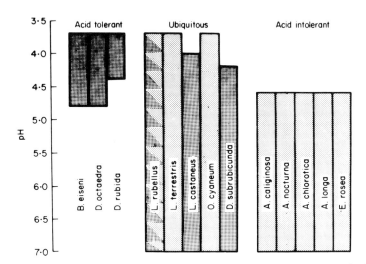

FIG. 6.9. Distribution of earthworms according to the pH of Lake District woodland soils (from Edwards & Lofty 1977, after Satchell 1955). The British Isles do not contain the full complement of European lumbricidae and these species would not necessarily show the same distribution in mainland Europe where more species are present.

(1955, 1967) who classified the British Lumbricidae according to their pH tolerance ranges (Fig. 6.9). Of the thirteen species he investigated, three surface inhabiting species were 'acid tolerant' (pH <4·5); five sub-soil inhabiting species were 'acid intolerant' (pH >4·5); five others were 'ubiquitous' and were found in soils ranging from more than pH 7·0 to less than pH 4·0. The ubiquitous group included both surface living and burrowing forms (*Lumbricus terrestris* and *Octolasium cyaneum*) which, although acid tolerant at the limits of their tolerance ranges, have their optima under more basic conditions. In Great Britain and northern Europe pH 4·5 approximately delimits the mor and mull forms of humus. In contrast, in Australia Wood (1974b) found extensive earthworm activity in alpine soils at low pH's which supported deep mull-type humus.

Finally, it is important to note that whilst local pH is the determinant of microbial activity the bulk pH may be useful as an indicator of distribution at a larger scale. The correlation of bacteria with more alkaline and fungi with acid soils has been shown so many times that it is almost safe to predict this inverse relationship on the basis of pH_B measurements. Table 6.3 shows

Table 6.3. The distribution of microbial populations and pH_B within a soil profile. The soil is a former sand dune now planted to a stand of *Pinus nigra laricia*. The Streptomycetes were grouped according to counts made on media with acid or alkali pH (from Goodfellow & Cross 1974).

		Number per g dry wt (x 10^4)			
		Streptomycetes			
Horizon	Mean pH	Acid medium	Neutral medium	Bacteria	Fungi
Litter	3·6	80	0·6	1083	225
A₁	4·2	0·6	8	1810	29·8
C	7·8	0	37·5	3080	1·3

this correlation in relation to the pH_B distribution of a soil profile. The foregoing account urges caution however in concluding that pH is the determining factor in view of the many other environmental and nutritional factors that are correlated with it.

6.5   Temperature

The temperature regime varies over the earth's surface on a latitudinal gradient modified locally by the aspect of the terrain. The amount of radiation entering the decomposer habitats of litter and soil is also regulated by the

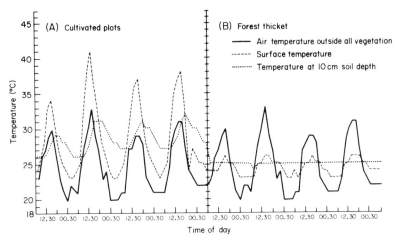

FIG. 6.10. Diurnal temperature fluctuations in tropical soils under different vegetation cover. (A) Light planting of cowpea (*Vigna unguiculata*). (B) Regrowth thicket from the natural rainforest vegetation of the area. The two plots arc adjacent so that soil temperature measurements were made within 10 m. Measurements were made at two hourly intervals for four days in August. Note that in the crop soil the surface temperature shows a greater amplitude than the air temperature. At 10 cm the amplitude is lower and there is a lag in development of the diurnal maximum so that the soil night temperatures are higher than those in air or at the surface. The thicket vegetation decreases the extent of thermal radiation reaching the soil so that the maximum temperatures and the temperature amplitude at the surface are much reduced. The temperature at 10 cm remains constant (unpublished data with permission from Perfect *et al.* 1978).

extent of vegetation cover (Fig. 6.10). Temperature also fluctuates over different time scales—notably seasonally and diurnally. Within decomposer habitats these major fluctuations are found but are modified by other features of the litter and soil environment. Whereas the transfer of heat energy in the atmosphere is brought about by radiation and convection, within the soil the transfer is by the process of conduction. Soil is a poor conductor of heat, so that the rate of penetration of heat from the surface into the body of soil is slow. Consequently diurnal fluctuations of temperature whilst very marked in the top centimetre or so of the soil decrease in amplitude with depth (Fig. 6.10). Due to the slow rate of conductance of heat there is also a lag in response time to surface fluctuations with depth. This can be as much as 2 hours per centimetre so that at a depth of 15 cm the surface maximum at noon is reflected in a maximum at midnight. The thermal characteristics of soil are the sum of the characteristics of the gaseous, solid and liquid components. As the moisture content is the most variable of these components and as water has a much higher thermal capacity than the components of the solid phase, the heat regime of a soil can be very significantly affected by its moisture content.

The response of all chemical reactions to temperature is very sensitive and enzyme-catalysed reactions are no exception. Arrhenius' equation predicts that the logarithm of reaction rate will be linear with the reciprocal of the absolute temperature. Many biological activities approximate to this relationship over a limited temperature range (Fig. 6.11A) defined by the cardinal points of maximum and minimum temperature for the activity in question. Between these points a fairly narrow optimum range is often evident and is usually closer to the maximum than the minimum. A convenient method often adopted of expressing response to temperature is by the temperature coefficient Q_{10}—the fold-increase in response to a 10°C change in temperature. For biological systems this is often assumed to approximate to 2·0 but deviations from this are fairly common. This is due both to the changing nature of the response as the limits of activity are approached (Fig. 6.11) and because other environmental and biological features vary with temperature range and may limit the biological response. Thus, whilst the Q_{10} of the decomposition processes shown in Fig. 6.11B over the temperature range 5° to 15° was 3·2 that of plant litter in the Antarctic over the range 0° to 10°C was reported as being 3·7 (Bunnell *et al.* 1975) and of dung in the tropics as 2·6 between 22° and 32°C (Anderson & Coe 1974).

Organisms also differ markedly in the range of temperature embraced by the cardinal points. Some micro-organisms are clearly adapted to extreme environments and are described variously as thermophiles (high temperature adapted) or psychrophiles (low temperature adapted) in comparison with the majority of mesophilic organisms. There have been many suggestions of the temperature limits that should be applied to these groups (e.g. see Fig. 6.12) but it is probable that in nature there is a continuous spectrum of overlapping temperature ranges. Animals show behavioural as well as physiological responses to temperature changes. They will seek out favourable thermoclimates by migration, e.g. by vertical movements within the soil and litter. Petrusewicz & Macfadyen (1970) have detailed some of the difficulties entailed in obtaining a realistic estimate of the respiratory activity (as an index of decomposition) of soil animals in relation to temperature. Laboratory experiments must take account of the effects of age structure, CO_2 levels, RH, population density and thigmotatic stimuli. The temperature-activity response also differs according to whether the other features of the environment are constant or fluctuating and whether the rate of temperature change is slow or rapid. This latter feature relates to the ability of many animals to acclimatize to temperature change. Thus whilst all invertebrates are poikilothermic, and theoretically directly responsive to environmental temperature, they do show a range of compensatory mechanisms to maintain a more stable metabolic level.

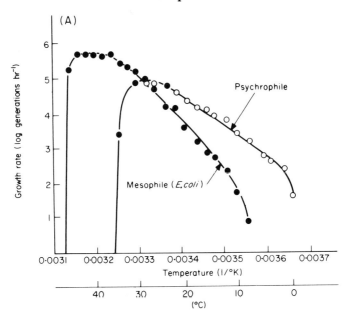

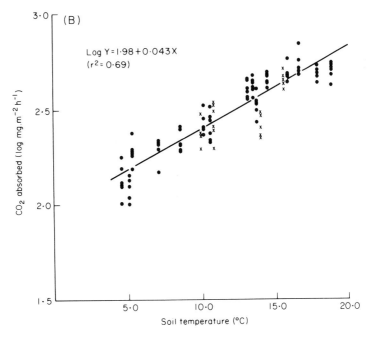

FIG. 6.11.

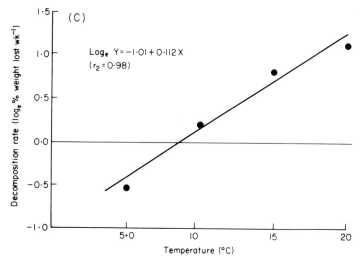

FIG. 6.11. Relation between the rate of decomposer activity and temperature plotted in a number of ways. (A) Growth rate of bacteria plotted in Arrhenius form (Ingraham 1962). (B) Soil respiration in a *Castanea* woodland. Individual values taken at different times of the year plotted against soil temperature taken from Anderson 1973c). (C) Rate of decomposition of *Castanea* sawdust by the basidiomycete *Coriolus versicolor* under laboratory conditions (Swift unpublished data).

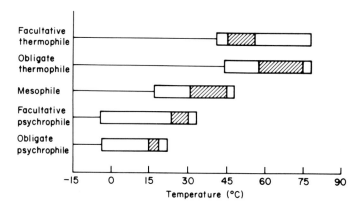

FIG. 6.12. The temperature limits for the growth of psychrophilic, mesophilic and thermophilic bacteria. The approximate optimum range for each group is shown shaded. Note however that for a mixed community a high level of activity may be expected over the full range of temperature from about 10°C to 70°C (from Doetsch & Cooke 1973).

Whilst temperature is a very important regulator of microbial and animal activity the importance of other environmental and biological factors in modifying the response makes prediction at the micro-habitat scale a hazardous business. Responses to temperature at larger scales are readily perceived however. Fig. 6.13 shows the close correlation between the diurnal pattern

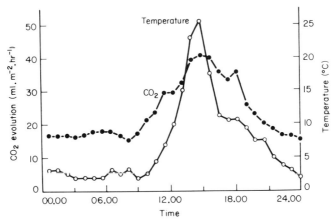

FIG. 6.13. Diurnal fluctuation in decomposition rate. The diurnal march of temperature ($-\bigcirc-$) is closely paralleled by the evolution of CO_2 ($-\bullet-$) from decomposing litter on the floor of a temperate deciduous woodland (from Witkamp & Frank 1969).

of CO_2 release from decomposing litter and the diurnal march of temperature. A seasonal correlation of temperature and soil respiration is also evident in Fig. 6.11B—an aspect we shall take up in more detail in a later section.

Interesting responses to temperatures beyond the range for normal mesophilic activity are observed during composting of plant detritus. The large mass of material in these artificial systems insulates the heap so that the loss of heat to the exterior is much slower than the rate of heat generation by the decomposition processes. As a result the internal temperature may reach levels approaching 70–80°C. A typical pattern for the march of temperature with time of a wheat straw compost is shown in Fig. 6.14, together with data on the rate of decomposition and the changes in microbial populations (Yung Chang & Hudson 1967). From this it is clear that the high temperatures and the period of maximum weight loss and depolymerase activity coincide. This period is also characterised by a thermophilic microflora of bacteria, actinomycetes and fungi with activity optimal typically in the 45–55°C range. Many of these organisms, particularly the fungi, are active decomposers which have been shown to have the capacity to degrade polysaccharides

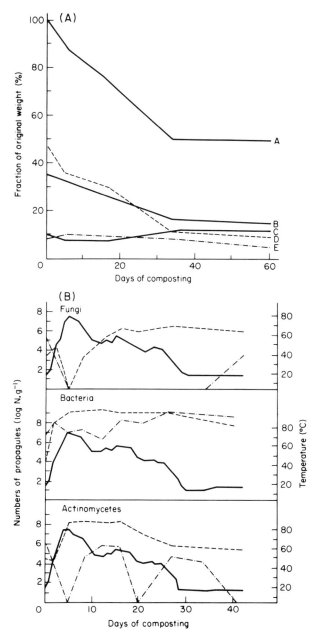

FIG. 6.14. Changes in wheat straw (A) and microbial populations (B) during composting. In (A) the following components are shown; A = total weight; B = hemicellulose; C = lignin; D = cellulose; E = total soluble matter. In (B) temperature is shown by the solid line, the thermophilic organisms by the dashes, and the mesophilic flora by the dots and dashes (from Yung Chang 1967, and Yung Chang & Hudson 1967).

(Yung Chang 1967, Fergus 1969). These activities are probably less wide-spread in other thermophilic microbes. The cellulases of thermophilic organisms appear to have temperature optima in the region of 65°C (Stutzen-berger *et al.* 1970). The capacity of enzymes to remain functionally stable at such temperatures has attracted a lot of attention but the population changes induced by the temperature change may be of greater ecological significance. The shifting population structure of the compost dominated by thermophilic organisms during high temperature periods and by mesophiles at lower temperatures may provide a model for the response of decomposer populations to fluctuating temperatures. The data of Fig. 6.10 indicate that the surface layers of exposed tropical soils may reach temperatures approaching the extreme of the mesophilic range—and indeed frequently do go beyond it up to 50°C or more. Similar high temperatures may occur locally when ambient is high and catabolic activity is generating heat within well-insulated microsites. Under such circumstances a thermophilic component of the microflora may be important in maintaining decomposer activity. There is evidence for analogous population fluctuations between mesophiles and psychrophiles in cold environments (Bunnell *et al.* 1975).

6.6 Leaching and other direct effects of climate

In Chapter 1 we identified leaching as a major contributory process to the weight loss during the decomposition of detrital resources and a number of examples of this have been quoted in subsequent sections (e.g. Fig. 4.3). We have discussed the role of leaching as a determinant of initial loss rates in fresh detritus and as a modifier of resource quality (Chapter 4). We shall now consider its quantitative role in decomposition processes and its relation-ship to climate.

Leaching may initially take place from the senescent resources before litter fall, the leachates reaching the soil by stemflow or drip from the canopy. As we have already seen the quantity of inorganic ions removed in this way may contribute significantly to the total input to the soil (Table 1.8) although different ions show marked variations in mobility. Regrettably few quantita-tive studies have been carried out so that it is not possible at present to determine a general relationship between the quantity of leachate and the rainfall. Tukey (1970) has reviewed the subject and concluded that whilst such a correlation may exist the area of highest rainfall—tropical rainforest— also showed the presence of adaptations to conserve losses of this kind. The most obvious of these is the drip-tips present on the leaves of many tropical species which probably speed up the rate of water run-off and minimise the extent of extraction of soluble materials. Organic materials may also be leached from the canopy but the quantities are usually relatively low. They may be enhanced markedly by herbivore activity in the canopy (Carlisle *et al.* 1966b) and in particular by aphid attack. The effect of such large inputs of soluble organics on the soil system has been studied by Dighton (1977).

There has been surprisingly little quantitative experimental work on the leaching of decomposing litter at the soil surface although the possible quantitative importance of the effect has often been remarked upon (see Section 4.3.1). Many decomposition curves show an initial weight loss which is much higher than subsequent decay rates (Fig. 4.1) and chemical analysis frequently reveals a rapid initial loss of soluble constituents. Both these phenomena are as readily explained by leaching losses as by microbial catabolism. The experimental differentiation of these two processes has however proved difficult. The potential 'competition' between leaching and microbial immobilisation of soluble materials was illustrated by Witkamp (1969). He studied the release of different isotopes from tree leaf litter tagged with a variety of heavy metal ions and found that the relative solubility of the ions was directly paralleled by the relative extent to which they were immobilised by the leaf microflora. He also demonstrated that the rate of leaching of the mobile ^{137}Cs ion increased with the intensity of simulated precipitation over a wide range, although the highest efficiency of extraction occurred at a medium intensity (equivalent to about 300 mm g^{-1}). Presumably this represents the point at which the microflora is unable to retain the ions against the extractive effect of the precipitation. Unfortunately no comparable experiments appear to have been done on the movement of the natural soluble organic and inorganic constituents of decaying litter. Attempts to quantify the extent of leaching loss by laboratory experiments have shown that deciduous leaf litter can lose up to 35% dry weight but much smaller losses occur from coniferous litters. Analysis of the organic fractions of *Eucalyptus* showed that under field conditions the loss of water soluble substances of a year represented a weight loss of 10–20%, the total weight loss due to microbial activity plus leaching was 29–60%. Thus, at most, about half of the observed weight loss could result from leaching because some of the water soluble compounds would be lost through microbial activity (Wood 1974a). Anderson (1973a) suggested that most of the 28% first year weight loss in beech (*Fagus*) and over 75% of the 37% weight loss of chestnut (*Castanea*) leaf litters could result from leaching in the absence of fauna. In contrast, Witkamp and Olson (1963) considered that less than 10% of the initial weight loss in white oak (*Quercus alba*) leaf litter resulted from leaching of soluble compounds.

None of the results for leaching under field conditions have measured leaching losses directly. Concurrent measurement of respiration and weight loss allow an indirect estimate of leaching losses by calculation of the difference between observed weight loss and weight loss due to catabolism. Again there are apparent contradictions in the results; Witkamp (1966) showed that only 16–30% of carbon lost from tree litter in a warm temperate forest in U.S.A. was recovered as CO_2, whilst Howard (1971) in a similar forest in U.K. and Bunnell *et al.* (1977b) in Alaskan tundra estimated that respiration accounted for virtually all the observed weight loss in litter. No general consensus of opinion on the magnitude of leaching losses can be gained and it seems clear that the position can only be clarified when experiments employing the same techniques are repeated in a variety of different climates. The concentration of water soluble components sets an upper limit to initial losses, but microbial decomposition of polysaccharides provides a continuing supply of water soluble sugars which may be leached from litter. Comminution also accelerates leaching losses—both Witkamp (1969) and Nykvist (1963) showed that reducing the particle size increases the amount of soluble material that can be leached by standard treatments. Although leaching may remove a considerable portion of the weight of the litter, the compounds removed are subject to microbial decomposition in lower parts of the soil profile and are not necessarily lost from the system. Exceptions to this may occur where, for example, there is rapid run-off through the surface layers during melt-off in tundra, flooding in areas of high rainfall or where there is high polyphenol or tannin content in the leachate. In general however the nett result is that within the soil the immobilisation of both organic and inorganic materials in solution is so efficient that only traces can be detected in run-off ground water.

The release of materials from detritus may be accelerated by other direct physical effects. The interaction of extreme conditions of temperature with the moisture environment resulting in desiccation or freezing may cause cellular disruption, with a consequent release of substrates and nutrients. Witkamp (1969) found that drying or freezing leaves resulted in a subsequent four-fold increase in the amount of leachable nuclide compared with untreated controls. He attributed this largely to the release of materials previously immobilised in microbial tissue. The enhancing effects of drying/ wetting cycles on the break-up of soil aggregates and consequent increased mineralisation have already been referred to in Chapter 5. The thawing of previously frozen detritus may result in the immediate release of large amounts of soluble material and contribute significantly to the burst of decomposer activity which occurs at the onset of snow-melt in tundra ecosystems (Bunnell *et al.* 1975).

6.7 Analysis of the effects of the physico-chemical environment on decomposition processes

The picture of the decomposition subsystem painted with a broad brush in Chapters 1 and 2 is of communities and processes largely regulated by the prevailing climate. If the climate sets upper and lower limits to the potential decay rate, the 'fine control' at the local level is determined by resource quality and factors of the edaphic complex. In Chapter 4 we analysed in some detail the resource quality factor and in the earlier part of this chapter we have looked at the effects of the main edaphic factors—soil moisture, aeration, pH. These have been examined as individual features although the strongly interactive character has constantly obtruded itself. In particular the dominating influence of moisture has been obvious—the moisture relations of the soil largely determine the aeration and strongly affect the pH and other ionic equilibria. We must therefore consider the ways in which climate and edaphic factors are interrelated, and to what extent the simple hypothesis of the 'hierarchy' of regulation advanced above can be maintained. This is the concern of this last part of the chapter. Most of the evidence on individual factors has come from experimental work employing some control over factors other than the one under investigation. In the field however the ecologist is concerned with analysing a multivariate situation and this is where we must start.

In Chapter 1 we saw how the multi-factorial analysis of Jenny (1961) on the formation of soils had come to the conclusion that the accumulation of organic carbon in the soil was predominantly determined by climate. The predominance of climatic factors is easily recognised when simple indices are used. Brady (1974) analysed the general trends of SOM accumulation in grassland soils of related type in the Mississippi Valley region of the U.S.A. (Fig. 6.15). This area, now largely converted to cultivated land, was mainly derived from the prairie lands to the west of the Eastern Deciduous Forest and to the south of the Boreal Forest. The North–South transect is associated with a marked increase in mean temperature and with a decreasing organic content of the soils. Correspondingly an increase in the annual rainfall from West to East—from the prairie grasslands to the edge of the deciduous forests—also shows an increased component of organic matter in the soils.

Other factors such as soil type also have to be separated from climatic factors. The most detailed multifactorial analysis yet conducted into the influence of abiotic factors on decomposition processes is that of the International Biological Programme for the Tundra Biome. Data were from a number of sites over a wide geographical area and were analysed for variation between sites in the annual weight loss of plant litters in medium mesh bags (Heal & French 1974). Although differences in resource quality between the litters were eliminated by comparing only similar materials (e.g. 'soft' leaves), significant differences in decay rate between sites were apparent. Fig. 6.16 shows the result of a Principal Component Analysis (PCA) defining two axes of site variation, broadly determined by 'climate' (I = temperature, and moisture, plus organic matter and nitrogen status of the soil which are correlated strongly with climate) and 'Soil' (IV = pH and P and Ca content). The decay rates were separated very clearly on the basis of these two axes showing trends of increasing losses from cold, dry, acid, mineral sites with

Chapter 6

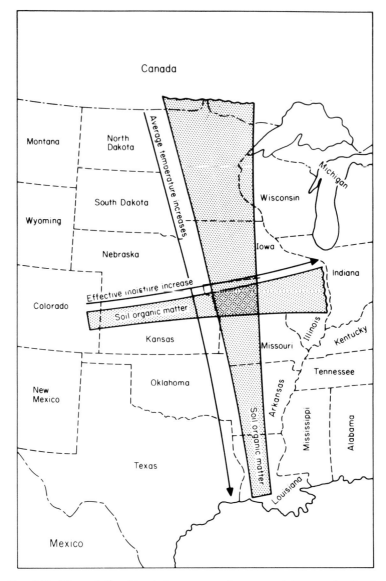

FIG. 6.15. The relationship between organic accumulation and climate. Gradients of soil organic matter accumulation in midwest-U.S.A. grassland soils are associated with gradients of temperature from north to south and moisture from west to east (from Brady 1974).

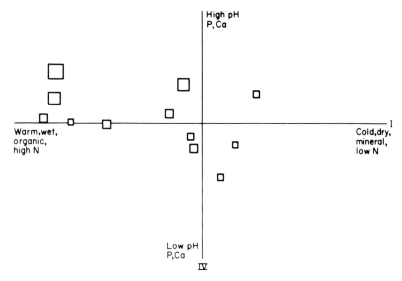

FIG. 6.16. Principal component analysis of the annual losses in soft leaves at ten geographically separated Tundra sites. The two axes broadly define climatic (I) and soil (IV) variables. The area of the boxes are proportional to the first year percent weight loss (from Heal & French 1974).

low P, Ca and N, to warm wet, nutrient rich sites with higher pH. With soft leaves climate explained more of the variation in loss rate than did soil, but other analyses showed that as resource quality was reduced soil factors became increasingly important. The interaction of temperature and moisture was explored further by regression analysis and showed (Fig. 6.17) that at most moisture levels the system responded linearly to increase in thermal energy but a defined optimum of soil moisture content occurred at about 400%. Unfortunately it is not possible to interpret this in terms of moisture potential as the relationship between potential and moisture content differs at the various sites. An interesting pointer is however given by the observation that when all litters (i.e. including more resistant resource types as well as soft leaves) are included in the analysis then the influence of moisture is even greater. This perhaps shows that the differing moisture-retention characteristics of different resources do have significant effects.

This general definition of the importance of climatic variables on decomposition rates in Tundra has been explored in detail by Bunnell *et al.* (1977a). They have described a model which defines the relationships between microbial respiration and the moisture and temperature environments of the resource being decomposed;

$$R(T, M) = \frac{M}{a_1 + M} \times \frac{a_2}{a_2 + M} \times a_3 \times a_4^{\frac{T-10}{10}}$$

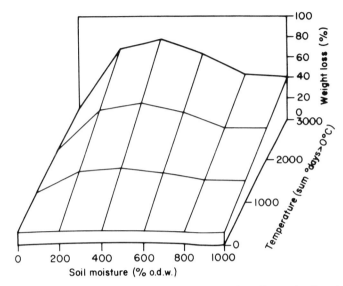

FIG. 6.17. A regression surface showing the interactive effects of soil moisture (% o.d.w.) and thermal environment (temperature sum = ∑ degree days at temperatures > 0°C). The regression equation for the relationship of the three factors was $Y = 11·62 + 0·0147.T.W - 0·00289.T.W^2 + 0·000152.T.W^3$ ($r = 0·769$), where Y = per cent weight loss, T = temperature sum and W = soil moisture (from Heal & French 1974).

where $R(T, M)$ = the respiration rate in $\mu l\ O_2\ g^{-1}\ hr^{-1}$ at temperature T (°C) and moisture M (% dry weight); a_1 = % moisture content at which the resource is half-saturated with water; a_2 = % moisture content at which half the pores are saturated or blocked with water; a_3 = the respiration rate that occurs at 10°C when neither oxygen nor moisture is limiting; a_4 = the Q_{10} coefficient.

The coefficient a_3 reflects resource quality by setting the level of the response surface of respiration to temperature and moisture; the derivation of this term has already been described (Chapter 4 page 138).

The authors have attempted to derive a model describing the influence of moisture and temperature in the simplest form possible. Thus they considered the desirability of using a measure of moisture potential rather than moisture content recognising the arguments of Griffin (1972) and others, as advanced in Section 6.2, that it is potential that regulates the availability of water to microbial cells. They concluded however that except at extremes moisture content was an equally acceptable basis. A second function of moisture content was included in the second term of the equation. Whereas the first term expresses the availability of water to the cells the second term expresses the limiting effect which increasing moisture content can have on gaseous exchange. The fourth term, the temperature effect is also admittedly simple in form using the Q_{10}, i.e. a simple exponential relationship between rate and temperature, rather than an Arrhenius' relationship. No limits to the temperature relationship are included.

The term a_3 establishes a base line for the decomposition of any resource as a resource specific (on the basis of chemical composition) rate at 10°C when neither moisture or O_2 availability are limiting. The Q_{10} (a_4) then gives a rate value for response to temperature values (T) and a_1 and a_2 establish responses to moisture contents (M) entered into the equation. Tests of goodness of fit of predicted respiration rates to measured rates for ten different resource type/micro-environment combinations were all significant. Predictions of weight loss in litters showed good agreement with litter-bag determinations (Table 4.9) even though the model ignores leaching and comminution. The realistic nature of the predictions is perhaps best shown by comparing the response surface of various litters to temperature and moisture fluctuation as predicted by the model (Fig. 6.18 right) with the actual responses derived from respirometric measurements made on the litters (Fig. 6.18 left).

This is the most extensively tested model relating decomposition rate to the regulatory variables of abiotic environment and resource quality. A number of other models have been published which differ in important respects. Nyhan (1976) for instance included terms for soil moisture potential and temperature based on an Arrhenius equation. This gave good predictions for carbon loss from grass litter buried in soil. He also included a time factor which allowed for a change in the relationships of the other factors as decomposition proceeded.

Witkamp (1966) derived a multiple regression equation for the field respiration (C) of four different tree leaves ($C = 46 \cdot 5 + 3 \cdot 2T + 26 \cdot 9M/D + 11 \cdot 4 \log B - 0 \cdot 6W$ ($r = 0 \cdot 71$, $n = 198$) where T = temperature, M/D = moisture content, B = bacterial population and, W = time from litter fall (i.e. equivalent to a resource quality factor). Temperature was the dominating factor in this equation explaining 64% of the variation as compared with only 5% for the moisture, but as the mean fortnightly moisture contents in these experiments never fell below 100% it is unlikely that moisture was a limiting factor. In laboratory experiments on the same leaf species at constant temperature but with more critical moisture contents the same author showed that CO_2 evolution was correlated with moisture content (Witkamp 1963). This is a good example of the caution that must be exercised in extrapolating a relationship, no matter how sophisticated, derived from particular circumstances, to a generalisation of the inter relationship between environmental variables. Meentmeyer (1971) used a different approach to modelling the relationship between abiotic variables and decomposition rates with a

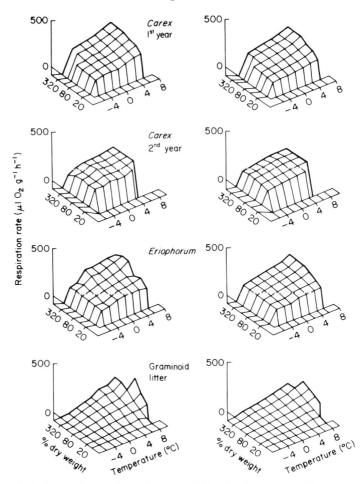

FIG. 6.18. Regression surfaces showing relationship between moisture content and temperature of litter and their rate of decomposition. Measured values (left) and values predicted from the model described in the text (right) are compared for litter of *Carex aquatilis* in the canopy during the first (top pair) and second (second pair) years of decomposition, for two-year old standing dead of *Eriophorum angustifolium* in the third pair and for mixed graminoid litter at the soil surface in the bottom pair (from Bunnell *et al.* 1977a).

single index combining moisture and temperature effects—that of the actual evapotranspiration. He was able to show a close correlation between the annual actual evapotranspiration and the weight loss of leaf litters for ten different forest sites in the U.S.A. (Fig. 6.19A). An equation derived from this relationship showed good predictive capacity for a number of tests against available data from forest systems (e.g. see Fig. 6.19B).

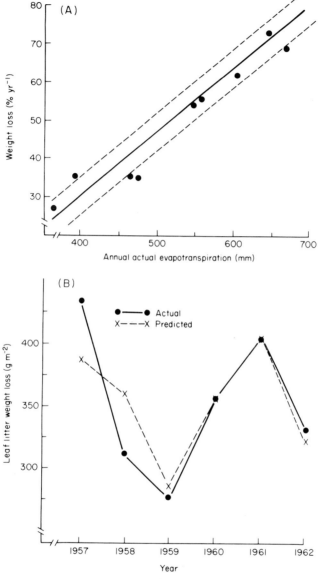

Fig. 6.19. The relationships between annual actual evapotranspiration and decomposition loss (A) plotted for first year weight loss and actual evapotranspiration for tree leaf litters in ten sites in the U.S.A. The regression equation is $Y = 0.166X - 36.3$. Standard errors of the regression line are indicated by the dotted line. (B) Yearly trends in the loss of weight by decomposition of leaf litter in deciduous forests of Eastern U.S.A. as observed (solid line) and predicted (broken line) by the above equation (from Meentmeyer 1971).

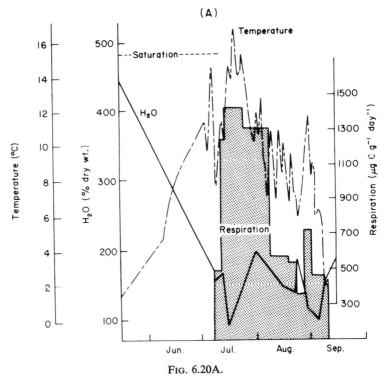

FIG. 6.20A.

FIG. 6.20. Relationship of seasonal climate and decomposition rate. (A) Relationship between respiratory loss of carbon from soil, air temperature and soil moisture in tundra at Hardangervidda, Norway (from Flanagan & Veum 1974). (B) Seasonal variation in decomposition losses from leaf standing crop in rainforest at Barro Colorado Island, Panama. Xo is the observed standing crop, Xe is the standing crop estimated to be present if litter fall occurred but with no decomposition. The shaded area thus gives an estimate of the decomposition taking place over the periods between sampling times (from Healey & Swift, unpublished data).

The changing nature of the relationship between temperature and moisture is perhaps best exemplified by reference to seasonal climatic effects, particularly the very marked ones which occur in extreme latitudes. In tundra, temperature is the predominant feature; in early June as snow melt is initiated, the radiation balance changes, resulting not only in heating of the soil but also significant evaporation which brings about a decrease in soil moisture content. The effect of this combination of events on soil respiration is shown in Fig. 6.20A. The development of maximal decomposer activity is dependent not only on the increase in temperature but also on the decline in soil moisture from saturation to a slightly sub-optimal level of around

(B)

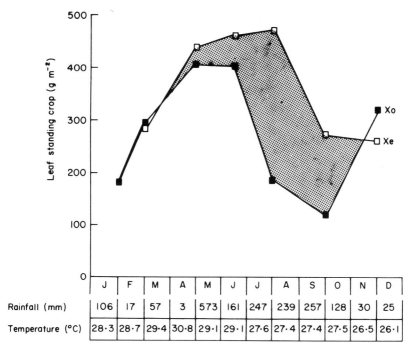

	J	F	M	A	M	J	J	A	S	O	N	D
Rainfall (mm)	106	17	57	3	573	161	247	239	257	128	30	25
Temperature (°C)	28·3	28·7	29·4	30·8	29·1	29·1	27·6	27·4	27·4	27·5	26·5	26·1

FIG. 6.20B.

200%. Seasonal tropical rainforest provides an example of extreme contrast (Fig. 6.20B). Temperature remains relatively constant but both litter fall and decomposition are related closely to rainfall. The same study (Healey and Swift, unpublished data) also showed that the availability of different resources to decomposers was a strongly seasonal factor with fruits and other reproductive material forming a dominant fraction of the litter decomposed at the start of the rainy season, leaves during the main early and middle stages and woody resources in the latter part.

Thus we can distinguish theoretically between two extremes (Fig. 6.21). In the first case (A) we have litter falling at a time most favourable to decomposition (e.g. at the onset of the rainy season in tropical rainforest as in Fig. 6.20B). The time of slowing of the decay rate due to resource quality may or may not coincide with the onset of unfavourable (dry) conditions, depending on the rate of decay and length of the rainy season. The second curve (B) is more typical of temperate forest; the curve is sigmoidal as litter input coincides with the onset of unfavourable (cold) winter conditions. Decomposition is initiated as temperature rises in the spring but the onset of

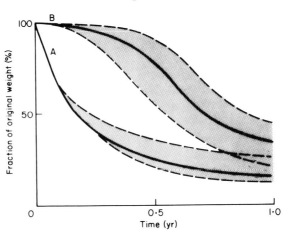

Fɪɢ. 6.21. Theoretical decay curves for first year weight loss of litter in different environments. (A) Litter fall coincides with a climatic period favourable to decomposition. (B) Litter fall coincides with a climatic period unfavourable to decomposition. The shaded areas round each curve indicate the variations consequent upon shortening (above) or lengthening (below) of the favourable season. Similar variations will occur due to lower or higher resource quality and unfavourable or favourable soil types.

resource quality limitation may in fact coincide with climatic conditions most favourable to decomposition. In Anderson's (1973a) litter-bag experiment, described in Chapter 3 the importance of the winter climatic inhibition was clearly demonstrated in a subsidiary experiment. A subsample of leaves collected at litter fall was placed out in the field in June alongside the litter-bags which had already been in the field for six months. By the following autumn the weight losses in the two samples were indistinguishable. Over-wintering is often pictured as a 'weathering' process in which the polyphenols and other modifiers are removed or denatured allowing decomposition to proceed in the spring. This experiment in contrast emphasises the dominating influence of seasonal climate on decomposition.

The experimental distinction of resource quality, edaphic, and climatic factors is not difficult, but few attempts have been made to unravel the complexities of the regulation of decomposition processes. It is to be hoped that this will become an important research objective in future for an understanding of regulation is an essential basis for the management of the decomposition subsystem.

7

THE DECOMPOSITION SUBSYSTEM; SYNTHESIS AND SUMMARY

We have analysed the processes and populations in the decomposition subsystem recognising that:

(1) decomposition is a combination of three major processes: catabolism, comminution and leaching;

(2) decomposition is controlled by three major groups of factors: the nature of the decomposer community, the resource quality and the physico-chemical environment;

(3) the decomposition subsystem can be conceived as a series of modules within each of which organic resources are decomposed according to the 'rules' defined in (1) and (2). The relative importance of the individual processes and factors varies in time and space so that the structure of the sub-system can be defined by combinations of modules arrayed in time and space;

(4) feedback mechanisms are involved in which the process of decomposition modifies both resource and environment within each module thus altering the regulatory status of succeeding modules;

(5) ultimately the composition and productivity of vegetation is regulated by the patterns of decomposition defined by the above structure. The character of the vegetation is in itself, however, one of the prime determinants of the nature of the decomposition subsystem.

In this last chapter we draw together these principles in two ways; first 'case-studies'—detailed analyses of ecosystems or parts of ecosystems. These have been selected to show the effects of different combinations of factors and to indicate some of the ways in which man influences and controls decomposition; second, we summarise the pattern of processes of decomposition in major types of ecosystems in a series of full-page charts. In the case-studies the emphasis is placed on the integration of dry matter, energy, carbon and nutrient flows. Where a single element is considered to be critical in the functioning of an ecosystem, the pathways of that element may be analysed in detail, but the transformations of element pools can, in general, be conceived in the same way as the decomposition module. So far we have dealt with ecosystems which are not intensively managed by man, but in two of the case studies, grassland and forest, we concentrate on systems in

267

which man manipulates, consciously or unconsciously, the decomposition subsystem.

The first two case-studies are confined to specific and specialised examples of the functioning of the decomposer subsystem.

(1) *The fungus/ant symbiosis*—is a case where a single fungal species acts as the main saprotrophic agent and its symbiotic partner, the ant, derives its energy and nutrient by microtrophic activity. The relationship is maintained by a series of complex biochemical interactions which may serve as a model of the more complex interactions of other decomposer communities.

(2) *The vertebrate rumen*—provides another example of a detailed analysis of the interaction of saprotrophic organisms in decomposition. The synergistic effects of serial utilisation of decomposition products are emphasised for a system which operates within a very stable environment.

(3) *Coastal tundra*—a site specific study at Point Barrow, Alaska, where the tundra ecosystem is dominated by decomposition pathways. The interactions within the system have been rigorously defined in a mathematical model.

(4) *Cold temperate peat bog*—a site specific study in northern England exemplifying the importance of events in geological and historical time on the current system, a blanket bog, in which water is the key factor.

(5) *Temperate cultivated grassland*—a grazing system in which the activities of the herbivore modify the rate and pathway of decomposition directly through dung, urine and trampling, and indirectly through modification of the vegetation. The circulation of N is the focus of attention in many cultivated ecosystems and here we examine the linking of N with organic matter decomposition and the effect of varying conditions and management practices on this intimate and complex cycle.

(6) *Temperate deciduous forest*—based mainly on the detailed catchment studies at the Hubbard Brook Experimental Forest in Eastern U.S.A. The balance of factors is disturbed by deforestation and the consequences of this practice are examined. The account concentrates on the interaction of nutrient and soil factors. The main features revealed by this experimental study are briefly developed in a discussion of the agricultural management of previously forested lands including the current development of tropical forest areas.

The selection of the case-studies emphasises the imbalance of ecosystem research; it is only in temperate regions that studies detailed enough to illustrate integrated ecosystem functioning have been carried out.

7.1 Fungus/ant symbiosis

The fungus growing ants of Central and South America have developed a symbiotic relationship with fungi which is only paralleled by that of the

Macrotermitinae (Termitidae, Isoptera, see Fig. 3.13). The Lower Attine genera utilise a range of plant materials for their fungus gardens including insect frass, rotten wood, plant litter and fresh leaves; some species also collect insect remains. The High Attines, *Sericomyrmex*, *Acromyrmex* and *Atta*, forage for fresh leaves which are cut into discs by the workers and carried back to the nest. Large colonies of these genera can defoliate trees and shrubs, and in many areas of South America they are serious agricultural pests particularly in citrus orchards. Several species inhabiting grasslands also exploit or specialise on Gramineae.

The nests are subterranean and colonies can be identified by heaps of excavated soil and discarded leaf discs which, in a large colony, may cover more than 50 m². The fungus gardens are located in cavities scattered throughout the colony and rarely exceed 20–30 cm in diameter. In most attine genera the garden consists of a loose-structured mass of chewed plant material penetrated by a dense fungal mycelium. The taxonomic identity of the fungus is uncertain, though it is believed to be one or more species of the Basidiomycete *Lepiota*. The fungus produces hyphal swellings, called gongylidia, at the surface of the leaves which are used by the ants as food. Chemical analysis of the fungus removed from the nest of an *Atta* species and cultured on dextrose medium (Table 7.1) has shown that, with the excep-

Table 7.1. Nutrients derived from the fungus cultivated by *Atta columbica tonsipes*. More than 50% of the fungus dry weight is soluble nutrient (Martin, Corman & MacConnell in Stradling 1977).

Nutrient	mg per g dry fungus
Trehalose	160
Glucose	47
Mannitol	51
Arabinitol	14
Protein-bound amino acid	130
Free amino acid	47
Lipid (major sterol, ergosterol)	2
Total weight identified	451

tion of a rather low lipid content, the fungus may represent a complete, high-quality food source for the ants. Additional nutrients and water are obtained directly by the ants from plant saps collected during foraging.

The fungus ants inhabiting the rain forests of South America forage for a wide variety of plant species. In one area of Guyana, 36 out of 72 plant species were attacked over a ten-week period in an area of 1194 m² around a large colony of *Atta cephalotus*. The selection of leaf material seems to be based on physical, morphological and chemical criteria. Plant leaves with a dense covering of trichomes are rejected and species containing resins or copious latex are also avoided. The preferred leaf materials are thin, soft, young leaves which are high in sap and soluble nutrients and readily cut by the ants. These characteristics are also possibly associated with low levels of plant secondary compounds, such as alkaloids and tannins, but detailed analyses of the leaves have not been carried out. The quantity of fresh leaf material cut by the ants and carried back to the nest along cleared foraging trails may be considerable. Leaf inputs as high as 290 g per hour have been recorded for large nests in Costa Rica.

The leaf discs undergo extensive pre-treatment by the ants before they are added to the fungus garden. After licking both surfaces of the leaves for several minutes, small fragments of material are removed from the edges of the discs and chewed to a pulp which is placed on the fungus garden. A small tuft of fungal hyphae is then transplanted to the fresh material. The exact nature of this preparative process is still not understood, but it appears to involve the removal of the protective epicuticular wax layer of the leaves, leaf surface micro-organisms and possibly surface active mycostatic agents. In addition, metathoracic gland secretions are added to the leaf material which have been shown not only to have broad spectrum bacteriostatic and mycostatic properties against possible competitors, but also stimulate growth of the ant fungus. The faecal secretions of the ants which are also added to the garden have further complex properties for the ant/fungus system. The faecal fluids contain pectinolytic enzymes which chemically macerate the plant tissues as well as a broad spectrum of enzymes which degrade starch, xylan, esters and chitin. In addition the rectal secretions contain significant protease activity which the fungus apparently lacks. The nitrogenous excretory products are utilised by the fungus and allantoin serves to disperse the leaf protein and render it more susceptible to hydrolysis. Both the allantoin and the chitinases may reinforce the antibiotic properties of the metathoracic gland secretions and assist the ants in maintaining a mono-species dominated culture. The fungus gardens of abandoned nests are rapidly overrun by bacterial and fungal contaminants. The interactions between the ants and their symbiotic fungus are summarised in Fig. 7.1.

In this example, as with the termites described in Section 3.2.6, we see a decomposer community condensed to two organisms. The ants initially perform as foraging herbivores but the main energy flow is maintained by the activity of the saprotrophic fungi with the ants obtaining energy and nutrient by microtrophic function. Colonial insects have a high energy demand for the foraging activities of the workers. The Macrotermitinae and Attini show a unique example of convergence in the cultivation of Basidio-mycetes for the degradation of plant structural carbohydrate. There is also a large demand for nutrients such as N, P, K, etc. for tissue production although recycling of nutrients in dead workers within the nest is important

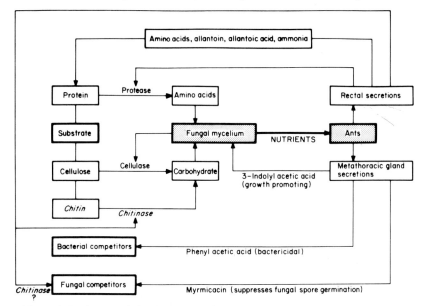

FIG. 7.1. Pathways of chemical interaction between attine ants, their symbiotic fungi and other possible competitors (from Stradling 1977).

in nutrient homeostasis. The production of sexuals (alatae) which disperse from the nest produces a significant drain on the colonies accumulated nutrient capital; large losses of worker termites may also occur by ant predation. In the termites the quality of the food resource (dead wood) is low, but nutrient accumulation is probably facilitated by the high rate of carbon flux through the termite/fungus system as well as by possible sources of nitrogen fixed in the termite gut. In the attines the development and maintenance of a very high ant biomass is made possible by the high resource quality of the living leaves which form the output to this unique saprotroph system.

This account is based on a general review of the food and feeding habits of ants by Stradling (1977) with further information from Cherrett (1968, 1972) on leaf selection and Quinlan and Cherrett (1977) on the significance of resource preparation for the maintenance of the fungus gardens.

7.2 The vertebrate rumen

The ant/fungus symbiosis provides an example of a decomposition system reduced to a few component variables which function in rather a constant manner. Resource quality is usually high, decomposition is carried out by a single fungus species which produces a high-quality food resource for the ants.

We turn now to the rumen as an example of a more complex decomposition system illustrating the synergistic interactions of a diverse bacterial flora which provide the nutrient and energy requirements of not only the ruminant but also large populations of commensal protozoa. The microbiota act in rather a constant, though specialised, physical environment but resource quality, especially in wild ruminants, may be extremely variable and produce major shifts in the relative importance of different groups of bacteria.

The ruminant digestive system is shown in Fig. 7.2. The rumen itself

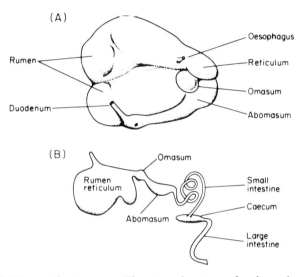

Fig. 7.2. The vertebrate rumen. The stomach system of a sheep shown as in the animal (A) and diagrammatically to show its relationship to other parts of the alimentary tract (B). Diagram (B) does not show the relative sizes of the organs (from Hobson 1976).

occupies a large proportion of the total gut volume and slows down the passage of food allowing the development of the bacterial flora and the fermentation of cellulose to take place. Plant material is swallowed with little chewing and passes into the rumen. Here it is mixed with copious saliva (100–190 litres per day in a cow) and kneaded by contractions of the rumen wall. At intervals a bolus of material is regurgitated and chewed (rumination). The food becomes comminuted to particles of a few cubic millimetres in size and the plant cell walls are exposed to attack by microbial enzymes. The products of catabolism are mainly absorbed through the rumen wall and the digesta then pass to the omasum where some water and salts are removed. Normal digestive processes start in the abomasum where bacterial and protozoal cells are broken down. Amino acids, lipids, some sugars and microbially produced vitamins are absorbed in the small intestine. Further microbial growth takes place in the caecum and much of the remaining undigested cellulose is catabolised. The products of this second fermentation are absorbed by the animal but no lysis of bacteria occurs and the cells pass out with the faeces.

The rumen forms a specialised microbial environment. Temperature is maintained close to 39°C by a combination of the fermentative processes and the ruminant's body temperature. The liquid contents of the rumen are in equilibrium with a gas phase consisting of approximately 65% CO_2, 35% methane (CH_4), up to 7% N, 0·2% H_2, 0·01% H_2S and 0·6% O_2. Oxygen entering the rumen with food or diffusing from the rumen wall is rapidly reduced and the liquid maintains highly anaerobic conditions with a redox potential of -400 mv or even lower. pH conditions in the rumen are buffered by a bicarbonate-phosphate mixture (pH c. 8·2) which neutralises the carboxylic acids from the fermentation processes and maintains a rumen liquid pH of about 6·5.

The reactions carried out by the mixed populations of rumen bacteria are shown in Fig. 7.3. The major products of the anaerobic fermentation process are acetic, propionic and butyric acids, CO_2 and methane. The gases are belched out (a cow produces 1500 litres of gas per day) and the acids are assimilated through the rumen wall to be aerobically oxidised by the animal as the major energy source from the feed. Ammonia is also produced by the rumen bacteria, absorbed into the blood stream and converted into urea in the liver. Part of this urea is recycled to the rumen in saliva or by diffusion through the rumen wall, and part is excreted in the urine. The recycled urea is available for synthesis of microbial tissues which are used as a source of protein by the animal. Under conditions when resource quality is low (high C/N ratio) the kidneys excrete less urea and N is conserved within the animal/rumen system. This mechanism allows ruminants to subsist on poor feed with low nitrogen.

A very large number of bacterial species are involved in rumen metabolism and it is impossible to consider their functional attributes in any detail within this review. However, the elegant experiments of Hobson and his colleagues, where the rumen microflora was built up in gnotobiotic lambs, provide an insight into the symbiotic association between some of the major groups of bacteria. The rumen is small and sterile in new born animals but grows rapidly after weaning under the combined influences of solid food and microbial metabolites. The bacterial and protozoan inocula come from contact with adult animals. Hobson *et al.* weaned lambs under aseptic conditions on to a straw, ground nut and grain feed and then introduced *Streptococcus bovis* and *Lactobacillus* sp. into the rumen. The *Streptococcus* is amylolytic (starch fermenting) and a facultative anaerobe. Its activities provide the maltose for the *Lactobacillus* and the anaerobic conditions

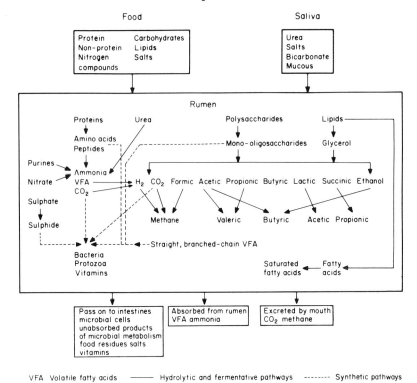

FIG. 7.3. The main reactions occurring in the rumen during digestion of food (from Hobson 1976).

which allow other bacteria to be introduced. These bacteria produced lactic acid, which accumulated in the rumen, and the pH fell from 6·5 to about 4·3. Two lactate-fermenting bacteria *Peptostreptococcus elsdenii* and *Veillonella gazogenes* were then introduced and the pH returned to approximately 6 within two days and the normal, negligible concentration of lactic acid in the rumen was re-established. The bacteria present in the rumen at this stage, and others which were added later, need amino acids for good growth and so hydrolysis of the feed protein was required. *Bacteroides amylophilus*, a starch fermenting and strongly proteolytic species, was then added together with the deaminative bacterium, *Bacteroides ruminicola*, to provide ammonia for the growth of *Bacteroides*. Both of these species produce succinic acid which is decarboxylated by the *Veillonella*. *Butyrivibrio fibrisolvens* was added to bring up the butyric acid concentration and assist in hemicellulose fermentation. Finally, *Ruminococcus albus* was added as a strongly cellulolytic species which favours hydrogen, ethanol, formate, acetate and lactate. This bacterium was supplied with its needs for ammonia and volatile fatty acids by the deaminating activity of *Bacteroides ruminicola*.

As inoculation proceeded over 30 days a closely similar microflora to the adult rumen was developed with bacteria counts of 2–3 $\times$ 10^{10} per ml and producing acetic, propionic and butyric acids in the proportions 39%:32%: 24% against the normal ratio of approximately 60%:30%:10%. This rumen flora allowed the lamb to digest food and gain weight in contrast to its still germ-free control. It is interesting, however, that the normal ratio of the three major acid products of the rumen is apparently not optimum for growth. Recent experiments have shown that the growth efficiency of shed-reared stock is improved by the addition of monezin to the diet. This results in a decrease in the proportion of acetic and butyric acids produced and a 10% increase in the production of propionic acid. Treatment with monezin was found to decrease food consumption by over 10% and save on the production of meat by about $0·10 per kg in 1976.

The rumen ciliate protozoa have adapted to their environment to such an extent that they are found nowhere else. The populations rarely exceed a few thousands per millilitre of rumen liquid and consist of two groups: the holotrichs and the entodiniomorphs. Both groups utilise not only low molecular weight plant compounds but also engulf small particles including starch granules and bacteria. Some species are predatory on other protozoa. Little is known about the role of rumen protozoa *in vivo*, but in cultures the bacteria and protozoa are mutually beneficial. *Enterodinium* species, which have been subject to more detailed investigation than other groups, digest starch and proteinacious material making sugars and amino acids available to bacteria in the medium as well as to bacteria in the protozoan endoplasm which function endosymbiotically for some time after they are engulfed. The protozoa digest the bacteria as a source of amino acids for protein synthesis. Some of the bacterial amino acids are released back into the medium where they are reutilised by the bacterial populations. The net effect of the protozoa is an increased turnover rate of bacterial carbon and nitrogen. Under some conditions up to 1% of the bacteria may be killed per minute in the rumen and experimental rumen developed without protozoa have significantly higher bacterial populations than those in which the normal complement of protozoa is present. The presence or absence of the protozoa, however, seems to make no difference to the health and growth of the ruminant.

In many non-ruminant, vertebrate herbivores, such as horses, microbial fermentation of cellulose occurs in the large intestine after digestion has taken place in the stomach. The bacterial flora is similar to that of the rumen, protozoa may also be present, and the plant residues are broken down under anaerobic conditions. The fermentation products and some amino acids from lysed cells are absorbed but a large proportion of the bacteria pass out with the faeces and the microbial protein is not available to these animals in the same way as ruminants. Rabbits, and some other herbivores, reingest a particular type of faecal pellet which they produce and this may permit the utilisation of microbial proteins and vitamins. This situation is analogous to that of most saprotrophic invertebrates which have a comparatively simple gut structure and utilise microbial protein and metabolic products of the soil microflora. This analogy has led to the concept of the soil as an 'external rumen' for soil animals.

The most comprehensive account of the rumen microflora is given by Hungate (1966) while Coleman (1975) reviews the role of protozoa in the rumen. This account is largely based on a monograph by Hobson (1976) which includes details of the experiments with gnotobiotic lambs.

7.3 Coastal tundra

On the coastal plain of Alaska, tundra ecosystems are developed in a climate in which air temperatures rise above freezing for only 80–90 days, rarely exceeding 10°C, and are below -20°C for much of the winter. In summer the surface layers of the soil thaw to a depth of about 40 cm below which the ground is perennially frozen. Thus there is a steep temperature gradient within the soil and the freeze-thaw cycle, acting in a flat landscape, produces characteristic ice-wedge polygons. Despite the underlying importance of temperature in tundra ecology, one of the main impressions in the field is of the wide variation in moisture. On the wet meadow the polygon formation produces dry rims surrounding low centres in which water is retained by the permafrost and the polygons are separated by wet troughs along which water can flow freely. The seasonal variation in moisture is also marked; the annual precipitation is only 124 mm, of which 40% falls as rain during the summer, but at thaw there is a considerable release of water resulting in flooding of much of the meadow. Winds produce drying and cooling conditions within the vegetation canopy and as a result of varying contents of organic matter and moisture in the soil the depth of thaw varies from 20 to 80 cm with temperatures at 10 cm depth in the polygon rims being 2–4°C higher than in the centres and troughs. Thus the flat coastal tundra shows marked regular gradients in microclimate over small vertical and horizontal distances, gradients which have a dominant influence on both the short- and long-term rates of decomposition.

The species composition and production of the vegetation varies over the area, mainly in relation to the moisture content, oxidation-reduction potential and soluble phosphate concentration of the soil. *Carex aquatilis*, *Eriophorum angustifolium* and *Dupontia fischeri* are most common in the moist and wet meadows but woody dicotyledons, mosses and lichens show increased biomass in the drier habitats. Net primary production ranges up to about 300 g m^{-2}

yr^{-1} on the wet sites, most of it above ground although the above:below ground biomass may be as high as 1:40.

The timing and quality of the input from primary production to the decomposition subsystem is markedly affected by the brown lemming (*Lemmus sibericus*) whose numbers vary in a cyclic fashion over a period of 3–5 years. During periods of high lemming density (up to 200 ha^{-1}) about 25% of the above ground primary production is consumed compared with less than 0·1% when densities are low. Much of the grazing occurs during the winter however when there is no plant growth and whilst the animals consume only a small proportion of the sedge and grass shoots they often completely sever them at their base. The combined effect of these two features is to fell the complete above-ground standing crop of vascular plants during times of high population density. The effect of this on the decomposition subsystem is four-fold; the leaf detritus contains higher concentrations of soluble components because translocation is restricted by severence; the leaves are transferred more rapidly from the canopy to the litter micro-environment; the nutrients in lemming faeces and urine which accumulate over winter are released as a spring flush at thaw; the soil temperatures during summer are higher because the insulating properties of the plant cover are reduced. The result is an increase in the rates of decomposition and nutrient cycling as the lemming numbers increase.

Decomposition rates in tundra are severely limited by the short summer period. Standing dead leaves of grasses and sedges lose about 4–5% of their weight in the first year but this increases to 7–16% when the leaves are in the moister environment of the litter layer. There is a differential loss of the chemical fractions of the leaves, ethanol soluble components decreasing by 49% in the first year in comparison with the ethanol insoluble fraction loss of only 11%. The weight losses from the standing dead leaves in the litter result from the combination of translocation by the plant, leaching, comminution and catabolism. In the studies at Point Barrow, only microbial respiration has been accurately measured for a range of resource types. From known relationships of respiration to temperature, moisture (Chapter 6, p. 259) and resource chemical composition (Chapter 4, p. 138) the weight loss to respiration throughout the year has been calculated using daily measurements of temperature and moisture and analyses of resource chemical composition. The computed weight loss is very close to measured weight loss in the field indicating the major contribution of microbial respiration to decomposition. The model also estimates well the broad chemical composition of the litter after one year's decomposition (Table 4.9) and thus expresses the combined effects of seasonal variation in temperature and moisture and the effects of chemical composition (i.e. resource quality).

Chemical definition of the later stages of decomposition is poor but using a response surface relating the respiration of soil organic matter to temperature and moisture, the seasonal accumulated output of carbon dioxide has been simulated along with the respiration of live roots (including the rhizosphere microflora). These simulations cannot be evaluated directly but the total system respiration is very close to measured field values and the results indicate that microbial respiration contributes 30–67% of the total respiration on any given day with CO_2 output equivalent to about 130 g dry matter m^{-2} over a summer period of 85 days (Fig. 2.10). Again the major influence of temperature and moisture is indicated but the results also raise conflicting conclusions; whether organic matter decomposition matches nett primary production giving a steady state system in which there is no further accumulation of organic matter; or whether the system is in a developing state with either nett gains or losses of organic matter.

The detailed analysis of the decomposition subsystem at Point Barrow has gone a long way towards providing an understanding of the factors influencing the processes of decomposition and of the ecology of the microbial populations involved, but the quantitative formulations are not yet sufficiently precise to determine the current balance between primary production and decomposition. For this we must rely on measurements at a broader scale. There is between 22 and 45 kg m^{-2} of dead organic matter in the top 20 cm of the soil in various meadow and polygon habitats; there is also an inverse correlation between primary production and accumulation. A reasonable hypothesis to account for these observations is that primary production decreases with increasing accumulation of organic matter through the lowering of soil temperatures and of the rates of nutrient circulation. Thus on sites with a high decomposition potential, accumulation of SOM with time will be small and primary production reduced to a small extent as the system tends towards a steady state (Fig. 7.4A). Where the decomposition potential is low there will be a greater fall in primary production as larger amounts of organic matter accumulate but again tending towards a steady state (Fig. 7.4B). An alternative hypothesis is that the rate of decomposition declines with depth in the soil profile, eventually becoming negligible, through low temperatures and eventual incorporation of the organic matter into permafrost. Accumulation would be a continuing process but would have little effect on primary production, as indicated in Fig. 7.4C. These three hypotheses show the interaction between production, decomposition and accumulation over time and could account for the observed negative relationship between production and accumulated organic matter.

Preliminary budgets of dry matter production and decomposition indicate that the systems are currently accumulating and in the absence of deep accumulations of organic matter (Fig. 7.4C) we must assume either that this tundra system is young and has not yet reached steady state, or that recurrent disturbances can reverse the pattern of accumulation in any habitat. The latter hypothesis is supported by the dynamic appearance of the landscape

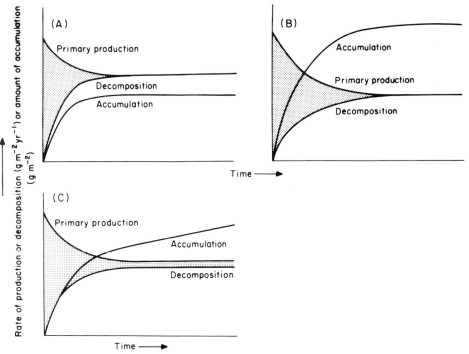

FIG. 7.4. Hypothesised relationships between primary production, decomposition and organic matter accumulation in tundra at Point Barrow. (A) model tending towards steady state, for a habitat of high decomposition potential; (B) model tending towards steady state, for a habitat of low decomposition potential; (C) non-steady state model, tending towards a condition of continuing accumulation (from Bunnell, MacLean & Brown 1975).

with the suggestion that on flat wet meadow tundra, polygons are formed by the action of the ground ice; polygons coalesce to form pools and lakes while aquatic habitats drain and form meadows which initially have low amounts of organic matter because of the lack of accumulation in aquatic habitats. This thaw-lake cycle as it has been called could maintain a non-steady system, initiated and maintained by the prevailing temperature and topography but with the vegetation and soil characteristics varying in relation to decomposition:

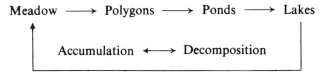

This summary of the tundra at Point Barrow is derived from Bunnell, MacLean & Brown (1975), Bunnell & Scoullar (1975), Bunnell, Tait, Flanagan

& Van Cleve (1977a), Bunnell, Tait & Flanagan (1977b), and Pitelka & Schultz (1964).

7.4 Cold temperate peat bog

The analysis of a moorland ecosystem in the northern Pennines of England provides a case study in which the balance between production and decomposition of organic matter is the product of a series of past and present environmental factors, particularly geology, climate and topography. Slight variations in time and space in one or other of these factors have resulted in a mosaic of blanket peat with *Calluna*, *Eriophorum* and *Sphagnum* interspersed with grassland swards on a variety of mineral soils.

The initial factor in the system is the geology, a series of horizontal alternating beds of Carboniferous limestone, sandstone and shale which, as a result of pre-glacial and glacial erosion, has produced a landform with a steep scarp rising from 300 m to a ridge at about 900 m and a gently sloping eastern dip slope falling to the River Tees at about 300 m. Glacial drift is the second factor, a layer of boulder clay deposited by the retreating ice about 13,000 BC. The drift was subjected to cryoturbation and solifluction and was largely removed from the steeper scarp and ridge but retained on the dip slope except on stream banks. Tundra scrub in the cool post-glacial Boreal period (10,000 BC) developed into birch woodland with an increasingly warm climate but at the Boreal-Atlantic transition (5,500 BC) a cooler wetter climate caused waterlogging of the soil especially where the boulder clay impeded drainage. Waterlogging plus the lower temperatures reduced the rate of decomposition and caused an increased accumulation of organic matter which gradually incorporated the remains of the birch woodland. The organic matter accumulation increased the water-holding capacity and acidity of the soil, and with the cool wet climate, a bog vegetation of *Calluna*, *Eriophorum* and *Sphagnum* developed. The litter from these species has an intrinsically low rate of decomposition (i.e. low resource quality) and thus contributed to the accumulation of peat to its present depth which varies from 0·5 to 4 m. Some peat accumulation probably occurred on the scarp but was unstable because of the slope and has mainly been lost through erosion.

Where the boulder clay was thin or absent in the early post-glacial period, the bedrock strongly influenced the soil development. Gley and peaty gley, peaty podzol and brown podzolic soils formed on the shale and sandstone with varying degrees of drainage. On the small areas of limestone less acid and well drained brown earths with mull humus were formed and now carry grasslands dominated by *Festuca* and *Agrostis*. Thus the combination of geology, glaciation, topography and climate has caused the current mosaic of soil and vegetation types within an area of 3,000 ha. Man's

and organic phosphorus. Initially nitrogen is fixed symbiotically giving increased primary production and there is a gradual build up of decomposing organic matter at a rate of about 200–300 g m^{-2} yr^{-1} with organic-N accumulating at about 5–6 g m^{-2} yr^{-1}. Phosphorus, initially present in inorganic form derived from the parent mineral rock, is assimilated by the vegetation and returned to the soil as organic-P. As P is not released rapidly from parent rock the proportion of organic-P increases as the vegetation develops until there is little inorganic-P available and it begins to limit primary production. As clover competes poorly with grass roots for P it tends to be eliminated from the sward and symbiotic N-fixation virtually ceases leaving a system approximating to a steady state with an organic matter content of about 1 kg m^{-2} (Fig. 7.7). This general sequence has been described by Walker (1956) for New Zealand soils and is probably applicable in U.K. (Floate 1971) where a change from arable to permanent grassland

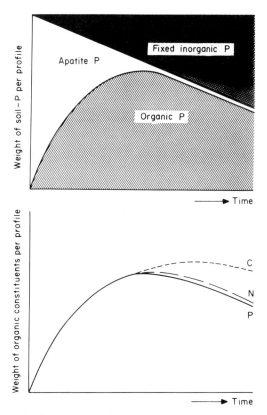

FIG. 7.7. Hypothetical pattern of accumulation of phosphorus, carbon and nitrogen in pasture (from Floate 1971 & Walker 1956).

at Rothamsted resulted in an accumulation of about 5 g N m^{-2} yr^{-1} for the first 40 years reaching equilibrium after about 100 years (Richardson 1938).

The general simplified pattern of nutrient circulation in a mature grassland ecosystem grazed by dairy cattle is shown in Fig. 7.8, the annual

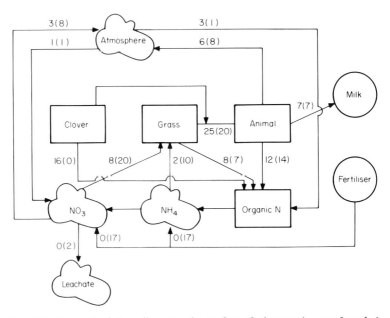

Fig. 7.8. Generalised flow diagram of transfers of nitrogen in g m^{-2} yr^{-1}, in cultivated temperate grassland grazed by dairy cattle. Transfers are for an unfertilised clover-grass sward and for a grass sward fertilised with about 34 g m^{-2} yr^{-1} of ammonium nitrate, the figures for the fertilised sward are in parentheses. The data are generalised figures from a wide literature and do not refer to specific sites (from Whitehead 1970).

transfers being given for fertilised and unfertilised systems. In unfertilised clover and grass swards producing a herbage yield of about 1000 g m^{-2} yr^{-1}, N-fixation by the legume can add up to 40 g N m^{-2} yr^{-1}, the rate being limited in some cases by availability to the clover of other nutrients especially phosphorus. The fixed N is released from the clover to the soil via grazing and defaecation, exudation of organic-N compounds from the living roots, leaching from leaves and by litter production or death. The fixed N is only available to the grass after mineralisation during decomposition but the clover residues with a high N content and low proportion of structural polysaccharides probably decompose rapidly. Despite the extensive research on fixation mechanisms there is remarkably little information on the processes of transfer to other parts of the ecosystem.

In a grazed sward consumption by livestock removed about 10–25 g N $m^{-2} yr^{-1}$ and output from the system is of the order of 2–7 g N for dairy milk, or 0·5–2 g N for beef cattle or sheep. Most of the ingested N is returned to the sward directly as faeces or urine. About half of the faecal N is bacterial material and only 10–20% is undigested plant residue. The faecal N is mineralised fairly rapidly because of its low C:N ratio, the predominance of amino-N, its comminuted state and the moist environment. Grazing gives a higher rate of N circulation through the cropping of N-rich live plant material and its return in faeces and urine compared to the input of plant litter, relatively low in N, in an ungrazed sward (Floate 1970c, 1971) and to a cut sward (Shaw, Brockman & Wolton 1966). Some of the mineralised N from faeces is released as ammonia, as is much of the N in urine. Under warm and moist conditions, where nitrification is inhibited by lack of oxygen, the ammonia may be volatilized and returned to the atmosphere if it is not assimilated by micro-organisms, the assimilation being dependent on the presence of a suitable source of carbon.

The N return to the decomposition subsystem through plant death and animal faeces is about 40 g $m^{-2} yr^{-1}$ but this may be an underestimate because of underestimation of primary production. The returned N represents most of the N which is circulating rapidly through the soil–plant chain of processes but it is only about 4%, and sometimes 1% or less, of the total organic-N in the soil. There is a fairly rapid mineralisation of much of the freshly added N which is available to plants but the incorporation into microbial tissues and into humic compounds causes short- and long-term immobilisation. With a nitrogen concentration of about 5% humus contains much of the soil N, releasing it at a very slow rate through slow decomposition. An annual turnover of less than 3% has been estimated for N in soil organic matter but with a standing crop of the order of 1 kg N m^{-2}, 10 g N m^{-2} is mobilised annually if the rate of mineralisation is only 1%. Thus the decomposition characteristics of the more recalcitrant fractions play a significant role in N circulation.

The complexities of mineralisation of organic N and the processes of nitrification and denitrification (see Section 5.4) have been extensively studied because of their importance in plant nutrition. Transfers in the grassland ecosystem (Fig. 7.8) are crude approximations and grossly simplified, ignoring particularly the cycling and immobilisation in the microflora, but accurate measurements of rates under field conditions are not available. However, the association of N mineralisation with resource quality is well established, as reflected in the C:N ratio which broadly expresses the extent to which N is available for the microbial catabolism of carbon or conversely, carbon is available for mineralisation of N. Thus although humus has a low C:N ratio it is resistant to decomposition and the lower C:N ratios in arable compared with grassland may result from a high proportion of humus in arable soils.

The microclimate in many pasture ecosystems allows nitrification and denitrification to proceed at high rates but there are wide variations as a result of local conditions. In heavy soils and in those which tend to be waterlogged, mineralisation of organic-N can proceed but nitrification may be restricted by the low levels of oxygen. Ammonium is readily absorbed on to cation exchange sites on organic matter and by clay minerals and thus varies in availability to plants. In addition the balance between nitrification and denitrification appears to be influenced by grass. Grassland soils tend to have higher concentrations of ammonium compared to nitrate than do arable soils and it has been suggested that grass roots inhibit nitrification possibly through bacteriostatic exudates or through a high root respiration causing oxygen deficiency. Nitrification may also be inhibited by pH values below about 5·5.

Gaseous losses from the soil through denitrification are readily observed under experimental conditions but are extremely difficult to measure in the field. Molecular N and nitrous oxide (N_2O) or ammonia (NH_3) are released particularly in response to waterlogging with the development of anaerobic conditions which may be temporary or restricted to microsites within the soil. Gaseous loss may also be stimulated by the presence of readily decomposable organic matter, faeces or urine and 20–30% of fertiliser N may be lost through ammonification. Losses of N through leaching are usually low in grasslands, partly associated with the continuous plant cover which reduces output of both water and nutrients. Even when inorganic N fertiliser is added, losses through leaching, mainly as NO_3', are similar to those from natural unfertilised ecosystems often about $1 \, g \, m^{-2} \, yr^{-1}$. Higher losses occur under heavy fertilisation, when the vegetation cover is not continuous in time or space (as in arable land), or when plant nutrient uptake is temporarily restricted (for example by climatic conditions). Losses are also related to the physical characteristics of the soil with a positive relationship to particle size and thus drainage capacity.

The rate and pattern of circulation of N in a temperate grassland ecosystem is thus the product of a mosaic of factors some of which are independent of the system (e.g. rainfall, temperature, soil structure) and some of which are a result of the functioning of the system itself (e.g. the plant species composition, the grazing intensity, the availability of organic carbon). Although the focus of attention may be on nitrogen, an understanding of its circulation requires consideration of other elements such as P and C and of the relative influence of a range of abiotic and biotic factors. The ecosystem needs to be considered as a functional unit within which changes in one part of the system can have repercussions throughout the system, for example

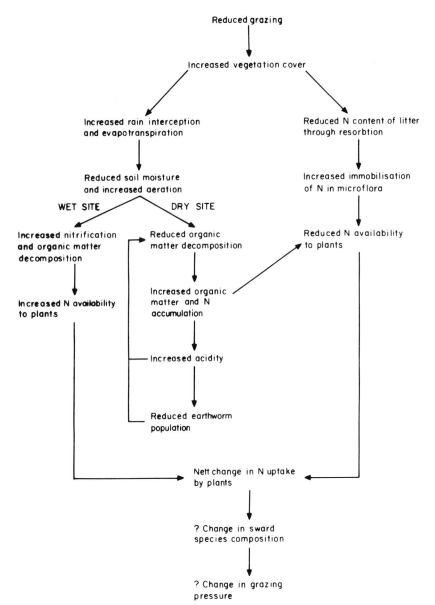

FIG. 7.9. Hypothetical sequence of changes in nitrogen transformations resulting from reduction in grazing on a pasture ecosystem.

a reduction in grazing could cause a sequence of changes within the de-composer subsystem (Fig. 7.9). The 'feedback' effect is obvious, the response will vary with site conditions, the nett result depends on the balance between the effects of resource quality and soil microclimate. Similar scenarios (Word Models) can be developed for other site conditions or management treatments, but the changes and interactions need to be quantified to provide a predictive understanding of the degree of change—mathematical models such as those for N circulation by Beek & Frissell (1975) and, for U.S. grasslands, by Innis (1978).

It is obvious that generalisation of the response of a grassland to the application of inorganic or organic fertilisers is likely to be misleading. The response will depend on the physical conditions of the site, the vegetation characteristics, the organic matter content of the soil, and the existing nutrient balance as well as the timing, concentration and chemical composition of the fertilisers. The capacity of the organic matter in the soil to retain fertiliser N which is in excess of that assimilated by the plants is probably a critical factor in controlling N output from the system, i.e. pollution. Nitrogen addition may itself cause an increase in the rate of organic matter decomposition resulting in N release from the N capital of the soils, a situation which may have contributed to the lower organic matter content in arable soils compared with grasslands. Conversely, attempts to maintain the soil organic matter content by addition of plant residues can result in immobilisation of soil nitrogen if the residues have a high $C:N$ ratio. Although research on soil N in agronomy has been extensive, there is now a need to apply this knowledge to the definition of the expected response of ecosystems to new management practices or to the application of existing practices to other systems as in developing countries. The comparison between expected and observed response provides an evaluation of the accuracy of the understanding and this could be an important focus for a major research effort in the next decade, stimulated also by the concern over environmental pollution. The vast increase in addition of inorganic fertiliser nitrogen, made available from industrial sources, has been linked with high levels of leached nitrate in streams, rivers and lakes causing eutrophication. Leached nitrate has also been identified as a potential cause of methaemoglobineamia in infants and cattle while increasing levels of nitrogen oxides in the atmosphere through denitrification in soil may lead to catalytic breakdown of ozone in the stratosphere and increased short-wave ultraviolet radiation which can cause skin cancer and reduce crop production. Increasing fertiliser cost, as well as environmental problems, may reduce the use of inorganic nitrogen and recent developments in the technology of symbiotic nitrogen fixation may provide an economic, and possibly more environmentally accept-able, substitute. Meanwhile the undoubted value of fertilisers to the ag-ronomist, and in particular the need to increase world food production in developing countries, will maintain a demand for sustained or increased soil fertility.

7.6   Temperate deciduous forest

The urban industrial societies of Western Europe and North America are very largely located in areas that were previously dominated by temperate deciduous forests. Indeed the history of western industrialised society can be said to have taken place by exploitation of this particular ecosystem-type and the neighbouring grasslands. The clearing of the forests goes back far beyond industrial times in Europe but in North America it has been a concomitant of the industrial and urban development that has taken place in the last two centuries. To serve this industrial society we also find the most intensive agricultural practice developed within this biome necessitated by the rapid expansion of urban populations. This pattern of land exploitation is now being repeated in the tropics. Tropical rain forest is being cleared at an increasingly high rate, either for the immediate cash benefit of timber, or to make way for extensive and intensive agriculture. In our last case-study we therefore consider some of the effects that clearing and management have on the decomposition subsystem in temperate forests and try to extrapolate the lessons to the development of tropical forest areas.

In our initial definition of ecosystems (Chapter 1) we suggested that the appropriate boundaries for a terrestrial ecosystem were those of the catchment* area for a drainage system. It is only by considering the gains and losses across these boundaries that the concept of an ecosystem as being essentially 'closed' with regard to the circulation of nutrients can be tested. Total catchment studies have been few in number and among the most extensive have been those at the Hubbard Brook Experimental Forest in New Hampshire, U.S.A. Detailed accounts of hydrologic and nutrient cycles in this forest have been given by Likens *et al.* (1967), Likens *et al.* (1971) and Likens & Bormann (1975).

The transfers of material that occur between terrestrial ecosystems largely depend on water movements. The study of the hydrologic budget for a catchment is dependent on having an impermeable bedrock so that drainage is confined to the stream system. This is the case in the Hubbard Brook Forest; the soil is derived from a bouldery glacial till overlying gneiss and quartz formations that appear to permit only minimal seepage. The altitude of the forest rises from 229 m at the lowest point of the main drainage stream, Hubbard Brook to just over 1000 m at the highest point of the mountain ridge (Fig. 7.10). Within the forest a series of primary and secondary streams drain into the main stream. A number of these have been the subject

---

* Note that in American terminology 'watershed' is the term used as equivalent to the English 'catchment'.

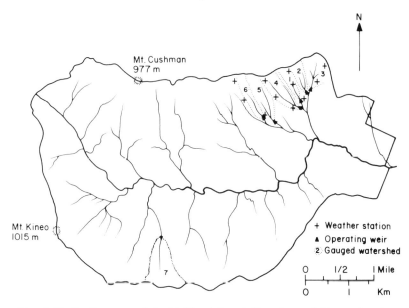

Fig. 7.10. Outline map of the Hubbard Brook Experimental Forest showing the gauged watersheds, weirs, weather stations and drainage streams tributary to Hubbard Brook (from Likens *et al.* 1971).

of catchment studies (Fig. 7.10). For each of these catchments a hydrologic budget was determined by weekly measurements of rainfall and stream flow measured at experimental weirs. For a varying number of catchments the mean annual precipitation between 1955 and 1969 was 1231 mm and stream flow was 730 mm. The difference between the two is evapotranspiration from forest and soil which thus has an annual value of 501 mm or 41% of the precipitation. The precipitation is evenly distributed through the year; the fourteen year monthly means varying from 8% of the annual total in March to 11% in November. Stream flow is more markedly seasonal, however, with over 35% occurring in April after the main snow-melt in contrast with only 0·7% in August during the evapotranspiration maximum.

Six catchment areas (numbered 1 to 6 on Fig. 7.10) were studied intensively during the period between 1963 and 1969 to determine the pattern of nutrient cycling. Inputs of nutrient elements to the catchments (other than those such as C and N which have gaseous reservoirs) were regarded as coming exclusively from rainfall, dryfall or weathering. Outputs occurred through carriage in streamflow either in ionic form or as particulate organic matter (Table 7.2). Particulate matter accounted for less than 10% of total nutrient output however and for the ions listed in Table 7.2 for less than 1%.

The pH of the rain and snow at Hubbard Brook was often below 4·0;

Table 7.2. Annual nutrient budgets for forested catchments (kg ha^{-1} yr^{-1}). Inputs are for dissolved materials in precipitation or as dryfall, outputs are losses in stream water as dissolved material plus particulate organic matter. The values are means for the period 1963–69 for catchments 1 to 6 (Ca, Mg, K and Na) or 2, 4 and 6 only (N and S). Data for catchment 2 were not included after 1965 (from Likens *et al.* 1971).

	NH$_4^+$	NO$_3^-$	SO$_4^{2+}$	K$^+$	Ca^{2+}	Mg^{2+}	Na$^+$
Input	2·7	16·3	38·3	1·1	2·6	0·7	1·5
Output	0·4	8·7	48·6	1·7	11·8	2·9	6·9
Nett change[1]	+2·3	+7·6	−10·3	−0·6	−9·2	−2·2	−5·4

[1] Nett change is positive when the ecosystem gains matter and negative when it loses it.

the sulphate concentration was also high (sometimes reaching 5 mg l^{-1}). Both these features are suggestive of fall-out of industrial pollution of the atmosphere. It was calculated that 50% of the incoming S may have originated from the burning of fossil fuels. This phenomenon of 'acid rain' is of some significance as H$^+$ may exchange with metallic cations on soil colloids and lead to increased loss from the system. Some 65–85% of the Ca^{2+}, Mg^{2+}, K$^+$ and Na$^+$ losses at Hubbard Brook may be due to this exchange process. The pH of the stream water varied from 5·0 to 6·0 and can be regarded as a dilute solution of sulphuric acid.

The concentrations of ions in stream water remained remarkably constant despite large fluctuations in stream flow within the year. The output is thus proportional to the water discharge. This is the first indication of the stabilising influence of the terrestrial part of the system. Water largely reaches the streams by percolation through the soil and the stream concentrations reflect the balance of biological and geological influences on the soil solution. The data of Table 7.2 show nett losses for all the cations and sulphate but the system has gained N in comparison with the input from precipitation. The losses by streamflow are assumed to be balanced by gains due to weathering of minerals in bedrock and soil. Based on the rates of loss of an element such as Na, which is probably available well in excess of biological demand, the rate of weathering was calculated as about 800 kg ha^{-1} of till and bedrock, or about 50 cm in the 13,000 years since the last glaciation. The significance of the ecosystem balance can only be seen however if the extent of nutrient movement between ecosystems is compared with the stock of nutrient within the ecosystem. The Hubbard Brook Forest is relatively immature and about 25% of the above ground NPP of 76 kg ha^{-1} yr^{-1} is added each year as growth increment. As with other temperate forests the proportion of nutrient in the dead organic matter. In Fig. 7.11 a total N budget for the system is given. This emphasises the insignificance of the inputs and outputs of the elements compared with the amounts cycling within the forest vegetation.

Chapter 7

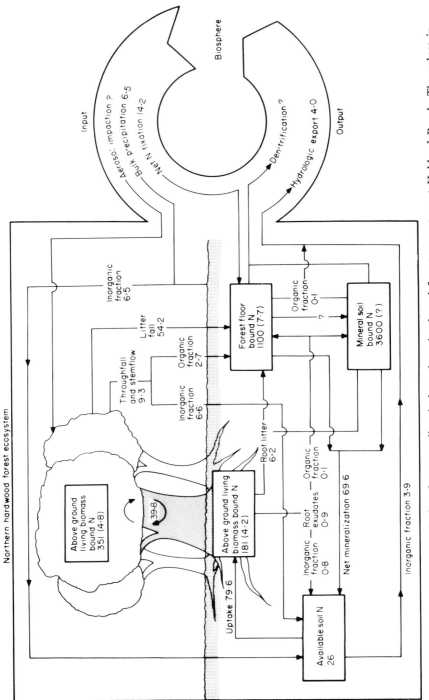

FIG. 7.11. Annual nitrogen budget for an undisturbed northern hardwood forest ecosystem at Hubbard Brook. The values in boxes are in kilograms of nitrogen per hectare. The rate of accretion of each pool (in parentheses) and all transfer rates are expressed in kilograms of nitrogen per hectare per year (from Bormann, Likens & Melillo 1977).

Thus the hydrologic loss of 4 kg N ha^{-1} each year is only 6% of that mineralised by decomposition processes and 5% of that taken up by the vegetation. In comparison with the standing crop of organic N the hydrologic output is completely negligible (less than 0·1%) emphasising the ability of the ecosystem to conserve this nutrient. The N budget shows that the nett gain of N from precipitation is also supplemented by N-fixation at a rate of about 14 kg ha^{-1} yr^{-1} but it should be noted that losses to denitrification have not been measured. The same ecosystem 'tightness' was found for all other macro-nutrients except S. Although details of mineralisation are not available the amount of S leaving the system annually (about 24 kg ha^{-1}) was far in excess of the amount in annual litter fall (5·5 kg ha^{-1}). This is again probably related to the large input of S from pollution which also exceeds the litter fall value.

Studies of nutrient release from decomposing litter on catchment 6 at Hubbard Brook Forest were reported by Gosz *et al.* (1973). Breakdown of leaves in 1 mm mesh bags, and twigs (< 5 mm diameter) attached to lines was studied over a twelve-month period from litter fall. During this period nett losses were recorded for K, Ca and Mg for all materials, (Table 7.3).

Table 7.3. Immobilisation and release of nutrients from decomposing litter at Hubbard Brook Forest. Percentage of original weight of nutrients remaining is shown after twelve months decay from October 1968 (recalculated from Gosz *et al.* 1973).

	Dry weight	N	P	S	K	Ca	Mg
Leaves							
Yellow birch	43	117	100	100	20	44	15
Sugar maple	60	164	300	36	16	73	69
Beech	69	142	253	107	33	54	38
Twigs[1] (< 5 mm diameter)	72	81	55	69	35	67	48

[1]   Mean of five species—the three deciduous plus red spruce and balsam fir.

For leaves these losses were principally attributable to leaching during the first month of decomposition (Fig. 7.12). In fact the later stages of leaf decomposition were characterised by nett immobilisation of K and Mg as C:N nutrient ratios fell to critical levels (see also Fig. 1.17). N, P and S typically show nett immobilisation in leaves over the twelve-month period although losses were recorded for the twigs. The major part of these nutrients in litter fall is thus not available to the plants until more than a year after litter fall. The decomposition study only continued for twelve months but it is likely that release would be delayed until the succeeding spring. It is thus probable that maximum release of critical nutrients from litter fall coincides

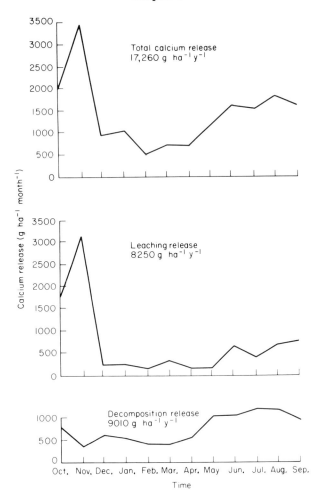

FIG. 7.12. Partitioning of nutrient loss between decomposition and leaching. Release of calcium from leaf and branch litter over a one-year period at Hubbard Brook Experimental Forest (from Gosz *et al.* 1973).

with maximum root growth in the second season after fall. This was the conclusion reached for the *Liriodendron* forest in Tennessee by Ausmus *et al.* (1976) where the period of maximum root growth in the early spring coincided with minimum microbial immobilisation of all nutrients (see Fig. 3.15). It may be a general feature of stabilisation in temperate forests where leaf litter takes something between twelve and twenty-four months to decompose. It emphasises the importance as a stabilising factor of the development of the O_2 and fermentation layers in forest systems.

No studies on the decomposer organisms have been reported from Hubbard Brook but the importance of the soil organisms in establishing a high degree of nutrient homeostasis is testified by many studies from elsewhere. For instance Satchell (1963) has shown that in European deciduous woodlands the demand for N by the earthworm population equals or even exceeds the N content of the annual leaf litter fall. Ausmus *et al.* (1976) and McBrayer *et al.* (1974) have similarly shown that in a North American forest lacking earthworms, the major part of all the nutrient elements available from litter fall are immobilised in microbial and animal tissues except for a very short period of the year (see Fig. 3.15).

In the normal temperate deciduous forest we may therefore picture an essentially closed cycle of nutrients in which the decomposer organisms play an important role in stabilising the system. We may now ask what happens when the system is disturbed.

Between 1965 and 1969 the Hubbard Brook Experimental Forest was used for an ecological experiment to explore the effects of dramatically changing the structure of an ecosystem (Likens *et al.* 1970). In the winter of 1965 all the vegetation on catchment 2 (Fig. 7.10) was felled and slashed. Herbicides were applied in the summers of the two following years to prevent any regrowth or fresh colonisation by plants. No detritus was removed from the site. Continuation of the monitoring of the hydrologic and nutrient budgets showed a marked change in the dynamics of the deforested catchment from that predicted on the basis of previous history and by comparison with catchment 6 which was used as a control.

The immediate effect was a change in the hydrologic budget. The outflow in the three succeeding years was 39, 28 and 26% higher than expected on the basis of undisturbed forest. This can be attributed to a reduction of evapotranspiration losses to below a third of their previous levels. The importance of evapotranspiration was further emphasised by the changed distribution of run-off; whereas the effect of deforestation on outflow was negligible in winter, the summer run-off was over four times as great as predicted on previous experience.

The effects on the nutrient budget were equally marked. Table 7.4 compares the inputs and outputs of nutrients for the control and experimental catchments over a two-year period following the felling operation. The inputs to the two catchments were virtually identical but the outputs on the deforested catchment showed marked deviation from those in the control. Whilst ammonium-N continued to show a nett gain to the terrestrial system, nitrate-N changed from a nett annual gain of about 3 kg ha^{-1} to a loss of about 120. Similarly dramatic increases in the loss of the major cations were also shown. Sulphate differed from other ions in showing a decrease in the amount lost. The patterns shown also continued into a third year (1968–69). The increased loss of elements from land could not be explained simply in

Chapter 7

Table 7.4. Nutrient budgets for control and deforested catchments (kg ha^{-1} yr^{-1}). Mean values for the two years 1966–67 and 1967–68 (June to May) (Likens *et al.* 1970).

		NH$_4$—N	NO$_3$—N	SO$_4$—S	K	Ca	Mg	Na
Input	Control	2·3	4·9	15·2	0·7	2·7	0·6	1·6
	Deforested	2·1	5·8	14·4	0·6	2·5	0·6	1·5
Output	Control	0·2	2·1	18·2	2·1	11·5	3·2	7·8
	Deforested	0·6	125·5	15·3	29·8	85·2	17·4	18·6
Net change	Control	+2·1	+2·8	−3·0	−1·4	−8·8	−2·6	−6·2
	Deforested	+1·5	−119·7	−0·9	−29·2	−82·7	−16·8	−17·1

terms of the increased water discharge. For instance the increase in water run-off in year one was only 34% whilst the NO$_3$—N increase was forty-six fold. The discrepancy is accounted for by massive changes in the concentration of the ions in the stream water which date from the spring following deforestation (Fig. 7.13). It should be remembered that run-off into the stream very closely resembles the ionic content of the soil solution. The massive losses of nitrate and major cations thus represent dramatic increases in the concentration of these materials in the soil solution.

The basic reason for this is obvious but the actual dynamics may be quite complex. The deforestation experiment effectively breaks the within-system nutrient cycle by uncoupling the link between the decomposition subsystem and the plant subsystem. In the absence of uptake of nutrients by plants in the spring the inorganic elements regenerated by decomposition processes are subject to leaching from the soil. The increased run-off of these nutrients is thus readily predictable but the balances between the individual ions merit further explanation.

The change is quantitatively dominated by the increase in NO$_3$—N. Under normal conditions about 90% of the N in the Hubbard Brook Forest is in soil organic matter, about 9·5% is in vegetation and 0·5% exists in available form largely as ammonium (Fig. 7.12). The N lost in the three years following deforestation was equivalent to about six times the annual litter fall, approximately equalled the N in the felled above-ground biomass but was only about 6·5% of the total organic-N of the ecosystem. This probably represents N released from readily-decomposed organic matter and the rate of release showed a marked decline five years after felling. No studies on decomposition on the deforested area have been reported but it is to be presumed that most of the leaf, herbaceous and fine twig litter would have decomposed in this time while the larger woody material still remained relatively intact. Without annual replenishment the character of the soil organic layers must have

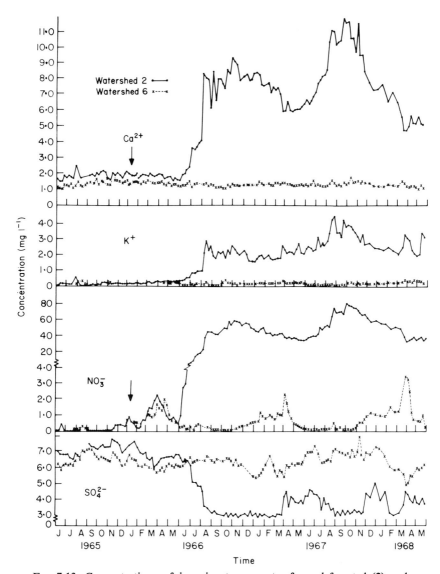

FIG. 7.13. Concentrations of ions in stream water from deforested (2) and control (6) catchments at Hubbard Brook Experimental Forest. Note that the nitrate graph has a break in the ordinate scale (from Likens & Bormann 1975).

changed markedly with the cellular component of SOM probably diminished as well as the leaf litter component.

The nutrient data also suggest a marked increase in the process of nitrification. That ammonium oxidation is relatively insignificant in the uncut ecosystem but is important within the soils on the deforested catchment was shown by Smith, Bormann & Likens (1968). They demonstrated that the number of *Nitrosomonas* and *Nitrobacter* increased eighteen and thirty-four-fold respectively following felling. Inhibition of nitrification under climax vegetation has been demonstrated in a large number of ecosystems. In some cases this has been thought to be due to the generally acid pH of forest soils because neutralisation of forest soils by liming can stimulate development of a nitrifying flora (Chase, Corke & Robinson 1968). More recent studies indicate that the inhibition occurs in climax systems other than forests and is also common at pH levels that should not in themselves inhibit nitrification, thus implying some direct toxicity perhaps involving tannins (Rice & Pancholy 1972). Whatever the immediate cause, the increased nitrification following deforestation is the main reason for the huge losses of N. The ammonium ion, having a positive charge, is attracted to the negatively charged soil colloids and is thus relatively immobile; nitrate has no such stabilising factor and is thus highly susceptible to leaching from the soil. The 'tightness' of the undisturbed ecosystem to N (Fig. 7.12; Table 7.2) is thus further explained by the assumption that nitrification is negligible and the N taken up by plants is largely in the form of ammonium.

After deforestation the pH of the stream in catchment 2 declined from a mean of 5·1 to 4·3. This probably reflects the increased acidity due to de-composition products. This increase in acidity has significance with regard to the increased mobility of cations such as K^+, Ca^{2+} and Mg^{2+} which will be displaced from colloidal exchange sites by H^+. Likens *et al.* (1970) suggested that the increased nitrification rate would also contribute to this due to the generation of H^+ during ammonium oxidation. They demonstrated that there was a highly significant correlation between the concentrations of NO_3' and exchangeable cation such as Ca^{2+}, Mg^{2+} and Al^{3+} in stream water. This could not be accounted for simply on the basis of water discharge variations as the concentrations of these ions varied independently of this factor.

We may thus conceive of a picture whereby the cations are released in increased amounts from the large amounts of decomposing litter, encounter a lowered capacity for base adsorption on the soil colloids and are leached from the soil. The importance of the organic detritus as a source of these cations is illustrated by a comparison with Na, an element of qualitatively lower biological significance (Ca:Na about 20:1 in litter). The Ca:Na ratio of the bedrock is 0·9:1 and thus weathering would tend to produce inputs in this ratio. In the undisturbed catchment the streamflow ratio was slightly higher than this (1·5:1, Table 7.2) indicating some biological origin of Ca. In the deforested system however Ca output increased sevenfold whereas that of Na only doubled. The ratio in catchment 2 was thus changed to 4·6:1. The differential increase in Ca is suggestive of organic origin as well as increased weathering due to higher acidity.

As already mentioned, sulphate behaves differently to all other ions, actually showing decreases in concentration where other ions show increases (Fig. 7.13). The undisturbed catchments show a high input to which a small component is added by mineralisation within the system so that the output exceeds input by 20–30% (Table 7.2 and 7.4). This difference is reduced to negligible proportions in the deforested system. If no change in internal mineralisation occurred then some decrease in sulphate concentration due to dilution by increased run-off could be expected but the nett sulphate balance should remain constant. Thus it must be supposed that S, in contrast to all other elements, is showing a reduction in nett mineralisation following deforestation. Likens *et al.* (1970) suggested two possible mechanisms to account for this; either inhibition of S-oxidation by nitrate toxicity or stimulation of S-reduction by increased anaerobiosis in the soil. For neither proposition however is it possible to make a good case—the relationship between NO_3' and SO_4'' concentrations is poor and S-reduction is inhibited at low pH—and so the actual mechanism remains uncertain.

The Hubbard Brook Experiment may represent an extreme case of the effects of forest clearance. The suppression of vegetational regrowth contributes to this as this will not normally be the case during conversion of land to agricultural practice. Even with only herbaceous cover we would expect ecosystem tightness to be increased by plant uptake of water and nutrient. Johnson & Swank (1973) compared the loss of nutrients from a mature hardwood catchment to those in three catchments which had been clearfelled, subjected to a good deal of disturbance but were then allowed to develop different types of vegetational cover. In the least mature system (a weed, grass and shrub, hardwood sere) stream discharge was greater by 6% than than the control, losses of Ca^{2+} and Mg^{2+} were six- and three-fold greater than the control, but K^+ and N^+ were only marginally higher. Catchments bearing coppiced hardwood and young pinewood plantations showed in contrast lower nett losses than the mature hardwood. The development of plant cover may thus rapidly restabilise the nutrient cycles but it must be remembered that in this example this has been shown only several years after the clearing operation and does not show to what extent the immediate dramatic changes might be modified by regrowth.

The planting of an annually harvested agricultural crop produces a situation somewhere between the clear ground and weed succession. This combined with the effects of yearly removal of a nutrient component in harvest, ploughing or burning of detritus, increase in nitrification, decrease in the input of materials of low resource quality—and thence increased mineralisation rates—must produce a situation of positive feedback in the nutrient cycling which may approximate to that of the deforested catchment. Maintenance of nutrient status by fertilisation and inhibition of nitrification by the use of such materials as N-serve are then necessary to achieve acceptable production. In this form of agriculture, productivity is maintained not by the natural internal regulation by the decomposition subsystem but by subsidy of energy and nutrient from elsewhere. Intensive agriculture is efficient in terms of meeting demand for food but highly inefficient in its use of resources.

This is of particular significance with regard to the development of agriculture in the tropics. Tropical ecosystems are particularly susceptible to nutrient depletion by the processes described above. The major part of their nutrient capital is stored in the plant biomass. This means that a greater proportion of the nutrient is immediately susceptible to loss after clearing. The heavy leaching of tropical rainstorms accelerates this process. The high temperatures and humid conditions promote decomposition processes so that the decline of soil organic status (Section 1.3, Fig. 1.18) is more rapid and more extensive than in temperate areas. Conversely, it is in these regions that the high cost of agricultural subsidy can be least afforded.

Our abilities to manipulate the decomposition subsystem to provide a more 'natural' agriculture are totally inadequate to meet these present demands. It is perhaps particularly depressing to realise that such manipulations as we are able to impose at present are purely chemical in basis—our understanding and hence our control of the decomposer community is still woefully poor. If it were possible to manipulate this part of ecosystems to the same extent that the plant and herbivore subsystems have been controlled to man's purpose then much of our fear about deteriorating agricultural potential would be overcome.

7.7   Summary: Decomposition processes in major ecosystem types

The charts on the following pages summarise the structure of the decomposition subsystems and its functional integration within major ecosystem types of the boreal, temperate and tropical zones. The triangle for each zone (p. 305) contains a modular synthesis of the driving variables of decomposition processes (physical environment, resource quality and saprotroph biota) acting in that region (see Chapter 2) and the contribution of the component processes: catabolism (K), comminution (C) and leaching (L). A forested and non-forested ecosystem type is described for each region in terms of its geographical distribution, climate, the contribution of the major resource types to plant biomass and litter fall, the characteristics of the major nutrient pools in the ecosystem, the soil biota and the dynamics of litter decomposition. The data are derived from various sources and emphasise the global patterns of decomposition processes rather than providing quantitative syntheses of individual ecosystems.

The following information is presented, with biomass pools represented as π diagrams in which the size of the circle is proportional to total biomass to provide a visual guide for comparison.

1. Seasonal patterns (as for the northern hemisphere) of mean monthly air temperatures and precipitation. An alternative presentation of these data, in the form of klima-diagrams, is given in Fig. 1.2.

2. Plant biomass is represented as a π-diagram divided into segments proportional to wood, leaf and root components. Nett primary production data are from Table 1.2.

3. Seasonal patterns of soil respiration are shown beneath the temperature and precipitation data to indicate the degree of regulation of decomposition rates by these physical environmental variables (see Chapter 6).

4. The standing crop of total N in the system with the allocation of N to major above- and below-ground pools. The distribution of N is typical of other nutrients which are potentially limiting to primary production and decomposition. C/N ratios for the soils are given (see Chapter 4).

5. Decomposer biota. The standing crop of total microflora is proportional to the diameter of the upper semi-circle. Active biomass is represented as a segment but the proportion is largely theoretical as these data are not available for most ecosystem studies.

The biomass of decomposer invertebrates (lower semi-circle) is usually much smaller than that of microbial standing crop and in some cases has been increased by an order of magnitude for presentation. The proportion of total biomass represented by micro-, meso- and macrofauna is shown to emphasise the contribution of the macrofauna to litter comminution.

6. Decomposition losses of wood and leaves are shown against a time basis of three years. Wood decay rates are largely hypothetical. The maximum and minimum rates of leaf decay are shown as a shaded area spanning the hypothetical seasonal patterns of decay. The decay constant 'k' and turnover constant '$3/k$' are taken from Table 1.1.

The following generalised latitudinal trends in decomposition processes are shown by these data.

The climatic regimes become progressively more favourable for decomposition from the poles to the equator both in terms of the upper limits of temperature and precipitation and the seasonal duration of favourable conditions. The non-forested systems, however, characteristically have soil moisture deficits at some point during the summer months. Plant biomass and NPP increases from the poles to the equator while SOM standing crop shows the reverse pattern. The strongly limiting effects of temperature on decomposition in the cold climates is emphasised by the high SOM accumulation with respect to low NPP. Plant resource quality entering the decomposition subsystem is generally low, particularly in forest ecosystems where modifiers significantly influence decomposition rates. The presence of organic soils in some tropical rain forests, where temperature and moisture regimes are optimum for saprotroph activities, emphasises the importance of resource quality in determining the characteristics of decomposition processes in ecosystems. The reduction of SOM standing crop is associated with a decrease in microbial standing crop, a decrease in the size of the soil nutrient pool, and an increase in above-ground plant biomass. Soil micro- and mesofauna populations decrease with decreasing SOM but total fauna biomass and macrofauna populations generally increase towards the tropics. Nutrient turnover times also increase from the poles to the equator and a large proportion of the total nutrient capital in tropical systems is located in plant and animal biomass. This has important consequences for the resilience of tropical ecosystems to major perturbations such as deforestation. Climax ecosystems with organic soils have high stability in a region since secondary succession is accelerated by the accumulated nutrient reserves.

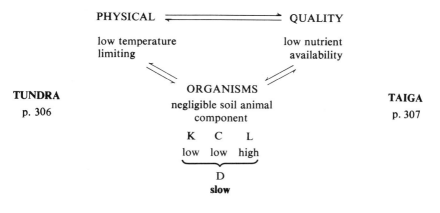

BOREAL ZONE

PHYSICAL ⇌ QUALITY

low temperature limiting   low nutrient availability

TUNDRA
p. 306

ORGANISMS
negligible soil animal component

K   C   L
low low high
_____
D
slow

TAIGA
p. 307

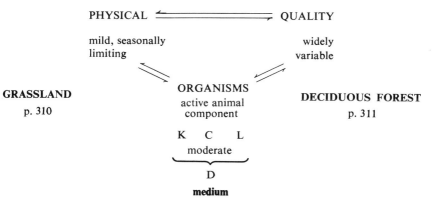

TEMPERATE ZONE

PHYSICAL ⇌ QUALITY

mild, seasonally limiting   widely variable

GRASSLAND
p. 310

ORGANISMS
active animal component

K   C   L
moderate
_____
D
medium

DECIDUOUS FOREST
p. 311

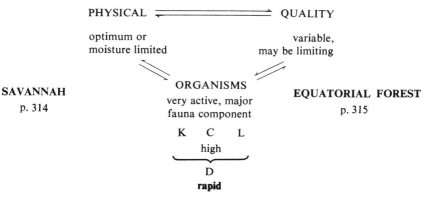

TROPICAL ZONE

PHYSICAL ⇌ QUALITY

optimum or moisture limited   variable, may be limiting

SAVANNAH
p. 314

ORGANISMS
very active, major fauna component

K   C   L
high
_____
D
rapid

EQUATORIAL FOREST
p. 315

BOREAL ZONE: TUNDRA

Major biome type around the Arctic
Circle north of the tree line. Exposed
habitat, with extreme climate, ranging
from stone deserts to waterlogged
moss meadows.

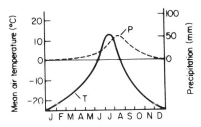

1. Air temperatures rise above
freezing for only 2 months of the
year; rarely exceeding 10°C.
Insolation more or less continuous in
summer. Winter temperatures
frequently below −30°C. Permafrost
at 0·5 m. Annual precipitation rarely
exceeds 200 mm of which 40% falls as
rain in summer. Drying winds and
elevated surface temperatures may
produce moisture deficit.

2. Nett primary production is limited
by the short growing season (when
free water is available) and exposed
conditions. Plants (shrubs and grasses)
have low growth form and a small
proportion of NPP is support tissues.
A large root biomass promotes
nutrient retention by the plants during
dormancy and rapid growth during
favourable periods. Litter input is
mainly in winter.

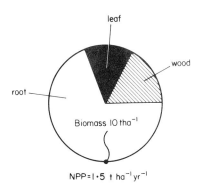

3. Low temperatures limit soil
animal and microbial respiration.
Psychrophilic micro-organisms are
active at −7·5°C but CO_2 evolution
from the soil is largely limited to the
brief summer of 70 to 80 days.
Microbial metabolism may attain a
$Q_{10} = 4$ relationship which capitalises
on the short season of elevated
temperatures.

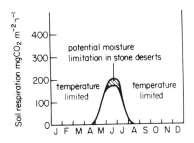

BOREAL ZONE: TAIGA

Taiga is the most extensive biome extending across N. America and Eurasia. Northern limit is the Arctic tree line. Trees are predominantly evergreen conifers. Soils podzols.

1. Winter temperatures are subzero and the soil is frozen to a depth of 2 metres but snow cover may keep soil temperatures as high as $-7°C$. Mean monthly temperatures are above freezing for 7–8 months rising to over $16°C$ in summer. Precipitation 370–620 mm falling mainly during the summer. Soil moisture deficits limited to soil surface as a result of higher rainfall than tundra and the ameliorating effect of forest on macroclimate.

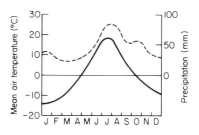

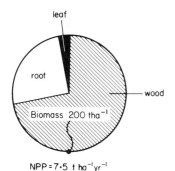

2. Photosynthesis can take place throughout the year in cold adapted foliage. Biomass up to 400 t ha^{-1} with woody tissues as dominant component. Conifers have shallow rooting system which exploit surface melt waters and spring nutrient flush. Litter input is more or less continuous.

3. Low temperatures limit saprotroph metabolism but CO_2 evolution is detectable throughout most of the year as a result of psychrophilic microbial activity. Q_{10} elevated as in tundra. Major proportion of CO_2 efflux from soil occurs over 300 days during summer.

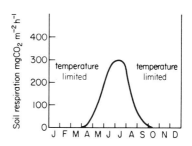

TUNDRA continued

4. A large standing crop of nitrogen is accumulated in organic soils but turnover is slow and it is largely unavailable to plants. The C/N ratio is 20–50:1 as a result of the accumulation of deep organic horizon of partially decomposed plant materials.

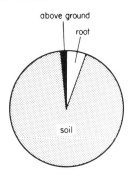

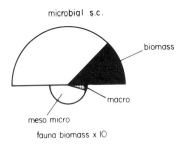

5. The microbial standing crop (predominantly fungal) is high but turnover times are low due to low temperature, moisture inhibition of lysis and low fauna activity. The invertebrate soil fauna is dominated by the microfauna although fly larvae and enchytraeid worms may be locally important in moss meadows. Litter comminution is negligible. Lemming cycles can introduce major perturbations in nutrient cycles.

6. Decomposition is slow regardless of resource quality as a result of the dominant physical limitations on soil biological processes. Anaerobic conditions impose further restraints to decomposition in organic, waterlogged soils. Leaching is high owing to the freeze/thaw cycle, catabolism is low and comminution is negligible. Weight losses from resources may follow a linear relationship over many years as a result of the small seasonal increment to weight losses on the soil surface and the physically limiting environment (cold and waterlogged).

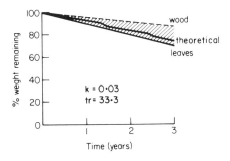

TAIGA continued

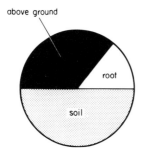

4. Nitrogen standing crop of above ground pool does not reflect the size of forest biomass due to low nitrogen content of wood and foliage. A large pool of nitrogen is located in soil organic matter but mycorrhizae are important in nutrient conservation during melt water flushes. C/N ratio lower than tundra (20–30:1) because of lower organic carbon accumulation.

5. Very large fungal standing crop up to 1 kg m^{-2} (excluding mycorrhizae) but with small biomass component. Nutrient turnover slow, but faster than tundra under influence of larger populations of microtrophic mesofauna (particularly mites, Collembola and enchytraeids). Macrofauna populations are low. Herbivore grazing is insignificant except during sporadic outbreaks of defoliating Lepidoptera larvae.

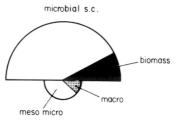

6. Decomposition primarily limited by temperature and secondarily by low nutrient status of litter. Oil and resin modifiers have biostatic properties. Gymnosperm litters have low water soluble (leachable) component. Fauna comminution is negligible. Some seasonality of weight losses may be detectable but overall decomposition rates are low and the general trend is for high levels of SOM accumulation.

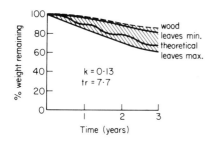

TEMPERATE ZONE: GRASSLAND

Rather homogeneous habitat derived by clearing, grazing, burning and/or low rainfall. Climax grasslands are mainly in the interior of continents away from maritime influence. Soils well drained. Czernozem soils characteristics of N. America and U.S.S.R.

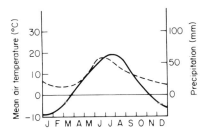

1. Mean air temperatures fall below 0°C for only 2–3 months of the year. Summer temperatures are high, owing to the absence of maritime influence, and soil moisture deficits, promoted by the thin vegetation cover, may last for 2–3 months. Rainfall low (250–750 mm) but occurs throughout the year, particularly in the summer months.

2. Woody component of biomass is absent. Roots of perennial grasses and dead leaves form matted turf which holds water near the surface and also helps to prevent tree seedling germination. Large root biomass is promoted by fire and grazing pressures as well as the demand for efficient moisture uptake. The shallow root system maintains grasses where tree seedlings are unable to survive. The large root component of NPP emphasises the importance of below ground decomposition processes.

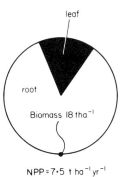

leaf

root

Biomass 18 t ha^{-1}

NPP = 7.5 t ha^{-1} yr^{-1}

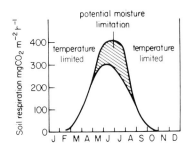

3. Soil CO_2 efflux is detectable throughout the year and follows the seasonal pattern of soil temperatures. Moisture may become limiting in the summer months. The active microflora is predominantly mesothermal, with a thermophilic component. $Q_{10} = 2$–3.

TEMPERATE ZONE: DECIDUOUS FOREST

Heterogeneous habitat as a result of clearing, forestry practice and on a wide variety of species assemblages, and dominants, on different parent soils. Predominantly located in areas with maritime influence. Soils vary from podzols (mor) to brown earths (mull).

1. Mean monthly temperatures rarely fall below 0°C. Maximum temperatures ameliorated by transpiration and microclimate. Moisture conditions are favourable throughout the year and are rarely limiting.

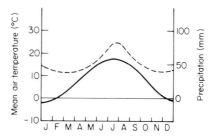

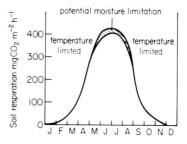

NPP=11·5 t ha^{-1} yr^{-1}

Biomass 350 tha^{-1}

2. Large wood biomass of variable quality (Q) according to species. Leaf litter represents a small biomass component but high concentrations of tannins and polyphenols in many species are important modifiers of decomposition rates and contribute to the formation of mull and mor humus forms under different tree species. Root standing crop has higher turnover rate than boreal forest and root production may be equivalent to annual above-ground production. Litter input, particularly of leaves, is highly seasonal and this strongly influences nutrient cycling and decomposition processes.

3. Soil CO_2 evolution follows seasonal temperature pattern with $Q_{10} = 2$–3. The soil microflora is predominantly mesothermal with a thermophilic component. Moisture rarely limiting to decomposition.

GRASSLAND continued

4. Total N pool is small and the root biomass component including VA mycorrhizae is important in nutrient retention. Nutrient turnover rates are high, particularly in heavily grazed pastures. C/N ratio is 15–20:1.

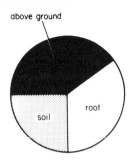

5. The soil microflora has a large bacterial component though fungal mycelium ramifies the root mat. The turnover of the microbial standing crop is rapid as a result of a very active soil fauna. Earthworm biomass is large in Eurasian grasslands (often one or two orders of magnitude greater than total herbivore biomass) and promote the development of deep, non-stratified brown soils. American prairies show similar structure despite the absence of burrowing earthworms but have lower productivity. Herbivores may graze 25% or more of NPP.

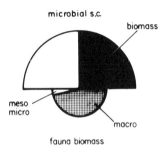

6. Grasses are not protected by secondary compounds and resource quality is higher than deciduous tree leaves. Decomposition is moisture limited in tussocks throughout most of the year but herbivore grazing and trampling brings about relocation of grass litter in favourable microclimates similar to large below-ground input. Decomposition is then rapid under the influence of high Q, high temperatures and an active soil biota; these characteristics are reflected by the low SOM content of grasslands. Earthworms and dung beetles are important in herbivore dung utilization and structuring of soil matrix.

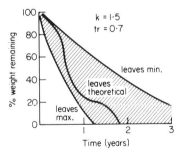

DECIDUOUS FOREST continued

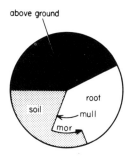

4. Nutrient content of above-ground biomass higher than in boreal forest. Nutrient standing crop in soil pool varies with humus form. In mor soils 60% of the total N is located in the soil organic matter (cf. tropical forest soils). The C/N ratio of mor soils is 20–30:1 while mull soils with low organic carbon content are similar to grasslands at 15–20:1.

5. Macrofauna, particularly earthworms and iulid millipedes, are important agents in mull humus formation on base rich soils. Litter high in phenolic acids may exclude many macrofauna groups and form mor humus where the mesofauna are the dominant elements. Herbivore consumption of leaves $\leqslant 5\%$ of above-ground production. Bacterial standing crop has rapid turnover and large biomass component in mull soils where fauna is active but a large fungal standing crop may accumulate in mor equivalent to boreal forest. Mycorrhiza important for nutrient conservation especially at leaf fall when leaching is high.

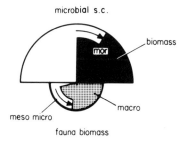

6. Moisture conditions are favourable throughout the year; decomposition is (a) temperature limited and (b) resource quality limited. In mull soils (high Q) decomposition is rapid under influence of active soil fauna, SOM accumulation is low and nutrient turnover is rapid. Mor soils are formed from litter with low Q (high modifiers, low N, P, K, Ca), litter decay takes 2–3 times longer than mull soils and k approaches 0·7 whereas mor soils are nearer to boreal conditions at 0·25. The important difference between boreal forest soils and temperate mor soils is that the former are formed 1° by physical restraints to decomposition while mor is largely determined by resource Q.

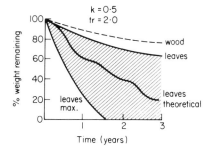

TROPICAL ZONE: SAVANNAH

The term 'savannah' refers to the grasses which are a major component of the vegetation. Savannah embraces much of the tropical zone which is not forested and includes thorn scrub and coarse grass where drought is lengthy, through park-like grasslands with scattered trees to savannah woodlands with tall elephant grasses. Vegetation forms are commonly affected by man, fire and grazing and many open areas are capable of regeneration to woodland. Soils lateritic.

1. Temperatures are high throughout the year particularly in the dry season when daytime maximum may exceed 40°C but fall to 15°C or less at night. Mean monthly temperatures show an annual range of 5–10°C. Amount and duration of precipitation varies widely from 1000–1500 mm, with only a short dry period, to central continental areas where less than 500 mm may fall over 1–2 months.

2. Growth of woody plants occurs throughout the year but mainly during wet season. Growth of grasses is confined to the wet season and the large root biomass is efficient in nutrient uptake during periods of high rainfall and in moisture absorption at the ends of the dry season. A very rapid growth response of grasses to rain is characteristic of savannah as nutrients are released from materials accumulated during the dry season. Resource quality of grasses varies from sun dried material at the end of the wet season, with high protein content, to highly silicious species.

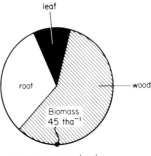

$NPP = 9.5 \ t \ ha^{-1} \ yr^{-1}$

3. Soil microbial activity is limited by moisture availability during the dry season but very high levels of CO_2 efflux are recorded during the rains when the non-woody NPP can be decomposed over a period of a few months. Soil respiration follows Q_{10} relationship 2. Thermophilic micro-organisms are important at midday when soil temperatures may exceed 35°C.

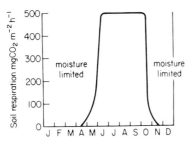

TROPICAL ZONE: EQUATORIAL FOREST

Tropical Rain Forest occurs in moist lowlands and up to about 900 m within 5°–10° of the equator. Short term 'slash and burn' agriculture settlements, with long periods of fallow, exist in equilibrium with forest regeneration but the exponential growth of timber felling and 'modern' agricultural practices has permanently destroyed much of the world's rain forest. Soils red earths to laterites.

1. High humidity (80–97%) and cloud cover prevent air temperatures exceeding 30°–35°C and temperatures near the forest floor are usually nearer 25°C. Night time temperatures seldom fall below 18°C. Mean annual soil temperatures 25°C. Rainfall over 2000 mm and may be seasonal (monsoon forest) or non-seasonal.

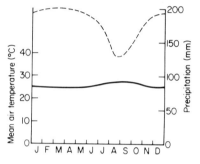

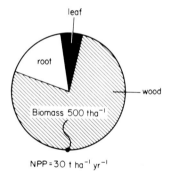

2. The forests are evergreen and there is no synchronous litter fall between or within tree species. The canopy is continuous and casts dense shade. A shrub stratum is absent but is represented by a tangled profile of immature tree species. Ground flora may be sparse. Growth is continuous (tree trunks do not show annual rings) and plant biomass is very high, dominated by a vast wood standing crop. Roots, with mycorrhiza, form a dense shallow mat for nutrient uptake. Trees supported by stilt roots and buttresses.

3. Soil CO_2 evolution not limited by moisture or temperature and shows little monthly variation. CO_2 evolution rates often lower than temperate forest as a result of continuous, non-seasonal, efflux. Thermophilic microflora less significant than savannah because of lower diurnal soil temperature fluctuations.

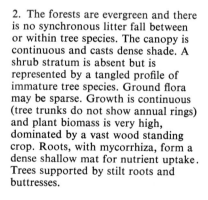

SAVANNAH continued

4. Nitrogen standing crop is low and is largely held in the plant (and animal) biomass. Nutrient turnover rates are high and the system has low stability if perennial grasses and woody shrubs are cleared. At the beginning of the dry season the SOM content is less than 1% with a C:N ratio c. 5:1 suggesting that residual organic matter is largely microbial tissues.

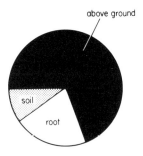

5. The soil fauna is dominated by the termites many of which can forage during the dry season by constructing sheets and galleries of soil over the resources. Almost all woody litter may be utilized by termites in West African savannah and the microbial component of wood decay is restricted to the fungus combs within termite mounds. Dung beetle burrowing activities and termite foraging produce equivalent effects to earthworms in temperate grasslands for soil turnover and structuring. Microbial standing crop has a high turnover rate. Bacteria and Actinomycetes are important components of the soil microflora.

6. Decomposition is primarily moisture limited and litter may accumulate on the soil surface during the dry season, although termite utilization of wood may be non-seasonal. When rain falls decomposition is rapid and temperature regulated. Grass litter has high nutrient content as a result of rapid drying at end of rains and non-woody NPP can be decomposed during rainy season. The effects of microclimate are similar but more extreme than temperate grassland and standing dead material has a long residence time in absence of termite attack.

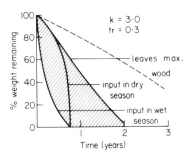

EQUATORIAL FOREST continued

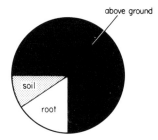

above ground
soil
root

4. Nitrogen standing crop is very large but is tightly conserved and has been accumulated in system over hundreds/thousands of years. Above ground pool is 60–80% of total. Standing crop of SOM is variable under different tree species but is generally low and does not represent a significant nutrient pool. Mycorrhizae are important for nutrient uptake and retention under high rainfall conditions and very little NPK, etc. is lost to ground water. C/N ratio similar to mull, and other mineral soils with low organic carbon content, at 10–15:1.

5. The saprotrophic fauna is active and not restricted to the forest floor owing to generally humid conditions within the forest. Wood feeding termites are abundant but not as important as in savannah. Earthworms and/or soil feeding termites are important for processing SOM. Microbial standing crop, particularly of fungi, is high with large active biomass and rapid turnover due to high temperatures, high moisture and active microtrophic fauna.

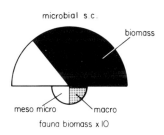

microbial s.c.
biomass
meso micro macro
fauna biomass x 10

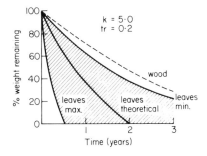

% weight remaining

$k = 5\cdot0$
$tr = 0\cdot2$

wood

leaves max.
leaves theoretical
leaves min.

Time (years)

6. Decomposition of wood and leaves is rapid under non-limiting conditions of moisture and temperature, and active saprotroph biota. Comminution is high, particularly by Diploda. Considerable resource weight loss may occur before reaching the ground. Leaf quality very variable and some tree species contain such high levels of tannins that decay is inhibited and SOM may accumulate locally to several kg m^{-2}.

Appendix I

METHODS IN DECOMPOSITION STUDIES

The previous sections show that decomposition is a complex and multi-disciplinary subject. Correspondingly the methods used in the study of decomposition are varied and a comprehensive review is beyond the scope of this book. In this Appendix we give a brief review of the main basic techniques, with selected references which are recent, comprehensive or detailed, but all of which provide an introduction to the literature. The choice of methods depends on the aims of the research, the level of resolution of detail required and the labour available. For this reason we have avoided recommending particular methods where a choice exists. An indication of relevant scale is given in Table 2.1 and we present this to emphasise the obvious but critical point of selecting methods appropriate to the scale and aims of the research. A combination of methods is often valuable, particularly when field measurements are used to evaluate the extrapolation of laboratory results. Methods of determining the relationships between decomposition and environmental variables are not discussed but can be approached by applying the measurements of decomposition to

(a)   naturally occurring environmental variation taking advantage of site heterogeneity and gradients

(b)   experimentally-induced variation in the field, e.g. fertiliser or transplant experiments, or

(c)   controlled variation in the laboratory.

The sections of the Appendix are broadly applicable to the chapters of the main text as follows:

1. Input of materials to the decomposition subsystem (Chapters 1 and 2)
2. Measurement of decomposition (Chapter 4)
3. Organism populations and activities (Chapter 3)
4. Physico-chemical factors in the environment (Chapter 6)
5. Chemical composition of plant and animal materials (Chapters 4 and 5)
6. Statistical analysis

1.   INPUT OF MATERIALS TO THE DECOMPOSITION SUBSYSTEM

Plant death is usually a gradual process and decomposition begins during senescence, at a time when withdrawal of soluble carbon compounds and nutrients is occurring. Radioactive tracers can help to apportion transloca-tion, heterotrophic catabolism and leaching, but most studies concentrate on plant material once it has obviously entered the standing dead or litter categories.

1.1 Input to standing dead

Inputs to dead material in the vegetation canopy (forest, dwarf shrub, herb or grassland) can be estimated from a change in standing dead biomass when death occurs at a particular period of the year. Where input continues throughout the year, as in many grasslands, marking of leaves or stems with paint or inert plastic tags can provide estimates of the death rate and of the residence time as standing dead which can be used to calculate input from biomass change. In dwarf shrub and woodland communities which are not in a seral state, the age composition of the plants determined from growth rings represents a mortality curve which, given the mean weight per individual in each age class, can be used to estimate production of standing dead.[1]

1.2 Input to litter

For woodlands a variety of designs have been used to measure litter fall.[2,3] The simplest traps are bags, boxes or plastic funnels supported clear of the ground with an aperture of $0 \cdot 25 - 1 \cdot 0$ m². In the tropics the material should be resistant to decomposition, termite-proof and lightweight. Woven polythene bags can be easily suspended, using D rings, to posts or trees. Exposed sites may require traps designed to catch wind-blown litter.

Traps should allow free drainage and, particularly in wet or humid conditions, should be cleared frequently to minimise decomposition and leaching losses. Birds may remove fruits, etc. from boxes or trays and may contaminate litter with faeces so the rim should be designed to deter perching. Trays on the ground surface can be used to measure litter fall in dwarf shrubs but animal activity, drainage and wind can be problems. Where there is little or no ground vegetation quadrats on the ground, either cleared of litter or covered with net, can be sampled repeatedly to estimate litter input.

Input from ground flora is difficult to measure. Above ground nett primary production may provide a reasonable estimate and, for species which are markedly seasonal in their growth, the peak standing crop represents about 50% of annual production.[4] In grasses, especially tussock species, leaves and stems may remain as standing dead for a number of years and standing dead and litter are continuous. Changes in biomass of dead material, estimated by clipping serial or paired quadrats[5] over time intervals can provide an approxi-mation to litter input, but where grazing occurs caged plots are necessary.[6]

Input of branch and trunk wood in woodland is irregular in time and space. Large quadrats, 25–400 m^2 or even 1 ha depending on the size of the litter component, can be sampled at intervals to determine input[2] but there are statistical problems related to their highly clumped distribution.[7] Woody materials should be categorised into diameter classes[8]; the proportion of bark to sapwood to heartwood varies with diameter and nutrients and modifiers are differentially distributed between these components.

1.3 Roots

Root standing crop can be estimated from soil core samples, excavation of trees, and from regression of root weight on above-ground weight, stem diameter or root base diameter, but the techniques are laborious.[3] The timing, rate and distribution of input of dead roots to the soil are very difficult to estimate. Change in standing crop over the season can provide reasonable estimates especially where the growth or death occurs in a short time period or the age of the roots can be detected from features such as colour. Peak standing crop gives a reasonable approximation for annual plants. Root observation chambers can be used to determine growth rate and the general time scale of death, especially for small roots, to give an approximation of input to dead roots when combined with standing crop of roots.

Root hairs, surface debris and exudates are further inputs to the soil which may have a considerable influence on decomposer activity in the rhizosphere and in the soil in general. ^{14}C labelling of photosynthate has proved useful here but the comprehensive research in the root zone has concentrated on microbiology and biochemistry and is at a high level of resolution which has poor links with the ecosystem scale. Methods are laboratory oriented and the ecology of root death and decomposition in the field is a major area of research potential.[9,10]

1.4 Other inputs of plant material and nutrients

Throughfall and stem flow of water are important pathways of soluble carbon and nutrient flux in forest ecosystems which can be measured by chemical analysis of samples from rain gauges and stem channels.[11] Similar pathways in grasslands and other ground vegetation have received less attention but may be approached experimentally and by determining concentration gradients in canopy and litter water.

Bird droppings may also be significant sources of nutrients.[12] Faecal production by other vertebrates can be assessed by measurement of change in faeces weight on sample areas, including paired plots,[5] but faeces are used by some species as terrestrial markers and normal behaviour patterns may be disturbed by their removal. Alternatively, faecal production can be calculated from known defaecation rates and population densities, as is the case for urine production.[13] Faeces of defoliating insects may be collected in litter traps[2] or derived from population parameters.

1.5 Secondary production input to the decomposition subsystem

The production of animal and microbial remains cannot be measured directly and estimates are derived from population parameters (see Section 3).

2. MEASUREMENT OF DECOMPOSITION

In this section we are concerned mainly with the decomposition of primary resources. No single method can elucidate the functioning of decomposer systems and all available techniques have limitations and/or introduce artefacts. Methods must be selected for particular purposes, an obvious but critical comment, and estimates of a process obtained by independent methods provide an assessment of methodological bias.

2.1 Weight loss of unconfined litter

The rate of disappearance of unconfined leaf litter can be estimated from the total annual input of litter (I) and the total amount of litter accumulated (X) on the soil surface.[14,15] The calculation of instantaneous decay constants ($k = I/X$) depends upon the assumption that a state of long-term equilibrium is maintained between litter input and decomposition. This condition is rarely achieved in the strict sense but the decay constant (k) nonetheless provides a useful comparison of decomposition processes in different ecosystem types (see p. 9).

Seasonal variation in k and actual measures of unconfined litter decomposition and SOM turnover may be obtained using the paired plot technique.[3,16] The standing crops of litter and SOM are measured at discrete time intervals on paired plots and decay rates calculated from the changes in standing crop. This technique is particularly useful in the tropics where the diversity of plant species and the difficulties of obtaining litter of known age limits the extent to which general statements can be made on decomposition processes from litter enclosure studies.

A major source of error in the calculation of decomposition rates using these methods is in the differentiation of age classes of litter, especially where there is a large standing crop of accumulated litter. This problem can be resolved by placing a marker layer, e.g. thin strips of metal foil, on the leaf fall or by covering the paired plots with fine netting to exclude litter input to the standing crop. Alternatively, differences between the observed litter standing crop (L_O) and the expected standing crop L_E (L_O plus the litter increment) may be compared over discrete time intervals as a measure of seasonal variation in decay rates (e.g. p. 265).

A micro-version of the paired plot technique can be employed where information is required on either variation in decomposition rates within a leaf cohort or on decomposition in litter micro-sites. Paired discs are punched out of the leaves at appropriate time intervals and decomposition rates calculated from the decrease in weights per unit area. The technique requires the recovery and identification of the original leaves with the attendant problems of introducing artefacts.

In a few situations the ages of dead plant material can be determined by its position on the plant, especially in perennial species, allowing measurement of the weight per unit length, area or volume of each age class to estimate rate of weight loss. Alternatively replicated marking of attached leaves or twigs with non-toxic paint, thread or inert tags can provide samples of known age for weight determination.[3]

2.2 Weight loss of confined litter

Accurate measurements of weight losses from litter and chemical changes during decomposition depend on retrieval of samples placed in the field. The weighed sample of a particular age or type of litter may be confined in a bag of terylene mesh or hairnet[17] or wooden box covered with wire mesh,[18] or in sieve-like containers,[19] or glass or polythene tube[20,21] or by a net covering the ground[22] or they may be simply restrained by attaching a nylon thread to a leaf petiole[23] or twig, or branch length.[24] A wide range of methods is possible, the choice depending on the type of litter and the habitat conditions.

Collection of a uniform litter is an important initial stage, especially where small quantities of material are used in the individual samples and variation in resource quality is high. Branchwood presents particular problems because of heterogeneity in the state of decay at the time of fall.[24] The weight of the sample should be determined after air drying (oven drying may cause chemical changes) but the moisture content of a subsample must be measured to calculate the oven dry weight. After placing the samples in the field, a number of replicates should be immediately retrieved to determine errors due to measurement and handling.

Litter bags are widely used[3] to measure weight loss but the weight of the sample and enclosing material should be selected to allow the test material to simulate the natural processes of decomposition and incorporation into the litter and soil profile. Fine mesh is sometimes necessary to prevent or reduce 'fall-out' from the bag when small-leaved litter is being studied, but variation in mesh size can allow estimation of the contribution to decomposition of microbial activity and of different size groups of soil animals.[25] In all studies with confined litter the effect of enclosure on microclimate of the litter,[26] on the availability to fauna, on compaction and on natural processes of fragmentation should be recognised and, preferably, measured.

Selection of mesh sizes depends partly on the size range of the fauna, but is often a compromise between providing maximum access and minimising 'fall-out'. In some tropical areas, millipedes up to 2 cm in diameter are major litter feeders, requiring large mesh bags but in many temperate sites mesh of 5–10 mm allows access of most decomposer fauna, while 1–2 mm allows detection of consumption by mesofauna by comparison with losses from samples in fine mesh (c. 50 μm) (see Fig. 3.3). The fine mesh material tends to be expensive and strongly influences moisture content of the sample.

Interpretation of results from litter bags and similar methods must be cautious because losses through fragmentation or removal by animals are quantified but not the subsequent fate of the material. In addition the sample is often increasingly contaminated by roots, leaves and mosses entering the bag. The accuracy of estimated loss rates can be improved by comparison of results from different methods, e.g. confined versus tethered leaves,[23] or by determination of the amount and type of material removed, e.g. by earthworms, or by assessment of losses by particular processes, e.g. abiotic removal in exposed sites.[27] Leaching may represent a significant part of total weight loss from litter, particularly in early stages of decomposition. Measurement of leaching has been made in many laboratory studies but seldom in the field and no particularly suitable methods are recognised (see p. 255). The magnitude of potential leaching loss can be obtained from determination of water-soluble content of the litter but experimental leaching under simulated field conditions[28] and measurement of the leachate appears to be the most useful approach.

2.3 Decomposition of woody litter

Decomposition of small woody plant litter, e.g. twigs, can be measured by the methods described above (Sections 2.1 and 2.2). Large wood may remain as standing dead for a number of years and, like wood on the forest floor, has a very discontinuous distribution which makes sampling difficult.[7,29] The annual change in weight of standing crop of dead wood is usually small in relation to total weight but the placing and retrieval of measured samples is reasonably convenient, with few of the difficulties associated with confining leaf litter.[24]

Actual weight of a wood sample can be used to measure decomposition but the weight per unit volume (relative density R.D.) is a particularly useful measure allowing subsampling of a log without sacrificing the whole sample.[7]

The extent of decay in naturally occurring wood in different habitats, e.g. standing dead and litter surface, can also be determined from the relative density when compared with live material.[24] Relative density is also useful in studies on decomposition of woody roots, but, as in other cases, the main difficulty is in age determination.

2.4   Decomposition of specific substrates

Chemical analysis of decomposing litter over time can provide a measure of the rate of decomposition of specific organic compounds within the litter[30] but refined pure substrates can be added to soil for comparison of the decomposition potential of different soil horizons, soil types, habitats, eco-systems or climatic regimes. Use of a variety of substrates provides assessment of their relative rates of decomposition in a site. The rate of decomposition however, cannot be assumed to be a measure of that substrate in naturally occurring plant remains because the latter is strongly influenced by the associated physical and chemical composition of the litter.

Cellulose is the most widely used pure substrate in the form of filter paper, wadding or board for which weight or cellulose loss can be measured. Loss of tensile strength of cotton strips,[31] or visual assessment of decomposition of films[32,33] are also used. Decomposition of lignin,[34] chitin,[35] keratin,[36] and other pure substrates have been used, particularly to determine the biochemical pathways of decomposition or to isolate groups of micro-organisms with specific catabolic capabilities.[37] Substrates which are not available in a convenient retrievable form may be mixed with an inert carrier such as kaolin.[38]

2.5   Radioisotopes

Tagging materials with radioisotopes is a convenient way to investigate losses of elements by leaching, catabolism and comminution[39,40] if the appropriate environmental safety requirements can be met. The limitations of this approach include the difficulties of obtaining isotopes, with suitable half-lives, which are biologically meaningful and which are naturally incorporated into the tissues. Labelling litter by injecting the isotopes into plants in the field requires doses with very high specific activity and isotopes with long half-lives. A number of classical studies have been carried out with ^{106}Pu, ^{60}Co, ^{85}Sr, ^{45}Ca, ^{137}Cs in deciduous woodlands[41-43] and ^{137}Cs in lichens, mosses and pine needles.[44]

Plants can be readily labelled with ^{14}C by growing them in an atmosphere containing $^{14}CO_2$ but only the end chains of cellulose are labelled by this method. Plant litter labelled with ^{14}C has been used to study the turnover of soil organic matter[45] and root decomposition[46] in the field. Atomic bomb

[14]C has also been used as a biological marker in studies of soil organic matter turnover.[47] Specifically labelled ^{14}C cellulose and ^{14}C lignin can be synthesised in the laboratory,[48,49] but the biochemical methodology is complex.

The role of soil animals in the breakdown of litter and release of nutrients has been investigated using radioisotopes[47,50–52] but the problems involved in the recycling of isotopes have not been generally recognised.[53] Pulse labelling using tagged fungal baits to detect fungal feeding soil animals[54] overcomes some of the problems of the long-term studies which build up high background levels of radiation. Analysis of feeding pathways can be qualitative or quantitative but there is always the danger that the label might be incorporated rapidly into organisms other than that initially labelled and ingested by fauna which do not feed on the test organism.

2.6 Respiration

There has been considerable interest in the evolution of CO_2 from the soil as a measure of total heterotroph activity and the literature has been extensively reviewed.[55–57]

Field measurements are made by covering the soil *in situ* with an open ended container or a cylinder permanently inserted into the soil, and which is capped for experimental readings. Carbon dioxide output is then measured in one of the following ways.[3,56]

(a) Estimation of increase in CO_2 content of the enclosed air by periodic analysis of small samples extracted at intervals.

(b) Continuous absorption of CO_2 in alkali and calculation of the amount absorbed by titration or gravimetrically.

(c) Continuous circulation of air from the container with absorption of CO_2 from the gas stream in alkali, followed by titration or gravimetric estimation.

Alternatively the sampling of the profile of soil atmosphere using diffusion wells[58] can be used to determine CO_2 flux rates and estimate soil respiration.

Recent methods include the use of infra-red gas analysis (IRGA) for the sensitive analysis of CO_2 in air streams[59] and the detection of CO_2 absorbed in alkali by conductimetric methods.[60]

The major limitation to the value of soil CO_2 evolution as a measure of soil heterotroph activity is the CO_2 output of living roots. This is a component of gross primary production and bears no direct relationship to soil heterotroph metabolism. Estimates of the proportion of total CO_2 represented by root respiration range from insignificant,[61] to 35%,[62] 40–50%[63] and even 69%.[64] Various methods of calculating root respiration have been proposed including the use of biocides to suppress soil organism metabolism[65] and establishing regressions between CO_2 evolution and root biomass.[64] Separation of roots from a soil sample and measuring their respiration for subtraction from total soil respiration has been used but the effect of disturbance is unknown. In view of the size and uncertainty of errors, expensive and highly sensitive methods of soil CO_2 measurement are not generally justified and absorption respirometry techniques are comparatively inexpensive and simple to use.[57]

Loss of gaseous carbon as CH_4 can be significant, particularly in water-logged soils, and can be measured along with CO_2 by IRGA or gas chromatographic techniques.

Measurement of CO_2 output from soil in the field should be related to soil temperature and moisture conditions since both factors are rate determinants, and although there may be a doubt about the value of soil respiration as an absolute measure of heterotroph activity it is useful in the comparison of sites and seasonal variation and the effects of experimental treatments of soils.

Measurement of uptake of O_2 or release of CO_2 has proved valuable in the analysis of microbial activity of litters. Warburg or Gilson respirometers have been widely used to measure microbial activity of samples of litter from litter-bags of known age or from other field or laboratory experiments.[3,20] The technique allows precise determination of the variation in activity between resources or with time and particularly the influence of temperature, moisture and other environmental variables. Disturbance caused by sample retrieval and preparation can affect respiration rate but this can be partially overcome by assessment of equilibration time or by placing undisturbed samples in the respirometer.[20]

Radiorespirometry is a particular approach to the use of ^{14}C labelled substrates added to soil to measure microbial activity, the [14]CO_2 released being continuously measured over time.[66] The method is very sensitive and can be used to analyse decay rates of various substrates, the effects of varying treatments and the role of specific microflora. As with most experimental methods the artificial conditions, with soil preparation and substrate introduction, make it difficult to relate results to actual field conditions but provide opportunities for unravelling the complexity of microbial processes.

2.7 Selective inhibition of organisms

The quantitative role of particular groups of organisms, particularly animals, is often assessed by calculation of their total annual respiration or food consumption, by prevention of access to litter (Section 2.2) or by experimental addition of organisms.[67] An additional method, used in both field and laboratory studies, is the application of more or less specific biocides or inhibitors. Naphtha has been used to inhibit animals activity in experimental plots,[68,69] and at high levels of application litter comminution by animals cannot be completely eliminated. DDT, dieldrin and other pesticides have been used for animals[70,71] with various fumigants, fungicides and bacteriocides to determine the contribution of microflora.[65,72]

2.8   Enzyme activity

Assay of the enzyme complement of soil has been used to determine the biochemical potential of the soil, measuring the activity of extracellular enzymes.[73,74] Thus oxidation of organic resources involves removal of hydrogen by dehydrogenase and the dehydrogenase activity in soil is measured by their action on tetrazolium dyes. These are converted to red coloured formozan compounds whose colour intensity is determined, the colour being related to a dehydrogenase standard.[75] The enzyme concentration is dependent on rates of production and breakdown and activity is dependent on inactivation by physical phenomena, as well as temperature, moisture, pH and other factors. Interpretation therefore can be difficult and the expected relationship of dehydrogenase activity to oxygen uptake or microbial numbers is often lacking. Methods of assay of a wide range of enzymes have been developed based on the principle of addition of a substrate to a soil sample and measurement of reaction products or substrate remaining after a period of incubation. Microbial growth is inhibited during incubation by addition of toluene or by radiation sterilisation[76,77] so that only free enzymes are measured, but the inhibitory treatment probably releases some intracellular enzymes by lysis of cells.

2.9   Nutrient mineralisation

The preceding sections have concentrated on decomposition of C compounds but in some studies it is necessary to focus on rates of mineralisation or transformation of N, P, S and other nutrients which are incorporated in the compounds of primary resources. Input of nutrients, either as total or as organic and inorganic forms, is usually measured by chemical analysis of the incoming litter, rain or leachate. Successive analyses of litter samples over time, using sampling methods given in Sections 2.1–2.4, can show the change in concentration and form of nutrients in organic matter. These measurements show only the nett change because transformations, e.g. from plant to microbial protein are not detected but they allow detection of main trends in concentrations. The *concentration* of an element may increase during decomposition, not through addition of the element but through a loss of C. Therefore to estimate the change in *absolute amount* of the element between t_0 and t_1 it is necessary to know the dry weight of the sample as well as the concentration. Measurement of nutrients in soil usually requires attempts to determine the component 'available' to plants or decomposers as well as the total amount of the nutrient.

Rates of transformation of forms of nutrients, e.g. amino-N to NO_3' or organic P to inorganic P can be estimated by addition of an initial form of the element and measurement of the change in concentration of alternative forms. This approach can be applied to analysis of the resource in the soil, to gaseous outputs using, for example, gas chromatography, to aqueous outputs from columns of soil, resin, glass beads or other suitable substrates for microbial activity, or in continuous flow systems. A vast range of incubation, enrichment, percolation and continuous flow systems has been developed and an introduction to these can be obtained from general textbooks on microbial ecology and physiology and on soil biochemistry.[37,78-80] These methods have very limited application in the field although resources labelled with ^{15}N, ^{32}P or other isotopes can be used while lysimeters and measurement of field drainage water can indicate nett transformations for soil systems or horizons.[81]

A valuable source of references to studies on the transformations and movement of plant nutrients in soil is Nye and Tinker.[82]

3. SOIL ORGANISM POPULATIONS AND ACTIVITIES

3.1 Estimation of populations, production and energy flow

The methods handbooks produced by the International Biological Programme provide detailed reviews of techniques for estimating faunal[83] and microbial[78,84] populations and activity, and there are many less specialised reviews of population sampling techniques, statistical analyses and activity measurements. The emphasis here is on a few selected approaches and recent advances.

3.2 Feeding relationships of soil fauna in the field

Gut content analysis: animals must be collected and fixed in the field or extracted from samples returned to the laboratory. In the latter case flotation, grease film or elutriation methods are preferred, allowing fixation of samples to prevent feeding after field sampling. Quantitative methods of gut analysis are available[85,86] and show considerable intra-specific variation in most fauna which can be overcome by pooling of gut contents.[87] Seasonal variation in feeding habits[86] necessitates repeated sampling and non-predatory fauna show marked differences in gut contents when taken from different microhabitats.[87] Embedding soil samples in gelatin allows sectioning and removal of arthropods from specific microhabitats for gut content analysis.[87,88]

Microscopic analysis of gut contents can show the main components of food but to determine the specificity of fungal-feeders, e.g. cryptostigmatid mites[89] isolation of fungi on agar has been used. Isolation, however, tends to favour spore-forming and rapidly growing fungi and techniques need to be developed to give more accurate estimates of other soil fungi, e.g. Basidiomycetes.

Analysis of the enzyme complement of animals from the field indicates the capacity to digest various substrates, but does not determine the source or amounts of substrates used (see Section 3.4).

Microbial bait: Microbial cultures contained in plastic side traps or in terylene net[90,91] have been placed in the field and later sampled for fauna. Colonisation by fauna of soil cores which have been sterilised by irradiation and inoculated with different species of fungi[92] or the addition to soil of radioisotope labelled mycelium followed by analysis of animals in the vicinity of the bait[54] are alternative methods of determining faunal feeding on microflora in the field. A range of other methods could be applied including serological analysis of gut contents as in predator–prey studies,[93] but production of specific antisera presents considerable technical problems. An ingenious technique has been used in soil predator–prey studies; DDT tolerant strains of the collembolan *Folsomia candida* have been bred, labelled with DDT, and released in the field. Predators were then detected by the presence of the pesticide in their tissues.[94]

The approaches to examination of feeding relationships described above can be applied to laboratory studies where a wide range of other methods are also applicable.

3.3 Microbial population estimates

The description of the species composition of the microflora of decomposer habitats depends on prior isolation with one of a variety of agar media or baits each of which has a selective effect for differing components of the microflora. Care must therefore be exercised in the choice of method. These features have been extensively reviewed as have the problems associated with the identification of soil micro-organisms.[84,95] Methods for the estimation of microbial standing crop and biomass have been reviewed,[84] but a number of developments in recent years have improved the range of available techniques. Because of the small size of micro-organisms and the opacity of soil environments direct methods of counting and observation are seldom possible although thin soil sections have been used for this purpose.[96] The second major problem is the distinction of the viable components (= biomass) of the total microbial mass (= standing crop). The most usual approach to standing crop estimates have been by microscopic counting of bacterial cells or measurements of fungal hyphal lengths in standard volumes of soil suspensions or litter macerates. These have most commonly been prepared in dried agar films,[97] although other methods have also been used.[84] The bacterial counts or lengths of hyphae are converted to weight by standard or independently measured conversion factors.[84,98] Viable components have been detected by phase contrast microscopy,[99] or a variety of staining techniques[100,101] but this still remains a major problem with the method.

Measurement of the active component of the mycoflora on the surface of leaves has been carried out by autoradiographic detection of the uptake of ^{14}C-labelled sugars by hyphae[102] but this technique ignores all penetrating mycelium. Chemical indices for standing crop have been used in recent years, such as chitin for fungi[103,104] and muramic acid[105] or diamino-pimelic acid[106] for bacteria. A detailed comparison of the chitin-assay and agar film methods for mycelial standing crop has been carried out[98] which includes a statistical analysis of appropriate experimental designs for the two methods. A major disadvantage of the chemical method is that its use is confined to situations free from contamination by arthropod chitin.

The standard method for counting viable populations of bacteria (which may then be converted to biomass) involves one of several methods of growing colonies from individual cell suspension prepared in a dilution series. All variants are subject to problems of selectivity and the choice of diluent, medium and experimental design depends on the experimental purpose.[84,107] A chemical assay of viable cells may be made by ATP estimation[108] but care must be taken to use this technique only where microbial ATP can be equivocally distinguished from other sources. The total microbial biomass in soil has been ingeniously estimated by measuring its breakdown, as a secondary resource, after fumigation.[109]

Methods of estimating microbial production include fluorescent-antibody (FAB) labelling of microbial cells to follow growth and mortality of bacterial cells in natural habitats[110] and the method is also applicable to studies of mycelial decay.[111] The balance between mycelial growth and decay on nylon meshes inserted in the soil has also been used to estimate fungal production,[112] although this technique may be criticised because of the introduction of an unnatural substratum for growth. Microbial production has also been estimated from the response of soil populations (as measured by total or viable counts) to added substrates.[113] In view of the major significance of fungi to energy flow within the decomposer community the development of techniques in this area must continue to be a major research priority.

3.4 Metabolic activities

A wide range of biochemical techniques exists for determining specific biochemical transformations either of organisms in pure culture or for the study of decomposer activities *in situ*. Prokaryotes (bacteria and actinomycetes) are commonly tested for a wide range of attributes such as C, energy and N source utilisation as part of the process of identification.[114,115] Studies of depolymerase activity in fungi commonly include cellulase (C_x), pectinase and amylase activity, the gallic acid test for phenolase which is taken as an indication of lignin degrading activity and a test for the ability to decolorise humic acids.[116] The recent increase in availability of a range of hemicellulose and lignin related substrates[117,118] should enable a widening of this type of testing. It is also sensible to test for C_1 activity as well as C_x[119] (see Chapter 5). Testing of polysaccharases may be facilitated by the use of a dye— Remazol Blue—which complexes with the polysaccharide and is only released upon hydrolysis of the glycosidic bonds.[33, 120] Enzyme tests can also be applied to intact guts or gut extracts from decomposer animals.[121] In this context it is sensible to attempt to distinguish whether the enzymes are of microbial or animal origin.[122] It should be remembered in relation to all these studies that the demonstration of enzyme capacity indicates only the potential for the participation in particular aspects of decomposition processes.[74] The strong regulatory influences of the chemical, physical and biological environment on enzyme activities discussed in Chapter 5 will all determine the functioning of specific enzyme systems within the decomposer community. Ideally enzyme studies should be combined with chemical analyses of resource or substrate change and decomposer growth. There is a great need for more detailed experiments of this kind particularly those including the interaction of various permutations of decomposer organisms.[123] Many specialised techniques exist for measurement of particular biochemical pathways and mention can be made of only a few; soil percolation for measurement of nitrification and other aspects of the N cycle;[124] acetylene reduction for N-fixation;[125] and the range of techniques applicable to anaerobic environments, and chemoautotrophic metabolism.[126]

4. PHYSICOCHEMICAL PARAMETERS

A number of excellent texts are available on soil physical[127,128] and chemical analysis,[129–131] the chemistry of soil organic matter[132] and general background information to the study of soil.[133–135] Ecological studies often involve the analysis of large numbers of samples and a recently published schedule of procedures simplifies the routine analysis of soluble salts, cation exchange, pH, soil organic matter and particle size.[136] The present section is intended to indicate important and/or recent papers which may not appear in the general texts.

4.1 Temperature

Metabolic functions of soil organisms tend to follow a logarithmic response to temperature. Thus geometric means of maximum–minimum temperatures provide only a broad approximation in calculating animal or microbial metabolism in the field from laboratory data. A simple method which provides an integrated mean temperature for use in biological studies depends upon the temperature-related inversion of sucrose solution. The original technique is inaccurate where the diurnal temperature variation exceeds 10–15°C but the addition of sodium chloride to the sucrose solution extends the useful range of the technique to all climatic conditions.[137] Other temperature integration methods include electrolysing a solution with a current which is proportional to the temperature of a thermistor inserted into the soil.[138]

Standard types of thermometer include liquid in glass, deformation, resistance wire, thermistor and thermocouple, data from which can be linked to a variety of data collection systems.[11,128,139]

4.2   pH

Bulk pH determinations (pH$_B$) are commonly measured with little critical evaluation of the methods used in their measurement and of the reproducibility of the results. Comparisons of pH$_B$ readings for woodland soils obtained using fresh soil with distilled water, fresh soil with KCl solution, dry soil with distilled water and bulked samples with distilled water[149] showed variation of nearly one pH unit from the same soils. It was concluded that composite samples left one hour in distilled water before testing provided the most consistent and assertable values. Other problems in measurement of soil pH are briefly discussed in Reference 131. However, it is not the precise and real pH which is often important but the correlations between objective results and standardised measurements. Relationships between bulk pH measurements and localised phenomena, such as microbial activity in different sub-horizons of a soil profile must always be interpreted with caution.

4.3   Soil moisture

The concepts, definitions and units applicable to the soil water regime will not be considered here as they are authoritatively reviewed elsewhere[141,135] (see also Chapter 6). The choice of methods, like pH, depends upon the scale and range of measurements which are required for correlation with particular biological phenomena. For simplicity, many soil ecologists resort to bulk, gravimetric measures of soil or litter water content where moisture losses on oven drying are expressed as a percentage of wet or dry weights, but diurnal variation in moisture in litter is often overlooked. The usefulness of these measurements can be considerably improved by expressing the water content as a percentage of ash-free SOM (determined by ignition losses)[57] but moisture potential measurements have greater validity in detailed studies of decomposition processes and soil microbial ecology.[141] A number of simple laboratory techniques are available for controlling and/or measuring soil moisture potential [142] but field apparatus tends to be complex and unreliable. A simple technique for soil moisture potential determination, suitable for field and expedition use, is based on the principle that moisture absorption by filter paper is controlled by matric potential. Filter paper is treated with $HgCl_2$ to inhibit microbial growth and then oven dried. A wad of three filter papers is packed between two layers of soil and stored under constant temperature for several days. The centre filter paper is then removed, oven dried to determine the water content and the corresponding matric potential is read from a calibration curve.[143]

4.4 Soil atmosphere

The carbon dioxide content of the soil atmosphere can be measured comparatively simply in the field and laboratory. The measurement of oxygen potential is more complex and involves polarographic [144] or gas exchange techniques.[145] Quantitative or semi-quantitative analyses of many gases may be simply carried out using Draeger tubes (Draeger Safety, Draeger House, Chesham, U.K.) but the volume of air which must be sampled precludes the use of this technique for studies of the soil atmosphere in microsites. All measurements of gases in soil cavities are complicated by the temporal exchanges between the soil atmosphere and interstitial water, and should be considered as relative to soil animal or microbial activity and not absolute measures of gas concentrations.

4.5 Redox potential

The relative degree of oxidation or reduction in a soil (E_h) is readily measured using a pH meter with a platinum electrode replacing the glass electrode. Care must be taken with cleanliness of the electrode and there are difficulties in obtaining reproducible results, especially in oxidising systems.[130] The ionic state of a particular ion can also be used to indicate the redox state, for example blackening of a surface by sulphide deposition when a silver sheet or wire is inserted into the soil shows the reducing zone or microhabitats.[146]

5. CHEMICAL ANALYSIS OF RESOURCES

Detailed and comprehensive accounts of analytical methods for ecological materials have recently been published [130,147] but two analytical procedures for litter analysis are of particular value in decomposition studies. One involves a scheme for the analysis of ash, sugars, lignin, polyphenols, total nitrogen and ether and methanol soluble fractions in only 100 mg of leaf material[148], while the other is a comparison of different methods of cellulose estimation.[149] It was found that the simplest method of analysis, when only the cellulose content of leaf and woody litter is required, was the total hydrolysis of the material with acids followed by the calculation of cellulose content from the hexoses in the hydrolysate. More detailed analyses using modern techniques of separation are given elsewhere.[117,131]

Humus chemistry is extremely complex and a wide range of physical and chemical techniques is necessary for the determination of its composition, activity, formation and breakdown.[150,151] Isotopic methods have been particularly useful for the elucidation of the origin and turnover of humus fractions.[152,153]

6. STATISTICAL ANALYSIS

We do not intend to discuss statistical analysis of data on decomposition, only to emphasise that much effort can be spent on data collection which can be wasted unless certain principles are adequately defined as follows:

(a)   The objectives of the work should be clear and realistic and any hypotheses tested should form part of a theory or model constructed to include any basic assumptions underlying the work.

(b)   The data collected must be relevant to the objectives of the work.

(c)   Sufficient data must be collected to allow an adequate statistical analysis and the collection itself must not be biased. The sample collection, survey techniques and/or experimental design must allow for these constraints.

(d)   In the laboratory and in the field the accuracy, precision and sensitivity of any methods used should be known.

These principles may appear obvious, but their implementation requires considerable care and effort.

REFERENCES

1 FORREST G.I. (1971) Structure and production of north Pennine blanket bog vegetation. *J. Ecol.* **59**, 453–479.

2 NEWBOULD P.J. (1967) *Methods for Estimating the Primary Production of Forests.* I.B.P. Handbook No. 2, Blackwell Scientific Publications, Oxford.

3 CHAPMAN S.B. (1976) Production ecology and nutrient budgets. In: *Methods in Plant Ecology*, ed. S.B. Chapman, pp. 157–228. Blackwell Scientific Publications, Oxford.

4 HUGHES M.K. (1970) Ground vegetation and forest litter production. In: *Methods of study in Soil Ecology*, ed. J. Phillipson, pp. 145–149. UNESCO, Paris.

5 WIEGERT R.G. & EVANS F.C. (1964) Primary production and the disappearance of dead vegetation on an Old Field in south eastern Michigan. *J. Ecol.* **45**, 49–63.

6 MILNER C. & HUGHES R.E. (1968). *Methods for the Measurement of the Primary Production of Grassland.* I.B.P. Handbook No. 6, Blackwell Scientific Publications, Oxford.

7 HEALEY I. N. & SWIFT M.J. (1971) Aspects of the accumulation and decomposition of wood in the litter layer of a coppiced beech-oak woodland. In: *Organismes du sol et production primaire, 4th Coll. Soil Zool.*, pp. 417–430. INRA, Paris.

8 CHRISTENSEN, D. (1975) Wood litter fall in relation to abscission, environmental factors and the decomposition cycle in a Danish oak forest. *Oikos*, **26**, 187–195.

9 COLEMAN D. (1976) A review of root production processes and their influence on soil biota on terrestrial ecosystems. In: *The role of terrestrial and aquatic organisms in decomposition processes*, eds. J.M. Anderson & A. Macfadyen, pp. 417–434. Blackwell Scientific Publications, Oxford.

10 MARSHALL J.K. (1978) (ed.) The belowground ecosystem: a synthesis of plant-associated processes. *Range Sci. Dept. Sci. Ser.* **26**, Colorado State Univ. Fort Collins.

11 PAINTER, R.B. (1976) Climatology and environmental measurement. In: *Methods in plant ecology*, ed. S.B. Chapman, pp. 369–410. Blackwell Scientific Publications, Oxford.

12 WEIR J.S. (1969) Importation of nutrients into woodland by rooks. *Nature, Lond.* **221**, 487–88.

13 WILKINSON S.R. & LOWRY R.W. (1973) Cycling of mineral nutrients in pasture ecosystems. In: *Chemistry and Biochemistry of Herbage.* Vol. 2, eds. G.W. Butler & R.W. Bailey, pp. 248–315. Academic Press, London.

14 JENNY H., GESSEL S.P. & BINGHAM F.T. (1949) Comparative study of decomposition rates of organic matter in temperate and tropical regions. *Soil Sci.*, **68**, 419–432.

15 OLSON J.S. (1963) Energy storage and the balance of producers and decomposers in ecological systems. *Ecology*, **44**, 327–332.

16 REINERS W.A. & REINERS N.M. (1970) Energy and nutrient dynamics of forest floors in three Minnesota forests. *J. Ecol.* **58**, 497–519.

17. BOCOCK K.L. & GILBERT O.J.W. (1957) The disappearance of leaf litter under different woodland conditions. *Plant Soil*, **9**, 179–185.

18 ATTIWILL P.M. (1968) The loss of elements from decomposing litter. *J. Ecol.* **49**, 142–145.

19 HERLITZIUS R. & HERLITZIUS H. (1977) Strenabban in laubwäldern. Untersuchungen in ĸalk- und Sanerhumusbuchenwäldern. *Oecologia (Berl.)*, **30**, 147–171.

20 HOWARD P.J.A. (1967) A method for studying respiration and decomposition of litter. In: *Progress in Soil Biology*, eds. O. Graff & J.E. Satchell, pp. 464–472. North Holland, Amsterdam.

21 CLYMO R.S. (1965) Experiments on breakdown of *Sphagnum* in two bogs. *J. Ecol.* **53**, 747–758.

22 WITTICH W. (1953) Untersuchungen uber den Verlauf der Stenzersetzung auf einem Boden mit starker Regenwurmfätigkeit. *Schriften r. Forstl. Fak. Gottingen*, **9**, 1–33.

23 WITKAMP M. & OLSON J. (1963) Breakdown of confined and non-confined oak litter. *Oikos*, **14**, 138–147.

24 SWIFT M.J., HEALEY I.N., HIBBERD J.K., SYKES J.M., BAMPOE V. & NESBITT M.E. (1976) The decomposition of branch-wood in the canopy and floor of a mixed deciduous woodland. *Oecologia (Berl.)*, **26**, 139–149.

25 EDWARDS C.A., & HEATH G.W. (1963) The role of soil animals in breakdown of leaf material. In: *Soil Organisms*, eds. J. J. Doeksen & J. van der Drift, pp. 76–84. North Holland, Amsterdam.

26 ANDERSON J.M. (1975) Succession, diversity and trophic relationships of some soil animals in decomposing leaf litter. *J. Anim. Ecol.* **44**, 475–495.

27 ANDERSON J.M. (1973) The breakdown and decomposition of sweet chestnut (*Castanea sativa* Mill) and beech (*Fagus sylvatica* L.) leaves in two deciduous woodland soils, I Breakdown, leaching and decomposition. *Oecologia (Berl.)*, **12**, 251–274.

28 NYKVIST N. (1959) Leaching and decomposition of litter. I Experiments on leaf litter of *Fraxinus excelsior*. *Oikos*, **10**, 190–209.

29 SOUTHWOOD T.R.E. (1966) *Ecological Methods*. Methuen, London.

30 MIKOLA P. (1954) Experiments on the rate of decomposition of forest litter. *Comm. Inst. Forest Fenn.* **43**, No. 1.

31 LATTER P.M. & HOWSON G. (1977) The use of cotton strips to indicate cellulose decomposition in the field. *Pedobiologia*, **17**, 145–155.

32 TRIBE H.T. (1957) Ecology of micro-organisms in soils as observed during their development upon buried cellulose film. In: *Microbial Ecology*, eds. C.C. Spicer & R.E.O. Williams. *Symp. Soc. Gen. Microbiol.* **7**, 287–98.

33 MOORE R.L., BASSET B.B. & SWIFT M.J. (1979) Developments in the Remazol Brilliant Blue dye-assay for studying the ecology of cellulose decomposition. *Soil Biol. Biochem.* (in press).

34 KIRK T.K. (1971) Effects of microorganisms on lignin. *Ann. Rev. Phytopath.* **9**, 185–210.

35 OKAFOR N. (1966) The ecology of micro-organisms on, and the decomposition of, insect wings in soil. *Plant Soil*, **25**, 211–237.

36 PUGH G.J.F. (1971) Factors which influence the early colonization of organic matter by fungi. In: *Organismes du sol et production primaire, 4th Coll. Soil Zool.* pp. 319–328. INRA, Paris.

37 ALEXANDER M. (1977) *Introduction to soil microbiology*. 2nd edition Wiley, New York.

38 HENDERSON M.E.K. (1961) The metabolism of aromatic compounds related to lignin by some hyphomycetes and yeast-like fungi of soil. *J. gen. Microbiol.* **26**, 155–665.

39 International Atomic Energy Agency (1972) *Isotopes and radiation in soil-plant relationships including forestry*. IAEA, Vienna.

40 PAUL E.A., BIDERBECK V.O. & ROSHA N.S. (1970) Investigation of the metabolism of soil micro-organisms by the use of C^{14}. In: *Methods of study in soil ecology*, ed. J. Phillipson, pp. 111–116. UNESCO, Paris.

41 OLSON J.S. & CROSSLEY D.A. (1963) Tracer studies of the breakdown of forest litter. In: *Radioecology*, eds. U. Schultz & A.W. Klement, pp. 411–416. Reinhold Publishing Co., New York.

42 THOMAS W.A. (1969) Accumulation and cycling of calcium by dogwood trees. *Ecol. Mongr.* **39**, 101–120.

43 AUSMUS B.S., EDWARDS N.T. & WITKAMP M. (1976) Microbial immobilization of carbon, nitrogen, phosphorus and potassium: implications for ecosystem processes. In: *The role of terrestrial and aquatic organisms in decomposition processes*, eds. J.M. Anderson & A. Macfadyen, pp. 397–416. Blackwell Scientific Publications, Oxford.

44 WITKAMP M. & FRANK M.L. (1967) Retention and loss of caesium-137 by components of the ground cover in a pine (*Pinus verginiana L.*) stand, *Health Phys.* **13**, 985–990.

45 JENKINSON D.A. (1971) Studies on the decomposition of ^{14}C labelled organic matter in soil. *Soil Sci.* **3**, 64–70.

46 DAHLMAN R.C. & KUCERA C.C. (1968) Tagging native grassland vegetation with carbon-14. *Ecology*, **49**, 1199–1203.

47 STOUT J.D., TATE K.R. & MOLLOY L.F. (1976) Decomposition processes in New Zealand soils with particular respect to rates and pathways of plant degradation. In: *The role of terrestrial and aquatic organisms in decomposition processes*, eds. J.M. Anderson & A. Macfadyen, pp. 97–144. Blackwell Scientific Publications, Oxford.

48 CRAWFORD D.L., CRAWFORD R.L. & POMETTO A.L. (1977) Preparation of specifically labelled ^{14}C lignin and ^{14}C cellulose—lignocelluloses and their decomposition by the microflora of soil. *Appl. Env. Microbiol.* **33**, 1247–1251.
49 HACKETT W.R., CONNORS W.T., KIRK T.K. & ZIEKNS J.T. (1977) Microbial decomposition of synthetic labelled ^{14}C lignins in nature: lignin biodegradation in a variety of natural materials. *Appl. Env. Microbiol.* **33**, 43–51.
50 CROSSLEY D.A. & WITKAMP M. (1964) Forest soil mites and mineral cycling. *Acarologia* spec. vol. (C. R. Ier Cong. Int. d'Acarologie, Fort Collins, 1963). 137–145.
51 REICHLE D.E. & CROSSLEY D.A. (1967) Investigations on heterotrophic productivity in forest insect communities. In: *Secondary Productivity of Terrestrial Ecosystems*, ed. K. Petrusewicz. 563–587. Panstowowe Wydawnictwo Naukowe, Warsaw.
52 GHILAROV M.S. & KRIVOLUTSKY D.A. (1971) Radioecological researches in Soil zoology. In: *Organismes du sol et production primaire, 4th Coll. Soil Zool.* pp. 537–544. INRA, Paris.
53 CONOVER R.J. & FRANCIS V. (1973) The use of radioisotopes to measure the transfer of materials in aquatic food chains. *Marine Biol.* **18**, 272–283.
54 COLEMAN D.C. & McGINNIS J.J. (1970) Quantification of fungus-small arthropod food chains in the soil. *Oikos*, **21**, 134–137.
55 MACFADYEN A. (1970) Simple methods for measuring and maintaining the proportion of carbon-dioxide in air, for use in ecological studies of soil respiration. *Soil Biol. Biochem.* **2**, 9–18.
56 MINDERMAN G. & VULTO J.C. (1973) Comparison of techniques for the measurement of carbon-dioxide evolution from soil. *Pedobiologia*, **13**, 73–80.
57 ANDERSON J.M. (1973) Carbon-dioxide evolution from two temperate, deciduous woodland soils. *J. Appl. Ecol.* **10**, 361–378.
58 VON BAVEL C.H.M. (1965) Composition of soil atmosphere. In: *Methods of soils analysis. Agronomy*, **9**, 315–318.
59 KUCERA C.L. & KIRKHAM D.R. (1971) Soil respiration studies in tall grass prairie in Missouri. *Ecology*, **52**, 912–915.
60 CHAPMAN S.B. (1971) A simple conductimetric soil respirometer for field use. *Oikos*, **22**, 348–353.
61 WITKAMP M. (1966) Rates of carbon-dioxide evolution from the forest floor. *Ecology*, **47**, 492–494.
62 EDWARDS N.T. & SOLLINS P. (1973) Continuous measurement of carbon dioxide evolution from partitioned forest floor compartments. *Ecology*, **54**, 406–412.
63 WITKAMP M. & FRANK M.L. (1969) Evolution of carbon-dioxide from litter, humus and sub-soil of a pine stand. *Pedobiologia*, **9**, 358–365.
64 MINDERMAN G. & VULTO J. (1963) Carbon-dioxide production by tree roots and microbes. *Pedobiologia*, **13**, 337–343.
65 ANDERSON J.P.E. & DOMSCH K.H. (1973) Quantification of bacterial and fungal contributions to soil respiration. *Arch. Microbiol.* **93**, 113–127.
66 MAYAUDON J. (1971) Use of radiorespirometry in soil microbiology and biochemistry. In: *Soil Biochemistry* Vol. 2, eds. A.D. McLaren & J. Skujins, pp. 202–256. Dekker, New York.
67 STANDEN V. (1978) The influence of soil fauna on decomposition by microorganisms on blanket bog litter. *J. Anim. Ecol.* **47**, 25–38.
68 KURCERA G.F. (1960) The role of invertebrates in the decomposition of leaf litter. *Pedology, Leningr.* **4**, 16–23.
69 EDWARDS C.A., REICHLE D.E. & CROSSLEY D.A. (1969) Experimental manipulation of soil invertebrate populations for trophic studies. *Ecology*, **50**, 495–498.

70 WEARY G.C. & MERRIAM H.G. (1978) Litter decomposition in a red maple wood lot under natural conditions and under insecticide treatment. *Ecology*, **59**, 180–184.

71 EDWARDS C.A., REICHLE D.E. & CROSSLEY D.A. (1970) The role of soil invertebrates in turnover or organic matter and nutrients. In: *Ecological Studies, Analysis and Synthesis* 1. *Analysis of Temperate Forest Ecosystems*, ed. D.E. Reichle, pp. 147–172. Springer-Verlag, Berlin.

72 JENKINSON D.S. & POWLSON D.S. (1970) Residual effects of soil fumigation on soil respiration and mineralization. *Soil Biol. Biochem.* **2**, 99–108.

73 KUPREVICH V.F. & SHCHERBAKOVA T.A. (1971) Comparative enzymatic activity in diverse types of soil. In: *Soil Biochemistry*, Vol. 2, eds. A.D. McLaren & J. Skujins, pp. 167–201. Dekker, New York.

74 SKUJINS J.J. (1967) Enzymes in soil. In: *Soil biochemistry*, eds. A.D. McLaren & G.H. Peterson, pp. 371–414. Edward Arnold, London.

75 HOWARD P.J.A. (1972) Problems in the estimation of biological activity in soil. *Oikos*, **23**, 235–240.

76 MCLAREN A.D. (1969) Radiation as a technique in soil biology and biochemistry. *Soil Biol. Biochem.* **1**, 63–74.

77 HOWARD P.J.A. & FRANKLAND J.C. (1974) Effects of certain full and partial sterilization treatments on leaf litter. *Soil Biol. Biochem.* **6**, 117–123.

78 ROSSWALL T. (1973) (ed.). Modern methods in the study of microbial ecology. *Ecol. Bull. (Stockholm)*, **17**.

79 GRAY T.R.G. & WILLIAMS S.T. (1971) *Soil Micro-organisms*. Oliver & Boyd, Edinburgh.

80 AARONSON S. (1970) *Experimental microbial ecology*, Academic Press, London and New York.

81 Ministry of Agriculture, Fisheries and Food (1976) Agriculture and water quality. *MAFF Tech. Bull.* **32**. HMSO, London.

82 NYE P.H. & TINKER P.B. (1977) *Solute movement in the soil-root system.* Blackwell Scientific Publications, Oxford.

83 PHILLIPSON J. (1971) (ed.) *Methods of study in quantitative soil ecology: Population, production and energy flow.* I.B.P. Handbook No. 18. Blackwell Scientific Publications, Oxford.

84 PARKINSON D., GRAY T.R.G. & WILLIAMS S.T. (1971) (eds.) *Methods for studying the ecology of soil micro-organisms.* I.B.P. Handbook No. 19. Blackwell Scientific Publications, Oxford.

85 ANDERSON J.M. & HEALEY I.N. (1972) Seasonal and inter-specific variation in major components of the gut contents of some woodland collembola. *J. Anim. Ecol.* **41**, 359–368.

86 ANDERSON J.M. (1975) Succession, diversity and trophic relationships of some soil animals in decomposing leaf litter. *J. Anim. Ecol.* **44**, 475–495.

87 ANDERSON J.M. (1978) Competition between two unrelated species of soil Cryptostigmata (Acari) in experimental microcosms. *J. Anim. Ecol.* **47**.

88 ANDERSON J.M. (in press). The preparation of gelatine embedded soil and litter sections and their application to some soil ecological studies. *Biol. Ed.*

89 STEFANIAK O. & SENICZAK S. (1976) The microflora of the alimentary canal of *Achipteria coleoptata* (Acarina, Oribatei). *Pedobiologia* **16**, 185–194.

90 DASH M.C. & CRAGG J.B. (1972) Selection of microfungi by Enchytraeidae (Oligochaeta) and other members of the soil fauna. *Pedobiologia*, **12**, 282–286.

91 SPRINGETT J.A. & LATTER P.M. (1977) Studies on the microfauna of blanket bog with particular reference to Enchytraeidae. 1. Field and laboratory tests of microorganisms as food. *J. Anim. Ecol.* **46**, 959–974.

92 COLEMAN D.C. & MACFADYEN A. (1966) The recolonization of gamma-irradiated soil by small arthropods. *Oikos*, **17**, 62–70.

93 DAVIES R.W. (1966) The production of antisera for detecting triclad antigens in the gut contents of predators. *Oikos*, 20, 248–260.

94 MANLEY G.U., BUTCHER J.W. & ZABIK M. (1976) DDT transfer and metabolism in a forest litter macro-arthropod food chain. *Pedobiologia*, **16**, 81–98.

95 BARRON G.L. (1971). Soil fungi. In: *Methods in Microbiology*, vol. 4, ed. C. Booth, pp. 405–428. Academic Press, London and New York.

96 BURGES A. & NICHOLAS D.P. (1961) Use of soil sections in studying amount of fungal hyphae in soil. *Soil Sci.* **92**, 25–29.

97 JONES P.C.T. & MOLLISON J.E. (1947) A technique for the quantitative estimation of soil micro-organisms. *J. gen. Microbiol.* **1**, 54–68.

98 FRANKLAND J.C. & LINDLEY D.K. with SWIFT M.J. (1978) An analysis of two methods for the estimation of mycelial biomass in leaf litter. *Soil Biol. Biochem.* **10**, 323–333.

99 FRANKLAND J.C. (1975) Estimation of live fungal biomass. *Soil Biol. Biochem.* **7**, 339–340.

100 BABIUK L.A. & PAUL E.A. (1970) The use of fluorescein isothiocyanate in the determination of the bacterial biomass of grassland soil. *Can. J. Microbiol.* **16**, 57–62.

101 SODERSTROM B.E. (1977) Vital staining of fungi in pure cultures and in soil with fluoroscein diacetate. *Soil Biol. Biochem.* **9**, 59–63.

102 WAID J.S., PRESTON K.J. & HARRIS P.J. (1971) A method to detect metabolically-active micro-organisms in leaf litter habitats. *Soil Biol. Biochem.* **3**, 235–241.

103 SWIFT M.J. (1973) The estimation of mycelial biomass by determination of the hexosamine content of wood tissue decayed by fungi. *Soil Biol. Biochem.* **5**, 321–332.

104 RIDE J.P. & DRYSDALE R.B. (1972) A rapid method for the chemical estimation of filamentous fungi in plant tissue. *Physiol. Plant Path.* **2**, 7–15.

105 MILLAR W.N. & CASIDA L.E. (1970) Evidence for muramic acid in soil. *Can. J. Microbiol.* **16**, 299–304.

106 STEUBING L. (1970) Chemische Methoden zur Bewertung des mangenmassigen Vorkommens von Bakterien und Algen im Boden. *Zbl. Bakt.* **124**, 245–249.

107 JENSEN V. (1968) The plate count technique. In: *The ecology of soil bacteria* eds. T.R.G. Gray & D. Parkinson, pp. 158–170. Liverpool University Press.

108 AUSMUS B.S. (1973) The use of the ATP assay in terrestrial decomposition studies. In: *Modern methods in the study of microbiol ecology*, ed. T. Rosswall. *Ecol. Bull. (Stockholm)*, **17**, 223–234.

109 JENKINSON D.S. & POWLSON D.S. (1976) The effects of biocidal treatments on metabolism in soil. V. A method of measuring soil biomass *Soil Biol. Biochem.* **8**, 209–213.

110 HILL I.R. & GRAY T.R.G. (1967) Application of the fluorescent antibody technique to an ecological study of bacteria in soil. *J. Bact.* **93**, 1888–1896.

111 PREECE T.F. (1971) Immunological techniques in mycology. In: *Methods in microbiology*. Vol. 4, ed. C. Booth, pp. 599–607. Academic Press, London and New York.

112 NAGEL-DE BOOIS H.M. & JANSEN E. (1971) The growth of fungal mycelium in forest soil layers. *Rev. Ecol. Biol. Sol.* **8**, 509–520.

113 SHIELDS J.A., PAUL E.A., LOWE W.E. & PARKINSON D. (1974) Turnover of microbial tissue in soil under field conditions. *Soil Biol. Biochem.* **5**, 753–764.

114 ROSSWALL T. & CLARHOLM M. (1974) Characteristics of tundra bacterial populations and a comparison with populations from forest and grassland soils. In: *Soil organisms and decomposition in tundra*, eds. A.J. Holding, O.W. Heal, S.F. MacLean Jr. & P.W. Flanagan, pp. 93–108, Tundra Biome Steering Committee, Stockholm.

115 SZABO I.M. (1974) *Microbial Communities in a forest-rendzina ecosystem; the pattern of microbial communities.* Akademiai Kiaro, Budapest.

116 FLANAGAN P.W. & SCARBOROUGH A.M. (1974) Physiological growth of decomposer fungi on Tundra plant remains. In: *Soil organisms and decomposition in Tundra*, eds. A.J. Holding, O.W. Heal, S.F. MacLean Jnr., P.W. Flanagan, pp. 159–181. Tundra Biome Steering Stockholm.

117 ALBERSHEIM P. & ANDERSON-PROUTY A.J. (1975) Carbohydrates, proteins, cell surfaces and the biochemistry of pathogenesis. *Ann. Rev. Plant Phys.* **26**, 31–66.

118 KIRK T.K., CONNORS W.J., BLEAM R.D., HACKETT W.F. & ZEIKUS J.G. (1975) Preparation and microbial decomposition of synthetic ^{14}C lignins. *Proc. Nat. Acad. Sci.* U.S.A. **72**, 2515–2519.

119 SELBY K. (1968) Mechanism of biodegradation of cellulose. In: *Biodeterioration of materials*, eds. A.H. Walters & J.S. Elphick, pp. 62–78. Elsevier, Amsterdam.

120 POINCELOT T.R.P. & DAY P.R. (1972) Simple dye release assay for determining cellulolytic activity of fungi. *Appl. Microbiol.* **23**, 875–879.

121 LUXTON M. (1972) Studies on the oribatid mites of a Danish beech wood Soil. I. Nutritional biology. *Pedobiologia*, **12**, 434–463.

122 HASSALL M. & JENNINGS J.B. (1975) Adaptive features of gut structure and digestive physiology in the terrestrial isopod *Philoscia muscorum* (Scop.) *Biol. Bull.* **149**, 348–364.

123 VISSER S.A. (1971) Investigations on the nitrogen flow in a model ecosystem. In: *Organismes du sol et production primaire, 4th Coll. Soil Zool.*, pp. 119–130. INRA, Paris.

124 GREENWOOD D.J. & LEES H. (1956) Studies on the decomposition of amino acids in soils. I. A preliminary survey of techniques. *Plant Soil*, **7**, 253–268.

125 POSTGATE J.R. (1971) The acetylene test for nitrogenase. In: *The chemistry and biochemistry of nitrogen fixation*, ed. J.R. Postgate, pp. 311–316. Academic Press, London and New York.

126 NORRIS J.R. & RIBBONS D.W. (1970) *Methods in microbiology.* Vol. 3B. Academic Press, London and New York.

127 BAVER L.D. (1972) *Soil Physics*, Wiley, New York.

128 MONTEITH J.L. (1973) *Principles of Environmental Physics.* Arnold, London.

129 HESSE P.R. (1971) *A textbook of soil chemical analysis*, Murray, London.

130 METSON O.J. (1956) *Methods of chemical analysis for soil survey samples.* New Zealand Soil Bureau Bulletin No. 12, DSIR, New Zealand.

131 ALLEN S.E., GRIMSHAW H.M., PARKINSON J.A. & QUARMBY C. (1974) *Chemical Analysis of Ecological Materials.* Blackwell Scientific Publications, Oxford.

132 KONONOVA M. (1961) *Soil Organic Matter.* Pergamon Press, Oxford.

133 TOWNSEND W.N. (1972) *An introduction to the scientific study of the soil.* Edward Arnold, London.

134 RUSSELL E.W. (1973) *Soil conditions and plant growth.* Longmans, London.

135 BALL D.F. (1976) Site and soils. In: *Methods on plant ecology*, ed. S.B. Chapman, pp. 297–367. Blackwell Scientific Publications, Oxford.

136 TOMLINSON P.R., BECKETT P.H.T., BANNISTER P. & MARSDEN R. (1977) Simplified procedures for routine soil analyses. *J. Appl. Ecol.* **14**, 253–260.

137 LEE R. (1969) Chemical temperature integration. *J. Appl. Meteorol.* **8**, 423–430.

138 MACFADYEN A. (1956) The use of a temperature integrator in the study of soil temperature. *Oikos*, **7**, 56–81.

139 RAFAREL C.R. & BRUNSDON G.P. (1976) Data collection systems. In: *Methods in plant ecology*, ed. S.B. Chapman, pp. 467–506. Blackwell Scientific Publications, Oxford.

140 VEZINA P.E. (1965) Methods of pH determination and seasonal pH fluctuations in Quebec forest humus. *Ecology*, **46**, 752–754.

141 GRIFFIN, D.M. (1972) *Ecology of soil fungi.* Chapman and Hall, London.

142 HOLMES J.W., TAYLOR S.A. & RICHARDS S.J. (1967) Measurement of soil water. In: *Irrigation of agricultural lands. Agricultural Monograph 11.*, eds. R.M. Hogan, H.W. Haise & T.W. Edminster, pp. 275–303. Academic Press, New York.

143 Fawcett R.G. & COLLIS-GEORGE N. (1967) A filter paper method of determining the moisture characteristics of soil. *Austr. J. Exp. Agric. and Anim. Husbandry*, 7, 162–7.

144 ARMSTRONG W. & WRIGHT E.J. (1976) A polarographic assembly for multiple sampling of soil oxygen flux in the field. *J. Appl. Ecol.* 13, 849–856.

145 HARLEY J.L. & BRIERLY J.K. (1953) A method of estimation of oxygen and carbon dioxide concentrations in the litter layer of beech woods. *J. Ecol.* 41, 385–387.

146 URQUHART C. & GORE A.J.P. (1973) The redox potential of four peat soils. *Soil Biol. Biochem.* 5, 659–672.

147 GOLTERMAN H.L. CLYMO R.S., & OHNSTAD M.A.M. (1978) *Methods for physical and chemical analysis of fresh waters* I.B.P. Handbook No. 8 (2nd Edition). Blackwell Scientific Publications, Oxford.

148 KING H.G.C. & HEATH G.W. (1967) The chemical analysis of small samples of leaf material and the relationship between the disappearance and composition of leaves. *Pedobiologia*, 7, 192–197.

149 MINDERMAN G. & BIERLING J. (1968) The determination of cellulose and sugars in forest litter. *Pedobiologia*, 8, 536–542.

150 SCHNITZER M. & KHAN S.U. (1978) *Soil organic matter.* Elsevier, Amsterdam.

151 SCHNITZER M. (1978). Humic substances: Chemistry and reactions. In: *Soil organic matter*, eds. M. Schnitzer & S.U. Khan, pp. 1–64. Elsevier, Amsterdam.

152 International Atomic Energy Agency (1968) *Isotopes and radiation in soil organic matter studies.* IAEA, Vienna.

153 International Atomic Energy Agency (1977) *Soil organic matter studies.* IAEA, Vienna.

REFERENCES

ADLER E. (1966) The chemical construction of lignin. *Sv. Kem. Tidskr.* **80**, 279–290.

AINSWORTH G.C., SPARROW F.K. & SUSSMAN A.S. (1973) *The Fungi*, Vols 4A–4B. Academic Press, London and New York.

ALBERSHEIM P. (1975) The wall of growing plant cells. *Scientific American*, **232** (4), 80–95.

ALBERSHEIM P., JONES T.M. & ENGLISH P.D. (1969) Biochemistry of the cell wall in relation to infective process. *Ann. Rev. Phytopath.* **7**, 171–194.

ALEXANDER M. (1977) *Introduction to Soil Microbiology* (2nd Edition). John Wiley, New York and London.

ALEXANDER M. (1971) *Microbial Ecology.* John Wiley, New York and London.

ALLEN S.E., GRIMSHAW H.M., PARKINSON J.A. & QUARMBY C. (1974) *Chemical Analysis of Ecological Materials.* Blackwell Scientific Publications, Oxford.

ALLISON F.E. (1968) Soil aggregation—some facts and fallacies as seen by a microbiologist. *Soil Sci.* **106**, 136–143.

ALLISON F.E. (1973) *Soil Organic Matter and its Role in Crop Production.* Elsevier Scientific Publishing Co, Amsterdam, London and New York.

ALLISON F.E. & COVER R.G. (1960) Rates of decomposition of short-leaf pine sawdust in soil at various levels of nitrogen and lime. *Soil Sci.* **89**, 194–201.

ANDERSON J.M. (1973a) The breakdown and decomposition of sweet chestnut (*Castanea sativa* Mill) and beech (*Fagus sylvatica* L). leaf litter in two deciduous woodland soils: I, Breakdown, leaching and decomposition. *Oecologia (Berl.)*, **12**, 251–274.

ANDERSON J.M. (1973b) The breakdown and decomposition of sweet chestnut (*Castanea sativa* Mill) and Beech (*Fagus sylvatica* L.) leaf litter in two deciduous woodland soils: II, Changes in carbon, hydrogen, nitrogen and polyphenol content. *Oecologia (Berl.)*, **12**, 275–288.

ANDERSON J.M. (1973c) Carbon dioxide evolution from two temperate, deciduous woodland soils. *J. appl. Ecol.* **10**, 361–378.

ANDERSON J.M. (1975) Succession, diversity and trophic relationships of some soil animals in decomposing leaf litter. *J. Anim. Ecol.* **44**, 475–495.

ANDERSON J.M. (1978a) The organisation of soil animal communities. In: *Soil organisms as components of ecosystems*, eds. U. Lohm & T. Persson, *Ecol. Bull.* **25**, pp. 15–23, Swedish Natural Science Research Council, Stockholm.

ANDERSON J.M. (1978b). Inter- and intra-habitat relationships between woodland Cryptostigmata species diversity and the diversity of soil and litter microhabitats. *Oecologia (Berl.)*, **32**, 341–348.

ANDERSON J.M. & COE M.J. (1974) Decomposition of elephant dung in an arid tropical environment. *Oecologia (Berl.)*, **14**, 111–125.

ANDERSON J.M. & HEALEY I.N. (1972) Seasonal and interspecific variation in major components of the gut contents of some woodland Collembola. *J. Anim. Ecol.* **41**, 359–368.

ARMS K., FEENY P. & LEDERHOUSE P.C. (1974) Sodium stimulus for puddling behaviour by Tiger Swallowtail Butterflies: *Papilio glaucus. Science*, **185**, 372–374.

AUSMUS B.A. (1977) Regulation of wood decomposition rates by arthropod and annelid populations. In: *Soil organisms as components of ecosystems*, eds. U. Lohm & T. Persson, *Ecol. Bull.* **25**, pp. 180–192, Swedish Natural Science Research Council, Stockholm.

AUSMUS B.S., EDWARDS N.T. & WITKAMP M. (1976) Microbial immobilisation of carbon, nitrogen, phosphorus and potassium: implications for forest ecosystem processes. In: *The role of terrestrial and aquatic organisms in decomposition processes*, eds. J.M. Anderson & A. Macfadyen, pp. 397–416. Blackwell Scientific Publications, Oxford.

BABEL U. (1975) Micromorphology of soil organic matter. In: *Soil Components Vol. 1 Organic Components*, ed. J.E. Gieseking, pp. 369–473. Springer-Verlag, New York.

BABIUK L.A. & PAUL E.A. (1970) The use of fluorescein isothiocyanate in the determination of the bacterial biomass of grassland soil. *Can. J. Microbiol.* **16**, 57–62.

BAILEY P.J., LIESE W. & ROSCH R. (1968) Some aspects of cellulose degradation in lignified cell walls. In: *Biodeterioration of materials*, eds. A.H. Walters & J.S. Elphick, pp. 546–557. Elsevier Scientific Publishing Co., Amsterdam.

BAVENDAMM W. & REICHERT H. (1938) Die Abhängigkeit des Wachstums holzzersetzenden Pilze von Wassergehalt des Nahrsubstrates. *Arch. Mikrobiol.* **9**, 486–544.

BEEK J. & FRISSELL M.J. (1975) *Simulation of nitrogen behaviour in soils.* CAPD, Wageningen.

BENOIT R.E. & STARKEY R.L. (1968) Inhibition of decomposition of cellulose and some other carbohydrates by tannin. *Soil Sci.* **105**, 291–296.

BERNAYS E. & CHAPMAN R.F. (1970) Experiments to determine the basis of food selection by *Chorthippus parallelus* (Zetterstedt) (Orthoptera Acrididae) in the field. *J. Anim. Ecol.* **39**, 761–776.

BERRIE A.D. (1976) Detritus, micro-organisms and animals in fresh water. In: *The role of terrestrial and aquatic organisms in decomposition processes*, eds. J.M. Anderson & A. Macfadyen, pp. 323–440. Blackwell Scientific Publications, Oxford.

BIRCH H.F. (1958) The effect of soil drying on humus decomposition and nitrogen availability. *Plant Soil*, **10**, 9–31.

BLACK C.A. (1968) *Soil Plant Relationships* (2nd edition). John Wiley, London.

BLACKITH R.E. & BLACKITH R.M. (1975) Zoogeographical and ecological determinants of collembolan distribution. *Proc. Roy. Irish Acad. B.* **75**, 345–68.

BLOOMFIELD B.J. & ALEXANDER M. (1967) Melanins and resistance of fungi to lysis. *J. Bact.* **93**, 1276–1280.

BOOTH R.G. & ANDERSON J.M. (1979) The influence of fungal food quality on the growth and fecundity of *Folsomia candida* (Collembola: Isotomidae). *Oecologia (Berl.)*, (in press).

BORMANN F.H., LIKENS G.E. & MELILLO J.M. (1977) Nitrogen budget for an aggrading northern hardwood forest ecosystem. *Science*, **196**, 981–983.

BORNEBUSCH C.H. (1930) The fauna of forest soil. *Forest. Fors Vaes. Danm.* **11**, 1–224.

BOUQUEL G., BRUCKERT S. & SAVIN L. (1970) Inhibition de la nitrification par les extraits aqueux de litière de hêtre (*Fagus sylvatica*). *Rev. Ecol. Biol. Sol.* **8**, 357–366.

BOYNTON D. & COMPTON O.C. (1944) Normal seasonal changes of oxygen and carbon dioxide percentages in gas from the larger pores of three orchard subsoils. *Soil Sci.* **57**, 107–117.

BRADY N.C. (1974) *The nature and properties of soils* (8th edition). Macmillan, New York.

BRASHER S. & PERKINS D.F. (1978) The grazing intensity and productivity of sheep in the grassland ecosystem. In: *Production ecology of British moors and montane grasslands*, eds. O.W. Heal & D.F. Perkins, pp. 354–374. Springer-Verlag, Berlin.

BREZNACK J.A., BRILL W.J., MERTINS J.W. & COPPEL H.C. (1973) Nitrogen fixation in termites. *Nature, Lond.* **244**, 577–580.

BRIERLEY J.K. (1955) Seasonal fluctuations in the oxygen and carbon dioxide concentrations in beech litter with reference to the salt uptake of beech mycorrhizas. *J. Ecol.* **43**, 404–408.

BROADFOOT W.M. & PIERRE W.H. (1939). Forest soil studies. I. Relation of rate of decomposition of tree leaves to their acid-base balance and other chemical properties. *Soil Sci.* **48**, 329–348.

BROCK T.D. (1966) *Principles of microbial ecology.* Prentice-Hall International, London.

BROWN A.H.F. (1974) Nutrient cycles in oakwood ecosystems in N.W. England. In: *The British Oak*, eds. M.G. Morris & F.H. Perring, pp. 141–161, Botanical Society British Isles & E.W. Classey, Faringdon.

BROWNING B.L. (1963) *The Chemistry of Wood.* John Wiley—Interscience, New York.

BROWNING B.L. (1967) *Methods of wood chemistry*, Vol. 1. John Wiley—Interscience, New York.

BUCHANAN R.E. & GIBBS N.E. (1974) *Bergey's manual of determinative bacteriology* (8th edition). The Williams and Wilkins Co., Baltimore.

BUNNELL F.L. (1973) Decomposition: Models and the real world. In: *Modern methods in the study of microbial ecology*, ed. T. Rosswall, *Ecol. Bull.* **17**, pp. 407–415, Swedish Natural Science Research Council, Stockholm.

BUNNELL F.L. & DOWDING P. (1974) ABISKO—a generalised decomposition model for comparison between Tundra sites. In: *Soil organisms and decomposition in Tundra*, eds. A.J. Holding, O.W. Heal, S.F. MacLean Jr. & P.W. Flanagan, pp. 227–248. Tundra Biome Steering Committee, Stockholm.

BUNNELL F.L., MACLEAN S.F. Jr. & BROWN J. (1975) Barrow, Alaska, U.S.A. In: *Structure and function of Tundra ecosystems*, eds. T. Rosswall & O.W. Heal, *Ecol. Bull.* **20**, pp. 73–124. Swedish Natural Science Research Council, Stockholm.

BUNNELL F.L. & SCOULLAR K.A. (1975) ABISKO II. A computer simulation model of carbon flux in tundra ecosystems. In: *Structure and function of tundra ecosystems*, eds. T. Rosswall & O.W. Heal, *Ecol. Bull.* **20**, pp. 425–448, Swedish Natural Science Research Council, Stockholm.

BUNNELL F.L. & TAIT D.E.N. (1974) Mathematical simulation models of decomposition processes. In: *Soil organisms and decomposition in Tundra*, eds. A.J. Holding, O.W. Heal, S.F. MacLean Jr. & P.W. Flanagan, pp. 207–226. Tundra Biome Steering Committee, Stockholm.

BUNNELL F.L., TAIT D.E.N. & FLANAGAN P.W. (1977b) Microbial respiration and substrate weight loss. II. A model of the influences of chemical composition. *Soil Biol. Biochem.* **9**, 41–47.

BUNNELL F.L., TAIT D.E.N., FLANAGAN P.W. & VAN CLEVE K. (1977a) Microbial respiration and substrate weight loss. I. A general model of the influences of abiotic variables. *Soil Biol. Biochem.* **9**, 33–40.

BURD J.S. & MARTIN J.C. (1924) Secular and seasonal changes in the soil solution. *Soil Sci.* **18**, 151–167.

BURGES N.A. & RAW F. (1967) *Soil Biology.* Academic Press, London and New York.

BURNETT J.H. (1968) *Fundamentals of mycology.* Edward Arnold, London.

BURNETT J.H. (1975) *Mycogenetics.* John Wiley, London.

CAMPBELL C.A., PAUL E.A., RENNIE D.A. & McCALLUM K.J. (1967) Applicability of the carbon-dating method of analysis to soil humus studies. *Soil Sci.* **104**, 217–224.

CAMPBELL N.E.R. & LEES H. (1967) The Nitrogen cycle. In: *Soil Biochemistry*, eds. A.D. McLaren & G.H. Peterson, Vol. I, pp. 194–215. Edward Arnold, London.

CAMPBELL W.G. (1929) The chemical aspect of the destruction of oakwood by powder-post and death-watch beetles, *Lyctus spp* and *Xestobium* sp. *Biochem. J.* **23**, 1290–1293.

CAMPBELL W.G. (1952) The biological decomposition of wood. In: *Wood chemistry*, (2nd edition), eds. L.E. Wise & E.C. Jahn, pp. 1061–1118. Reinhold Publishing Corporation, New York.

CAMPBELL W.G. & BRYANT S.A. (1941) Determination of pH in wood. *Nature, Lond.* **147**, 357.

CARLISLE A., BROWN A.H.F. & WHITE E.J. (1966a) Litter fall, leaf production and the effects of defoliation by *Tortrix viridiana* in a sessile oak (*Quercus petraea*) woodland. *J. Ecol.* **54**, 75–85.

CARLISLE A., BROWN A.H.F. & WHITE E.J. (1966b) The organic matter and nutrient elements in the precipitation beneath a sessile oak (*Quercus petraea*) canopy. *J. Ecol.* **54**, 87–98.

CHANG YUNG (1967) The fungi of wheat straw compost. II. Biochemical and physiological studies. *Trans. Br. mycol. Soc.* **50**, 667–677.

CHANG YUNG & HUDSON H. (1967) The fungi of wheat straw compost. I. Ecological studies. *Trans. Br. mycol. Soc.* **50**, 649–666.

CHAPMAN H.D. (1965) Chemical factors of the soil as they affect microorganisms. In: *Ecology of soil-borne plant pathogens: prelude to biological control*, eds. K.F. Baker & W.C. Snyder, pp. 120–139. John Murray, London.

CHASE F.E., CORKE C.T. & ROBINSON J.B. (1968) Nitrifying bacteria in soil. In: *The ecology of soil bacteria*, eds. T.R.G. Gray & D. Parkinson, pp. 593–611. Liverpool University Press.

CHEN A.W. & GRIFFIN D.M. (1966) Soil physical factors and the ecology of fungi. V. Further studies in relatively dry soils. *Trans. Br. mycol. Soc.* **49**, 419–426.

CHENG C. & LEWIN R.A. (1974) Fluidization as a feeding mechanism in beach flies. *Nature, Lond.* **250**, 167–8.

CHERRETT J.H. (1968) The foraging behaviour of *Atta cephalotes* L. (Hymenoptera, Formicidae). I. The foraging pattern and plant species attacked in tropical rain forest. *J. Anim. Ecol.* **37**, 387–404.

CHERRETT J.M. (1972) Some factors involved in the selection of vegetable substrate by *Atta cephalotes* L. (Hymenoptera: Formicidae). *J. Anim. Ecol.* **41**, 647–660.

CHRISTENSEN M. (1969) Soil microfungi of dry to mesic conifer-hardwood forests in Northern Wisconsin. *Ecology*, **50**, 9–27.

CHRISTIAN J.H.B. & INGRAM M. (1959) The freezing points of bacterial cells in relation to halophilism. *J. gen. Microbiol.* **20**, 27–31.

CHRISTIANSEN K. (1964) Bionomics of Collembola. *Ann. Rev. Ent.* **9**, 147–178.

CLARK F.E. (1967) Bacteria in soil. In: *Soil Biology*, eds. A. Burges & F. Raw, pp. 15–49. Academic Press, London and New York.

CLARK F.E. (1968) The growth of bacteria in soil. In: *The ecology of soil bacteria*, eds. T.R.G. Gray & D. Parkinson, pp. 441–457. Liverpool University Press.

CLARK F.E. & PAUL E.A. (1970) The microflora of grassland. *Adv. Agron.* **22**, 375–436.

COCHRANE V.W. (1958) *Physiology of fungi.* John Wiley, London and New York.

COLE M. & WOOD R.K.S. (1961) Pectic enzymes and phenolic substances in apples rotted by fungi. *Ann. Bot. N.S.* **25**, 435–452.

COLEMAN G.S. (1975) The role of bacteria in the metabolism of rumen into diniomorphid protozoa. In: *Symbiosis*, eds. D.H. Jennings & D.L. Lee, *Symp. Soc. Exp. Biol.* **29**, pp. 533–558. Cambridge University Press, Cambridge.

COOK S.F. (1932) The respiratory gas exchange in *Termopsis nevadensis. Biol. Bull. mar. biol. Lab., Woods Hole*, **63**, 246–257.

COOKE R. (1977) *The biology of symbiotic fungi.* John Wiley, London and New York.

COULSON C.B., DAVIES R.I. & LEWIS D.A. (1960) Polyphenols in plant, humus and soil. I. Polyphenols of leaves, litter and superficial humus from mull and mor sites. *J. Soil Sci.* **11**, 20–29.

COWLING E.B. (1961) Comparative biochemistry of the decay of sweetgum sapwood by white-rot and brown-rot fungi. *Tech. Bull.* **1258**, USDA, Washington.

CRITCHLEY B.R., COOK A.G., PERFECT T.J., RUSSELL-SMITH A. & YEADON R. (1979) Response of soil invertebrates to bush clearing and soil cultivation in the humid tropics. *J. appl. Ecol.* (in press).

CROMACK K. (1973) *Litter production and decomposition in a mixed hardwood watershed and a white pine watershed at Coweeta hydrologic station, North Carolina.* PhD Thesis, University of Georgia, Athens.

CROMACK K., SOLLINS P., TODD R.L., FOGEL R., TODD A.W., FENDER W.M. & CROSSLEY D.A. (1977) The role of oxalic acid and bicarbonate in calcium cycling by fungi and bacteria: some possible implications for soil animals. In: *Soil organisms as components of ecosystems*, eds. U. Lohm & T. Persson, *Ecol. Bull.* **25**, pp. 246–252. Swedish Natural Science Research Council, Stockholm.

CROMACK K., TODD R.L. & MONK C.D. (1975) Patterns of basidiomycete nutrient accumulation in conifer and deciduous forest litter. *Soil. Biol. Biochem.* **7**, 275–268.

CROSSLEY D.A. & WITKAMP M. (1964) Forest soil mites and mineral cycling. *Acarologia Fasc. h.s.* 137–145.

DAGLEY S. (1967) The microbial metabolism of phenolics. In: *Soil Biochemistry*, eds. A.D. McLaren & G.D. Peterson, pp. 287–317. Edward Arnold, London.

DASH M.C. & CRAGG J.B. (1972) Selection of microfungi by Enchytraeidae (Oligochaeta) and other members of the soil fauna. *Pedobiologia*, **12**, 282–86.

DAUBENMIRE R. & PRUSSO D.C. (1963) Studies of the decomposition rates of tree litter. *Ecology*, **44**, 589–592.

DICKINSON C.H. & PUGH G.J.F. (1974) (eds.) *Biology of plant litter decomposition*. (2 vols). Academic Press, London and New York.

DICKINSON S. (1960) The mechanical ability to breach the host barriers. In: *Plant Pathology*, Vol. 2, eds. J.G. Horsfall & A.E. Dimond, pp. 203–232. Academic Press, London and New York.

DICKSON A. & CROCKER R.L. (1953a). A chronosequence of soils and vegetation near Mt. Shasta, California. I. Definition of the ecosystem investigated and features of the plant succession. *J. Soil Sci.* **4**, 123–141.

DICKSON B.A. & CROCKER R.L. (1953b) A chronosequence of soils and vegetation near Mt. Shasta, California. II. The development of the forest floors and the carbon and nitrogen profiles of the soils. *J. Soil. Sci.* **4**, 142–154.

DIGHTON J. (1977) *In vitro* experiments simulating the possible fates of aphid honeydew sugars in soil. *Soil Biol. Biochem.* **10**, 53–57.

DOETSCH R.N. & COOKE T.M. (1973) *Introduction to bacteria and their ecobiology*. Medical and Technical Publishing Co. Ltd., Lancaster.

DOUGLAS L.A. & TEDROW J.C.F. (1959) Organic matter decomposition rates in Arctic soils. *Soil Sci.* **88**, 305–312.

DOWDING P. (1976) Allocation of resources; nutrient uptake and utilisation by decomposer organisms. In: *The role of terrestrial and aquatic organisms in decomposition processes*, eds. J.M. Anderson & A. MacFadyen, pp. 169–183. Blackwell Scientific Publications, Oxford.

DOWDING V.M. (1967). The function and ecological significance of the pharyngeal ridges occurring in some cyclorhaphous Diptera. *Parasitology*, **57**, 371–88.

DRIEL, W. VAN (1961). Studies on the conversions of amino acids in soil. *Acta Bot. Neerl.* **10**, 209–247.

DRIFT J. VAN DER (1951) Analysis of the animal community in a beech forest floor. *Tijdschr. Ent.* **94**, 1–168.

DRIFT J. VAN DER & WITKAMP M. (1959) The significance of the breakdown of oak litter by *Enoicyla pusilla* Burm. *Arch. neerl. Zool.* **13**, 486–492.

DUNGER S. (1958) *Tiere im Boden*. Ziemsen Verlag, Wittenberg Lutherstadt.

DUVIGNEAUD P & DENAEYER DE SMET S. (1970) Biological cycling of minerals in temperate deciduous forests. In: *Analysis of temperate forests ecosystems*, ed. D. Reichle, pp. 199–225. Ecological studies I. Chapman and Hall, London.

EDWARDS C.A. (1970) *Persistent pesticides in the environment*. Chemical Rubber Co., Cleveland.

EDWARDS C.S. & HEATH G. (1963) The role of soil animals in breakdown of leaf material. In: *Soil organisms*, eds. J. Doeksen & J. van der Drift, pp. 76–84. North Holland, Amsterdam.

EDWARDS C.A. & LOFTY J.R. (1977) *Biology of earthworms* (2nd edition). Chapman & Hall, London.

ENDE G. VAN DEN. & LINSKENS H.F. (1974) Cutinolytic enzymes in relation to pathogenesis. *Ann. Rev. Phytopath.* **12**, 247–258.

ENEBO L. (1963) Symbiosis in thermophilic cellulose fermentation. *Nature, Lond.* **163**, 805.

ENGLISH M.P. (1965) The saprophytic growth of non-keratinophilic fungi on keratinised substrates and a comparison with keratinophilic fungi. *Trans. Br. mycol. Soc.* **48**, 219–235.

EYRE S.R. (1968) *Vegetation and soils: a world picture* (2nd edition). Edward Arnold, London.

EYRE S.R. (1971) (ed.) *World vegetation types*. Macmillan, London.

FELBECK G.T. (1971) Structural hypotheses of soil humic acids. *Soil Sci.* **111**, 42–48.

FERGUS C.L. (1969) The cellulolytic activity of thermophilic fungi and actinomycetes. *Mycologia*, **61**, 120–129.

FIENNES R.N. (1972) (ed). *The Biology of Nutrition*. I.E.F.N. 18, Pergamon Press, Oxford.

FINCH P., HAYES M.H.B. & STACEY M. (1971) The biochemistry of soil polysaccharides. In: *Soil biochemistry*, eds. A.D. McLaren & J. Skujins, Vol. 2, pp. 257–319. Marcel Dekker, New York.

FINDLAY W.P.K. (1934) Studies in the physiology of wood-decay fungi. I. The effect of nitrogen content upon the rate of decay. *Ann. Bot.* **46**, 109–117.

FITZPATRICK E.A. (1971) *Pedology. A systematic approach to soil science*. Oliver & Boyd, Edinburgh.

FLAIG W. (1966) The chemistry of humic substances. In: *The use of isotopes in soil organic matter studies*, ed. R.A. Silow, pp. 103–127. Pergamon Press, Oxford.

FLANAGAN P.W. & VEUM A.K. (1974) Relationships between respiration, weight loss, temperature and moisture in organic residues in tundra. In: *Soil organisms and decomposition in tundra*, eds. A.J. Holding, O.W. Heal, S.F. MacLean, Jr. & P.W. Flanagan, pp. 249–278. Tundra Biome Steering Committee, Stockholm.

FLOATE M.J.S. (1970a) Decomposition of organic materials from hill soils and pastures. II. Comparative studies of the mineralisation of carbon, nitrogen, and phosphorus from plant materials and sheep faeces. *Soil Biol. Biochem.* **2**, 173–185.

FLOATE M.J.S. (1970b) Decomposition of organic material from hill soils and pastures. IV. The effects of moisture content on the mineralisation of carbon, nitrogen and phosphorus from plant materials and faeces. *Soil Biol. Biochem.* **2**, 275–283.

FLOATE M.J.S. (1970c) Mineralisation of nitrogen and phosphorus from organic materials of plant and animal origin and its significance in the nutrient cycle of grazed upland and hill soils. *J. Br. Grassland Soc.* **25**, 295–302.

FLOATE M.J.S. (1971) Plant nutrient cycling in hill land. *Hill Farming Research Organization*. Report No. 5, 14–34.

FORD G.W., GREENLAND D.J. & OADES J.M. (1969) Separation of the light fraction from soils by ultrasonic dispersion in halogenated hydrocarbons containing a surfactant. *J. Soil Sci.* **20**, 291–300.

FORTESCUE J.A.C. & MARTEN G.G. (1967) Micronutrients: forest ecology and systems analysis. In: *Analysis of temperate forest ecosystems*, ed. D.E. Reichle, pp. 173–198. Ecological Studies I, Chapman & Hall, London.

FRANKLAND J.C. (1966) Succession of fungi on decaying petioles of *Pteridium aquilinum*. *J. Ecol.* **54**, 41–63.

FRANKLAND J.C., LINDLEY D.K. & SWIFT M.J. (1978) An analysis of two methods for the estimation of mycelial biomass in leaf litter. *Soil Biol. Biochem.* **10**, 323–333.

FRANKLAND J.C., OVINGTON J.D. & MACRAE C. (1963) Spatial and seasonal variations in soil, litter and ground vegetation in some Lake District woodlands. *J. Ecol.* **51**, 97–112.

FRENCH J.R.J. (1975) The role of termite hindgut bacteria in wood decomposition. *Materie u. Organismen*, **10**, 1–13.

FULLER W.H. & NORMAN A.G. (1943) Cellulose decomposition by aerobic mesophilic bacteria from soil. III. The effect of lignin. *J. Bact.* **46**, 291–297.

GARRETT S.D. (1951) Ecological groups of soil fungi; a survey of substrate relationships. *New Phytol.* **50**, 149–166.

GARRETT S.D. (1956) *Biology of the root-infecting fungi*. Cambridge University Press.

GARRETT S.D. (1963) *Soil fungi and soil fertility*. Pergamon Press, Oxford.

GARRETT S.D. (1970) *Pathogenic root-infecting fungi*. Cambridge University Press.

GATES D.M. (1962) *Energy exchange in the biosphere*. Harper-Row, London.

GATES G.E. (1966) Requiem for Megadrile Utopias. A contribution towards the understanding of the earthworm fauna of North America. *Proc. Biol. Soc. Wash.* **79**, 239–54.

GHILAROV M.S. (1977) Why so many species and so many individuals can coexist in the soil. In: *Soil organisms as components of ecosystems*, eds. U. Lohm & T. Persson, *Ecol. Bull.* **25**, pp. 593–597. Swedish Natural Science Research Council, Stockholm.

GIESE A.C. (1962) *Cell physiology* (2nd edition). W.B. Saunders, Philadelphia and London.

GOKSKOYR J., EIDSA G., ERIKSEN J. & OSMUNDSVAG K. (1975) A comparison of cellulases from different microorganisms. In: *Enzymatic hydrolysis of cellulose*, eds. M. Bailey, T.M. Enari & M. Linko, pp. 217–230. SITRA, Aulanko, Finland.

GOODFELLOW M. & CROSS T. (1974) Actinomycetes. In: *Biology of plant litter decomposition*, eds. C.H. Dickinson & G.J.F. Pugh, Vol. 2, pp. 269–302. Academic Press, London and New York.

GOSZ J.R., LIKENS G.E. & BORMANN F.H. (1973) Nutrient release from decomposing leaf and branch litter in the Hubbard Brook Forest, New Hampshire. *Ecol. Monogr.* **47**, 173–191.

GRADUSOV B.P. (1958) Effect of forest litters on chemical properties of soils in the subzone of the southern Taiga subzone. *Soviet Soil Science*, **8**, 914–919.

GRAY T.R.G. & WILLIAMS S.T. (1971) Microbial productivity in soil. *Symp. Soc. Gen. Microbiol.* **21**, 255–286.

GREENWOOD D.J. (1968) Measurement of microbial metabolism in soil. In: *Ecology of soil bacteria*, eds. T.R.G. Gray & D. Parkinson, pp. 138–151. Liverpool University Press.

GREENWOOD D.J. & LEES H. (1956) Studies on the decomposition of amino acids in soils. I. A preliminary survey of techniques. *Plant Soil*, **7**, 253–268.

GREENWOOD D.J. & LEES H. (1960) Studies in the decomposition of amino acids in soils. 3. Aerobic metabolism. *Plant Soil*, **12**, 175–194.

GRIFFIN D.M. (1963) Soil moisture and the ecology of soil fungi. *Biol. Rev.* **38**, 141–166.

GRIFFIN D.M. (1972) *Ecology of soil fungi*. Chapman and Hall, London.

HAIDER K., MARTIN J.P. & FILIP Z. (1975) Humus biochemistry. In: *Soil biochemistry*, eds. E.A. Paul & A.D. McLaren. Vol. 4, pp. 195–244. Marcel Dekker, New York.

HANDLEY W.R.C. (1954) Mull and Mor formation in relation to forest soils. *Bull. For. Comm.* **23**, 1–115.

HANLON D. (1978) Soil animal/microbial interactions during litter decomposition. *Unpublished Ph.D. Thesis*, University of Exeter.

HANLON R.D.G. & ANDERSON J.M. (1979a) The influence of macroarthropods feeding activities on fungi and bacteria in decomposing oak leaves. *Soil Biol. Biochem.* (in press).

HANLON R.D.G. & ANDERSON J.M. (1979b) The effects of collembola grazing on microbial activity in decomposing leaf litter. *Oecologia (Berl.)*, **38**, 93–99.

HARBORNE J.B. (1977) *Introduction to ecological biochemistry*. Academic Press, London and New York.

HARDING D.J.L. & STUTTARD R.A. (1974) Microarthropods. In: *Biology of plant litter decomposition*, eds. C.H. Dickinson & G.J.F. Pugh, Vol. 2, pp. 489–532. Academic Press, London and New York.

HARGRAVE B.T. (1976) The central role of invertebrate faeces in sediment decomposition. In: *The role of terrestrial and aquatic organisms in decomposition processes*, eds. J.M. Anderson & A. Macfadyen, pp. 301–322. Blackwell Scientific Publications, Oxford.

HARLEY J.L. (1969) *The biology of mycorrhiza* (2nd edition) Leonard Hill, London.

HARLEY J.L. (1972) Fungi in ecosystems. *J. appl. Ecol.* **8**, 627–642.

HARLEY-MASON J. (1965) Melanins In: *Comprehensive biochemistry*, eds. M. Florkin & E.H. Stotz, Vol. 24, pp. 254–257. Elsevier Scientific Publishing Co., Amsterdam, London and New York.

HARRIS W.V. (1971) *Termites, their recognition and control.* Tropical Agriculture Series, Longman Group, London.

HARRISON A.F. (1971) The inhibitory effect of oak leaf litter tannins on the growth of fungi in relation to litter decomposition. *Soil. Biol. Biochem.* **3**, 167–172.

HASSALL M. & JENNINGS J.B. (1975) Adaptive features of gut structure and digestive physiology in the terrestrial isopod *Philoscia muscorum* (Scop.) 1963 *Biol. Bull.* **149**, 348–364.

HATCH A.B. (1937) The physical basis of mycotrophy in the genus *Pinus. Black Rock For. Bull.* **6**, 1–168.

HAWKER L.E. (1950) *Physiology of fungi.* University of London Press.

HAWORTH R.D. (1971) The chemical nature of humic acid. *Soil Sci.* **111**, 71–79.

HEAL O.W., FLANAGAN P.W., FRENCH D.D. & MACLEAN S.F. (in press) Decomposition and accumulation of organic matter in Tundra. In: *Analysis of Ecosystems: Tundra Biome*, L.C. Bliss, O.W. Heal & J.J. Moore, Cambridge University Press.

HEAL O.W. & FRENCH D.D. (1974) Decomposition of organic matter in tundra. In: *Soil organisms and decomposition in tundra*, eds. A.J. Holding, O.W. Heal, S.F. MacLean Jr. & P.W. Flanagan, pp. 279–310. Tundra Biome Steering Committee, Stockholm.

HEAL O.W., LATTER P.M. & HOWSON J. (1978) A study of the rates of decomposition of organic matter. In: *Production ecology of British moors and montane grassland*, eds. O.W. Heal & D.F. Perkins, pp. 136–159. Springer-Verlag, Berlin.

HEAL O.W. & MACLEAN S.F. Jr. (1975) Comparative productivity in ecosystems— secondary productivity. In: *Unifying concepts in ecology*, eds. W.H. van Dobben & R.H. Lowe-McConnell, pp. 89–108. Dr. W. Junk bv., The Hague and Pudoc, Wageningen.

HEAL O.W. & PERKINS D.F. (1978) *Production ecology of British moors and montane grasslands.* Springer-Verlag, Berlin.

HEALEY I.N. & RUSSELL-SMITH A. (1971) Abundance and feeding preferences of fly larvae in two woodland soils. In: *Organismes du sol et production primaire, Proc. 4th Coll. Soil Zool.*, pp. 177–192. INRA, Paris.

HEALEY I.N. & SWIFT M.J. (1971) Aspects of the accumulation and decomposition of wood in the litter of a coppiced Beech-Oak woodland. In: *Organismes du sol et production primaire, Proc. 4th Coll. Soil Zool.*, pp. 417–430. INRA, Paris.

HEATH G.W. & ARNOLD M.K. (1966) Studies in leaf-litter breakdown. II. Breakdown rate of 'sun' and 'shade' leaves. *Pedobiologia*, **6**, 238–243.

HEATH G.W. & KING H.G.C. (1964) Litter breakdown in deciduous forest soils. *4th Int. Congr. Soil Sci.* **3**, 979–987.

HEINEN W. & DE VRIES H. (1966) Stages during the breakdown of plant cutin by soil micro-organisms. *Arch. Mikrobiol.* **54**, 331–338.

HELLINGS C.S., KEARNEY P.C. & ALEXANDER M. (1971) Behaviour of pesticides in soil. *Adv. Agronomy*, **23**, 147–240.

HENDERSON M.E.K. (1955) Release of aromatic compounds from birch and spruce sawdusts during decomposition by white-rot fungi. *Nature, Lond.* **175**, 634–635.

HENDERSON M.E.K. (1960) Studies on the physiology of lignin decomposition by soil fungi. In: *The ecology of soil fungi*, eds. D. Parkinson & J.S. Waid, pp. 286–296. Liverpool University Press.

HENDERSON M.E.K. (1968) Fungal metabolism of certain aromatic compounds related to lignin. *Pure appl. Chem.* **7**, 589–602.

HENZELL E.F. (1973) The nitrogen cycle of pasture ecosystems. In: *Chemistry and biochemistry of herbage*, eds. G.W. Butler & R.W. Bailey. Vol. 2, pp. 228–247. Academic Press, London and New York.

HERING T.F. (1965) Succession of fungi in the litter of a Lake District Oakwood. *Trans. Br. mycol. Soc.* **48**, 391–408.

HERRERA R., MERIDA T., STARK N. & JORDAN C.F. (1978) Direct phosphorus transfer from leaf litter to roots. *Naturwissenschaft.* **65**, 208–209.

HINTIKKA V. & KORHONEN K. (1970) Effects of carbon dioxide on the growth of lignicolous and soil inhabiting Basidiomycetes. *Comm. Inst. For. Fenn.* **69**, 1–29.

HOBBS J.A. & BROWN P.L. (1957) Nitrogen changes in cultivated dry land soils. *Agronom. J.* **49**, 257–260.

HOBSON P.N. (1976) *The microflora of the rumen.* Meadowfield Press, Shildon.

HOWARD P.J.A. (1969) The classification of humus types in relation to soil ecosystems. In: *The soil ecosystem*, ed. J.G. Sheals, pp. 41–54. The Systematics Association, London.

HOWARD P.J.A. (1971) Relationships between activity of organisms and temperature and the computation of the annual respiration of micro-organisms decomposing leaf litter. In: *Organisms du sol et production primaire*, Proc. 4th Coll. Soil. Zool., pp. 137–144. INRA, Paris.

HOWARD P.J.A. & HOWARD D.M. (1974) Microbial decomposition of tree and shrub leaf litter. I. Weight loss and chemical composition of decomposing litter. *Oikos*, **25**, 341–352.

HUDSON H.J. (1968) The ecology of fungi on plant remains above the soil. *New Phytol.* **67**, 837–874.

HUGHES R.D. & WALKER J. (1970) The role of food in the population dynamics of the Australian Bush Fly. In: *Animal populations in relation to their food resources*, ed. A. Watson, pp. 255–270. Blackwell Scientific Publications, Oxford.

HULME M.A. & STRANKS D.W. (1971) Regulation of cellulase production by *Myrothecium verrucaria* grown on non-cellulosic substrates. *J. gen. Microbiol.* **69**, 145–155.

HUNGATE R.E. (1966) *The rumen and its microbes.* Academic Press, London and New York.

HUNSLEY D. & BURNETT J.H. (1970) The ultrastructural architecture of the walls of some hyphal fungi. *J. gen. Microbiol.* **62**, 203–218.

HUNT H.W. (1978) Decomposition submodel. In: *Grassland simulation model*, ed. G.S. Innis, pp. 257–303. Ecological Studies 26. Springer-Verlag, New York.

HURST H.M. & BURGES N.A. (1967) Lignin and Humic Acids. In: *Soil Biochemistry*, eds. A.D. MacLaren & G.H. Peterson, pp. 260–286. Edward Arnold, London.

HURST H.M., BURGES A. & LATTER P. (1962) Some aspects of the biochemistry of humic acid decomposition by fungi. *Phytochemistry*, **1**, 227–231.

HUTCHINSON G.E. (1957) *A treatise on limnology* Vol. I. John Wiley, New York.

IMMERGUT E.H. (1963) Cellulose. In: *The Chemistry of wood*, ed. B.L. Browning, pp. 103–190. John Wiley-Interscience, New York.

INGRAHAM J.L. (1962) Temperature relationships. In: *The Bacteria Vol. IV. The physiology of growth*, eds. I.C. Gunsalus & R.Y. Stanier, pp. 265–298. Academic Press, London and New York.

INNIS G.S. (1978) *Grassland simulation model.* Ecological Studies 26, Springer-Verlag, New York.

IRITANI W.M. & ARNOLD C.Y. (1960) Nitrogen release of vegetable crop residues during incubation as related to their chemical composition. *Soil Sci.* **89**, 74–82.

ISHIKAWA H., SCHUBERT W.J. & NORD F.F. (1963) Investigations on lignin and lignification. XVIII. The degradation by *Polyporus versicolor* and *Fomes fomentarius* of aromatic compounds structurally related to soft-wood lignin. *Arch. Biochem. Biophys.* **100**, 140–149.

JEFFREYS E.G., BRIAN P.W., HEMMING H.G. & LOWE D. (1953) Antibiotic production by the microfungi of acid heath soils. *J. gen. Microbiol.* **9**, 314–341.

JENKINSON D.S. (1968) Studies on the decomposition of plant material in soil. III. The distribution of labelled and unlabelled carbon in soil incubated with ^{14}C-labelled ryegrass. *J. Soil Sci.* **19**, 25–39.

JENKINSON D.S. (1977) Studies on the decomposition of plant material in soil. V. The effects of plant cover and soil type on the loss of carbon from ^{14}C-labelled ryegrass decomposing under field conditions. *J. Soil Sci.* **28**, 424–434.

JENKINSON D.S. & POWLSON D.S. (1976) The effects of biocidal treatments on metabolism in soil. V. A method for measuring soil biomass. *Soil. Biol. Biochem.* **8**, 209–213.

JENNY H. (1961) Derivation of state factor equations of soils and ecosystems. *Soil. Sci. Soc. Amer. Proc.* **25**, 385–388.

JENNY H., GESSEL S.P. & BINGHAM F.T. (1949) Comparative study of decomposition rates of organic matter in temperate and tropical regions. *Soil Sci.* **68**, 419–432.

JENNY H., SALEM A.E. & WALLIS J.R. (1968) Interplay of soil organic matter and soil fertility with state factors and soil properties. In: *Organic matter and soil fertility.* Pontificae Academiae Scientiarum Scripta Varia 32, pp. 6–33. North Holland, Amsterdam and John Wiley-Interscience, New York.

JEUNIAUX C. (1971) Chitinous structures. In: *Comprehensive biochemistry. Vol. 26C. Extracellular and supporting structures*, eds. M. Florkin & E.H. Stotz, pp. 597–625. Elsevier, Amsterdam, London and New York.

JOHNSON P.L. & SWANK W.T. (1973) Studies of cation budgets in the Southern Appalachians on four experimental watersheds with contrasting vegetation. *Ecology*, **54**, 70–80.

KAPUSTA L.A. & RICE E.L. (1976) Acetylene reduction (N_2-fixation) in soil and old field succession in Central Oklahoma. *Soil Biol. Biochem.* **8**, 497–503.

KEEGSTRA K., TALMADGE K.W., BAUER W.D. & ALBERSHEIM P. (1973) The structure of plant cell walls. III. A model of the walls of suspension-cultured sycamore cells based on the inter-connections of the macromolecular components. *Plant Physiol.* **51**, 188–196.

KHAN S.U. & SOWDEN F.J. (1971) Distribution of nitrogen in the Black Solonetzic and Black Chernozemic soils of Alberta. *Can. J. Soil Sci.* **51**, 185–193.

KHAN S.U. & SOWDEN F.J. (1972) Distribution of nitrogen in fulvic acid fraction extracted from the Black Solonetzic and Black Chernozemic soils of Alberta. *Can. J. Soil Sci.* **52**, 116–118.

KING H.G.C. & HEATH G.W. (1967) The chemical analyses of small samples of leaf material and the relationship between the disappearance and composition of leaves. *Pedobiologia*, 7, 192–197.

KING N.J. (1966) The extra-cellular enzymes of *Coniophora cerebella*. *Biochem. J.* **100**, 784–792.

KING N.J. & FULLER D.B. (1968) The xylanase system of *Coniophora cerebella*. *Biochem. J.* **108**, 571–576.

KIRK T.K. (1971) The effects of micro-organisms on lignin. *Ann. Rev. Phytopath.* **9**, 185–210.

KIRK T.K. (1973) Chemistry and Biochemistry of decay. In: *Wood deterioration and its prevention by preservative treatments. Vol. 1. Degradation and protection of wood*, ed. D.D. Nicholas, pp. 149–181, Syracuse University Press, New York.

KIRK T.K., CONNORS W.J., BLEAM R.D., HACKETT W.F. & ZEIKUS J.G. (1975) Preparation and microbial decomposition of synthetic 14C lignins. *Proc. Nat. Acad. Sci. (USA)*, **72**, 2515–2519.

KIRK T.K., HARKIN J.M. & COWLING E.B. (1968) Degradation of the lignin model compound syringylglycol-β-guaiacyl ether by *Polyporus versicolor* and *Stereum frustulatum. Biochem. Biophys. Acta.* **165**, 145–163.

KONONOVA M.M. (1966) *Soil organic matter* (2nd edition), Pergamon Press, Oxford.

KONONOVA M.M. (1968) Humus of the main soil types and soil fertility. In: *Organic matter and soil fertility*. Pontificae Academiae Scientiatrum Scripta Vara **32**, pp. 362–379. North-Holland, Amsterdam and John Wiley-Interscience, New York.

KOOPMANS J.J.C. (1970) Cellulases in molluscs. I. The nature of the cellulases in *Helix pomatia* and *Cardium edule. Neth. J. Zool.* **20**, 445–463.

KOUYEAS V. (1964) An approach to the study of moisture relations of soil fungi. *Plant Soil*, **20**, 351–364.

LANGE N.A. (1956) *Handbook of Chemistry* (9th edition). McGraw-Hill, New York.

LANIGAN G.W. (1963) Silage bacteriology. I. Water activity and temperature relationships of silage strains of *Lactobacillus plantarum*, *Lactobacillus brevis* and *Pediococcus cerevisiae. Austr. J. Biol. Sci.* **16**, 606–615.

LAVERACK M.S. (1963) *The physiology of earthworms*. Pergamon Press, Oxford.

LEE K.E. & WOOD T.G. (1971) *Termites and Soils*. Academic Press, London and New York.

LEVENS D.A. (1973) The role of trichomes in plant defence. *Quarterly Rev. Biol.* **48**, 3–15.

LEVI M.P. & COWLING E.B. (1969) Role of nitrogen in wood deterioration. VII. Physiological adaptation of wood-destroying and other fungi to substrates deficient in nitrogen. *Phytopathology*, **59**, 460–468.

LEWIS D.H. (1973) Concepts in fungal nutrition and the origin of biotrophy. *Biol. Rev.* **48**, 261–278.

LEWIS D.H. (1974) Micro-organisms and plants: the evolution of parasitism and mutualism. In: *Evolution in the microbial world*, eds. M.J. Carlile & J.J. Skehel, *Symp. Soc. Gen. Microbiol.* **24**, pp. 367–392. Cambridge University Press.

LEWIS J.G.E. (1965) The food and reproductive cycles of the centipedes *Lithobius variegatus* and *L. forticatus* in a Yorkshire woodland. *Proc. zool. Soc. Lond.* **144**, 269–283.

LIETH H. (1976) The use of correlation models to predict primary productivity from precipitation or evapotranspiration. In: *Water and plant life: Problems and modern approaches*, eds. O.L. Lange, L. Kappen & E.D. Schulze, pp. 392–407. Chapman & Hall, London and Springer-Verlag, Berlin, Heidelberg and New York.

LIKENS G.E. & BORMANN F.H. (1975) An experimental approach to New England landscapes. In: *Coupling of land and water systems*, ed. A.D. Hasler, pp. 7–30. Chapman & Hall, London and Springer-Verlag, Berlin, Heidelberg and New York.

LIKENS G.E., BORMANN F.H., JOHNSON N.M. & PIERCE R.S. (1967) The calcium, magnesium, potassium and sodium budgets for a small forested ecosystem. *Ecology*, **48**, 772–785.

LIKENS G.E., BORMANN F.H., JOHNSON N.M., FISHER D.W. & PIERCE R.S. (1970) Effects of forest cutting and herbicide treatment on nutrient budgets in the Hubbard Brook watershed ecosystem. *Ecol. Monogr.* **40**, 23–47.

LIKENS G.E., BORMANN F.G., PIERCE R.S. & FISHER D.W. (1971). Nutrient-hydrologic cycle interaction in small forested watershed-ecosystems. In: *Productivity of forest ecosystems*, ed. P. Duvigneaud, pp. 553–563. UNESCO, Paris.

LINDEBERG G. (1948) On the occurrence of polyphenol oxidases in soil-inhabiting Basidiomycetes. *Physiol. Plantarum*, **1**, 196–205.

LINDEN M.J.H.A. VAN DER (1971) Availability of protein in leaf litter—An enzymological approach. In: *Organismes du sol et production primaire. Proc. 4th Coll. Soil Zool.*, pp. 337–348. INRA, Paris.

LINDERSTRÖM-LANG K. & DUSPIVA F. (1935) Beitrage zur enzymatischen Histochemie. XVI. Die Verdauung von Keratin durch die Laruen der Kleidermolte. *Hoppe-Seyl. Z.* **237**, 131–158.

LOPEZ-REAL J.M. & SWIFT M.J. (1975) The formation of pseudosclerotia ('zone-lines') in wood decayed by *Armillaria mellea* and *Stereum hirsutum*. II. Formation in relation to the moisture content of wood. *Trans. Br. mycol. Soc.* **64**, 473–481.

LOPEZ-REAL J.M. & SWIFT M.J. (1977) Formation of pseudosclerotia ('zone-lines') in wood decayed by *Armillaria mellea* and *Stereum hirsutum*. III. Formation in relation to the composition of the gaseous atmosphere in wood. *Trans. Br. mycol. Soc.* **68**, 321–325.

LURIA S.E. (1960) The bacterial protoplasm: composition and organisation. In: *The Bacteria: A treatise on structure and function*, ed. I.C. Gunsalus & R.Y. Stanier, Vol. 1, pp. 1–34. Academic Press, New York and London.

LUXTON M. (1972) Studies on the oribatid mites of a Danish beech wood soil. I. Nutritional biology. *Pedobiologia*, **12**, 434–463.

MACARTHUR R. (1968) The theory of the niche. In: *Population biology and evolution*, ed. R.C. Lewontin, pp. 159–176. Syracuse University Press, Syracuse.

MANDELS M. & REESE E.T. (1960) Induction of cellulase in fungi by cellobiose. *J. Bact.* **79**, 816–826.

MANDELS M. & REESE E.T. (1965) Inhibition of cellulases. *Ann. Rev. Phytopath.* **3**, 85–102.

MARKS T.C. & TAYLOR K. (1972) The mineral nutrient status of *Rubus chamaemorus* L. in relation to burning and sheep grazing. *J. appl. Ecol.* **9**, 501–511.

MARSHALL K.C. (1971) Sorptive interactions between soil particles and micro-organisms. In: *Soil biochemistry*, eds. A.D. McLaren & J. Skujins, Vol. 2 pp. 409–445. Marcel Dekker, New York.

MARTIN J.T. (1964) Role of the cuticle in the defense against plant disease. *Ann. Rev. Phytopath.* **2**, 81–100.

MARTIN J.T. & JUNIPER B.E. (1970) *The cuticle of plants*. Edward Arnold, London.

MASON C.F. (1970) Snail populations, beech litter production and the role of snails in litter decomposition. *Oecologia (Berl.)*, **5**, 215–239.

MASON W.H. & ODUM E.P. (1969) The effect of coprophagy in retention and bio-elimination of radionuclides of detritus-feeding animals. In: *Proc. Second National Symposium on Radioecology*, eds. D.J. Nelson & F.C. Evans, pp. 721–724. Clearinghouse Fed. Sci. Tech. Inf., U.S. Dept. Commerce, Springfield.

MATHISON G.E. (1964) The microbiological decomposition of keratin. *Ann. Soc. belge. Med. trop.* **44**, 767–792.

MATHUR S.P. & PAUL E.A. (1967) Microbial utilization of soil humic acids. *Can. J. Microbiol.* **13**, 573–580.

MAYAUDON J. & SIMONART P. (1959a) Etude de la décomposition de la matière organique dans le sol au moyen de carbone radioactif. III. Décomposition des substances solubles dialysables des proteines et des humicelluloses. *Plant Soil*, **11**, 170–175.

MAYAUDON J. & SIMONART P. (1959b) Etude de la décomposition de la matière organique dans le sol au moyen de carbone radioactif. V. Décomposition de cellulose et de lignine. *Plant Soil*, **11**, 181–192.

MAYAUDON J. & SIMONART P. (1963) Humification des microorganismes marqués par 14C dans le sol. *Ann. Inst. Pasteur*, **105**, 257–266.

MAYAUDON J. & SIMONART P. (1965) Humification dans le sol d'un complexe poly-saccharidique 14C d'origine microbienne. *Med. Landbonw. Opzoek.* (*Gent*). **30**, 941–955.

MAYBERRY W.R., PROCHAZKA G.J. & PAYNE W.J. (1967) Growth yields of bacteria on selected organic compounds. *Appl. Microbiol.* **15**, 1342–1338.

MCBRAYER J.F. (1973) Exploitation of deciduous litter by *Apheloria montana* (Diplopoda: Eurydesmidae). *Pedobiologia*, **13**, 90–98.

MCBRAYER J.F., REICHLE D.E. & WITKAMP M. (1974) *Energy flow and nutrient cycling in a cryptozoan food-web.* Oak Ridge National Laboratory, Oak Ridge.

MCLAREN A.D. (1960) Enzyme activity in structurally restricted systems. *Enzymologia*, **21**, 356–364.

MCLAREN A.D. (1962) Use of mole fractions in enzyme kinetics. *Arch. Biochem. Biophys.* **97**, 1–16.

MCLAREN A.D. & PETERSON G.H. (1967) Introduction to the biochemistry of terrestrial soils. In: *Soil biochemistry*, ed. A.D. McLaren and G.H. Peterson, pp. 1–18. Edward Arnold, London.

MCLAREN A.D. & SKUJINS J.J. (1963) Nitrification by *Nitrobacter agilis* on surface and in soil with respect to hydrogen ion concentration. *Can. J. Microbiol.* **9**, 729–731.

MEENTEMEYER V. (1971) *A climatic approach to the prediction of regional differences in decomposition rate of organic debris in forests.* PhD Thesis. Southern Illinois University.

MEENTEMEYER V. (1978) Macroclimate and lignin control of litter decomposition rates. *Ecology*, **59**, 465–472.

MERRILL W. & COWLING E.B. (1966) Role of nitrogen in wood deterioration: amounts and distribution of nitrogen in tree stems. *Can. J. Bot.* **44**, 1555–1580.

MIKOLA P. (1955) Experiments on the rate of decomposition of forest litter. *Comm. Inst. for. Fenn.* **43**, 1–50.

MILLAR C.S. (1974) Decomposition of coniferous leaf litter. In: *Biology of plant litter decomposition*, eds. C.H. Dickinson & G.J.P. Pugh, Vol. 2, pp. 105–128. Academic Press, London.

MILLER R.D. & JOHNSON D.D. (1964) The effect of soil moisture tension on carbon dioxide evolution, nitrification and nitrogen mineralization. *Soil Sci. Soc. Amer. Proc.* **28**, 644–647.

MINDERMAN G. (1960) Mull and mor (Muller-Hesselman) in relation to the soil water regime of a forest. *Plant Soil*, **13**, 1–27.

MINDERMAN G. (1968) Addition, decomposition and accumulation of organic matter in forests. *J. Ecol.* **56**, 355–362.

MOMMAERTS-BILLIET F. (1971) Aspects dynamiques de la partition de la litière de feuilles. *Bull. Soc. R. Bot. Belg.* **104**, 181–195.

MONEY D.C. (1972) *Climate, soils and vegetation* (2nd edition). University Tutorial Press, Foxton.

MOORE R.L., BASSET B.B. & SWIFT M.J. (1979) A dye-release technique for studying the ecology of cellulose decomposition. *Soil Biol. Biochem.* (in press).

MOROWITZ H.J. (1968) *Energy flow in biology.* Academic Press, London and New York.

MORTON J. (1967) *Guts: the form and function of the digestive system.* Studies in Biology 7, Edward Arnold, London.

MOSSE B., TINKER P.B. & SANDERS F.E. (1976) (eds.) *Endomycorrhizas.* Academic Press, London and New York.

MULDER E.G. (1975) Physiology and ecology of free-living, nitrogen-fixing bacteria. In: *Nitrogen fixation by free-living micro-organisms*, ed. W.D.P. Stewart, pp. 3–28. Cambridge University Press.

NICHOLAS D.J.D. (1965) Influence of the rhizosphere on the mineral nutrition of the plant. In: *Ecology of soil-borne plant pathogens*; *prelude to biological control*, eds. K.F. Baker & W.C. Snyder, pp. 210–217. John Murray, London.

NICHOLSON P.B., BOCOCK K.L. & HEAL O.W. (1966) Studies on the decomposition of the faecal pellets of a millipede (*Glomeris marginata* Villers). *J. Ecol.* **54**, 755–766.

NIELSEN C.O. (1962) Carbohydrases in soil and litter invertebrates. *Oikos*, **13**, 200–215.

NIHLGARD B. (1971) Pedological influence of spruce planted on former beech forest in Scania, South Sweden. *Oikos*, **22**, 302–314.

NIHLGARD B. (1972) Plant biomass, primary production and distribution of chemical elements in a beech and a planted spruce forest in South Sweden. *Oikos*, **23**, 69–81.

NORTHCOTE D.H. (1972) Chemistry of the plant cell wall. *Ann. Rev. Plant Physiol.* **23**, 113–32.

NYHAN J.W. (1976) Influence of soil temperature and water tension on the decomposition rate of carbon-14 labelled herbage. *Soil Sci.* **121**, 288–293.

NYKVIST N. (1959a) Leaching and decomposition of litter. I. Experiments on leaf litter of *Fraxinus excelsior*. *Oikos*, **10**, 190–211.

NYKVIST N. (1959b) Leaching and decomposition of litter. II. Experiments on needle litter of *Pinus sylvestris*. *Oikos*, **10**, 212–224.

NYKVIST N. (1961a) Leaching and decomposition of litter. III. Experiments on leaf litter of *Betula verrucosa*. *Oikos*, **12**, 249–263.

NYKVIST N. (1961b) Leaching and decomposition of litter. IV. Experiments on needle litter of *Picea abies*. *Oikos*, **12**, 264–279.

NYKVIST N. (1963) Leaching and decomposition of water soluble organic substances from different types of leaf and needle litter. *Studia Forest. Suedica.* **3**, 1–31.

O'CONNOR F.B. (1967) The Enchytraeidae. In: *Soil Biology*, eds. N.A. Burges & F. Raw, pp. 213–257. Academic Press, London and New York.

ODUM E.P. (1969) The strategy of ecosystem development. *Science*, **164**, 262–270.

OLSON J.S. (1963) Energy storage and the balance of producers and decomposers in ecological systems. *Ecology*, **44**, 322–331.

OSBORNE L.D. & THROWER L.B. (1964) Thiamine requirement of some wood-rotting fungi and its relation to natural durability of timber. *Trans. Br. mycol. Soc.* **47**, 601–611.

OVINGTON J.D. & MADGWICK H.A.I. (1957) Afforestation and soil reaction. *J. Soil Sci.* **8**, 141–149.

PAIM U. & BECKEL W.E. (1963) Seasonal oxygen and carbon dioxide content of decaying wood as a component of the micro-environment of *Orthosana brunneum* (Forster) (Coleoptera: Cerambycidae). *Can. J. Zool.* **41**, 1133–1147.

PARINKINA O.M. (1974) Bacterial production in tundra soils. In: *Soil organisms and decomposition in Tundra*, eds. A.J. Holding, O.W. Heal, S.F. MacLean Jr. & P.W. Flanagan, pp. 65–78. Tundra Biome Steering Committee, Stockholm.

PARK D. (1975) A cellulolytic Pythiaceous fungus. *Trans. Br. mycol. Soc.* **65**, 249–257.

PARK D. (1976) Carbon and nitrogen levels as factors influencing fungal decomposers. In: *The role of terrestrial and aquatic organisms in decomposition processes*, eds. J.M. Anderson & A. Macfadyen, pp. 4–59. Blackwell Scientific Publications, Oxford.

PARKINSON D. & COUPS E. (1963) Microbial activity in a podzol. In: *Soil Organisms*, eds. J. Doeksen & J. van der Drift, pp. 167–175. North Holland, Amsterdam.

PARR J.F., PARKINSON D. & NORMAN A.G. (1967) Growth and activity of soil micro-organisms in glass micro-beads. II. Oxygen uptake and direct observations. *Soil Sci.* **103**, 303–310.

PAUL E.A. (1975) Recent studies using the acetylene-reduction technique as an assay for field nitrogen fixation levels. In: *Nitrogen fixation by free-living micro-organisms*, ed. W.D.P. Stewart, pp. 259–270. Cambridge University Press.

PAYNE W.J. (1970) Energy yield and growth of heterotrophs. *Ann. Rev. Microbiol.* **24**, 17–52.

PERFECT T.J., COOK A.G., CRITCHLEY B.R., CRITCHLEY U., MOORE R.L., RUSSELL-SMITH A., SWIFT M.J. & YEADON R. (1978) The effects of DDT on the populations of soil organisms and the processes of decomposition on a cultivated soil in Nigeria. In *Soil organisms as components of ecosystems*, eds. U. Lohm & T. Persson, *Ecol. Bull.* **25**, pp. 565–568. Swedish Natural Science Research Council, Stockholm.

PERKINS D.F., JONES V., MILLAR R.O. & NEEP P. (1978) Primary production, mineral nutrients and litter decomposition in the grassland ecosystem. In: *Production ecology of British moors and montane grasslands*, eds. O.W. Heal & D.F. Perkins, pp. 304–331. Springer-Verlag, Berlin.

PETERSON C.A. & COWLING E.B. (1973) Influence of various initial moisture contents on decay of Sitka spruce and Sweetgum sapwood by *Polyporus versicolour* in the soil block test. *Phytopathology*, **63**, 235–237.

PETRUSEWICZ K. & MACFADYEN A. (1970) *Productivity of terrestrial animals—principles and methods*. IBP Handbook No. 14. Blackwell Scientific Publications, Oxford.

PEW J.C. & WEYNA P. (1962) Fine grinding, enzyme digestion, and the lignin-cellulose band in wood. *Tappi*, **45**, 247–256.

PEYRONEL B. (1956) Considerazione sulle micocenosi e sui metodi per studiarle. *Allionia*, **3**, 85–109.

PITELKA F.A. & SCHULTZ A.M. (1964) The nutrient recovery hypothesis for microtine rodents. In: *Grazing in terrestrial and marine environments*, ed. D.J. Crisp, pp. 55–68. Blackwell Scientific Publications, Oxford.

PLATT W.D., COWLING E.B. & HODGES C.S. (1965) Comparative resistance of coniferous root wood and stem wood to decay by isolates of *Fomes annosus*. *Phytopathology*, **55**, 1347–1353.

PLICE M.J. (1934) Acidity, antacid buffering, and nutrient content of forest litter in relation to humus and soil. *Mem. Cornell Univ. agric. Exp. Stn.* **166**, 1–32.

POSTGATE J.R. (1971a) Relevant aspects of the physiological chemistry of nitrogen fixation. *Symp. Soc. gen. Microbiol.* **21**, 287–307.

POSTGATE J.R. (1971b) *The chemistry and biochemistry of nitrogen fixation*. Plenum, London and New York.

PUGH G.J.F. (1971) Factors which influence the early colonisation of organic matter by fungi. In: *Organismes du sol et production primaire, Proc. 4th Coll. Soil Zool.* pp. 319–328. INRA, Paris.

PUGH G.J.F. & BUCKLEY N.G. (1971) *Aureobasideum pullulans*; an endophyte in sycamore and other trees. *Trans. Br. mycol. Soc.* **57**, 227–231.

QUASTEL J.H. (1965) Soil metabolism. *Ann. Rev. Plant Physiol.* **16**, 217–240.

QUINLAN R.J. & CHERRETT J.M. (1977). The role of substrate preparation in the symbrosis between the leaf cutting and *Acromyrmex octospinosus* (Reich) and its food fungus. *Ecological Entomology*, **2**, 161–170.

READER R.J. & STEWART J.M. (1972) The relationship between net primary production and accumulation for a peatland in south eastern Manitoba. *Ecology*, **53**, 1024–1037.

REESE E.T. (1968) Microbial transformation of soil polysaccharides. In: *Organic matter and soil fertility*, Pontificae Academiae Scientiarum Scripta Varia 32, pp. 535–577. North Holland, Amsterdam and John Wiley-Interscience, New York.

REESE E.T. & MANDELS M. (1971) Enzymatic degradation. In: *Cellulose and cellulose derivatives*, eds. N.M. Bikales & L. Segal, Part V, pp. 1079–1094. John Wiley, London and New York.

REESE E.T., SEGAL L. & TRIPP V.W. (1957) The effect of cellulase on the degree of polymerisation of cellulose and hydrocellulose. *Textile Res. J.* **27**, 626–632.

REICHLE D. (1968) Relation of body size to food intake, O_2 consumption and trace element metabolism in forest floor arthropods. *Ecology*, **49**, 538–542.

REMEZOV N.P. & POGREBNYAK P.S. (1969) *Forest soil science*. Israel Program for Scientific Translations, Jerusalem.

RICE E.L. & PANCHOLY S.K. (1972) Inhibition of nitrification by climax ecosystems. *Amer. J. Bot.* **59**, 1033–1040.

RICHARDSON H.L. (1938) The nitrogen cycle in grassland with special reference to the Rothamsted Park grass experiment. *J. agric. Sci., Camb.* **28**, 73–121.

RODIN L.E. & BASILEVIC N.I. (1967) *Production and mineral cycling in terrestrial vegetation*. Oliver & Boyd, Edinburgh and London.

RODIN L.E. & BASILEVIC N.I. (1968) World distribution of plant biomass. In: *Functioning of terrestrial ecosystems at the primary production level*, ed. F.E. Eckardt, pp. 45–66. UNESCO, Paris.

ROGERS H.J. & PERKINS H.R. (1968) *Cell walls and membranes*. E. & F.N. Spon, London,

ROSENZWEIG M.L. (1968) Net primary productivity of terrestrial communities: prediction from climatic data. *Amer. Nat.* **102**, 67–74.

ROSSWALL T., FLOWER-ELLIS J.G.K., JOHANSSON L.G., JONSSON B.E., RYDEN B.E. & SONESSON M. (1975) Stordalen (Abisko), Sweden. In: *Structure and function of Tundra ecosystems*, eds. T. Rosswall & O.W. Heal, *Ecol. Bull.* **20**, pp. 265–294. Swedish Natural Science Research Council, Stockholm.

ROUATT J.W. (1967) Nutritional classifications of soil bacteria and their value in ecological studies. In: *The ecology of soil bacteria*, eds. T.R.G. Gray & D. Parkinson, pp. 360–369. Liverpool University Press.

RUSSELL E.W. (1973) *Soil condition and plant growth*. (10th edition). Longman, London.

SALTER R.M. & GREEN T.C. (1933) Factors affecting the accumulation and loss of nitrogen and organic carbon in cropped soil. *J. Amer. Soc. Agron.* **25**, 622–638.

SANDS W.A. (1969) The association of termites and fungi. In: *Biology of termites*, eds. K. Krishna & F.M. Weeser, Vol. I pp. 495–524. Academic Press, London and New York.

SATCHELL J.E. (1955) Some aspects of earthworm ecology. In: *Soil Zoology*, ed. D.K. McE. Kevan, pp. 180–201. Butterworths, London.

SATCHELL J.E. (1963) Nitrogen turnover by a woodland population of *Lumbricus terrestris*. In: *Soil organisms*, ed. J. Doeksen & J. van der Drift, pp. 60–66. North Holland, Amsterdam.

SATCHELL J.E. (1967) Lumbricidae. In: *Soil Biology*, eds. N.A. Burges & F. Raw, pp. 259–322. Academic Press, London and New York.

SATCHELL J.E. & LOWE D.G. (1967) Selection of leaf litter by *Lumbricus terrestris*. In: *Progress in soil biology*, eds. O. Graff & J.E. Satchell, pp. 102–119. North Holland, Amsterdam.

SCHEFFER T.C. (1936) Progressive effects of *Polyporus versicolour* on the physical and chemical properties of Red Gum sapwood. *Bull. U.S. Dept. Agric. No.* 527.

SCHEFFER T.C. & COWLING E.B. (1966) Natural resistance of wood to microbial deterioration. *Ann. Rev. Phytopathology*, **4**, 147–170.

SCHMIDT E.L. & STARKEY R.L. (1951) Soil micro-organisms and plant growth substances. II. Transformation of certain B-vitamins in soil. *Soil Sci.* **71**, 221–231.

SCHNITZER M. & KHAN S.U. (1972) *Humic substances in the environment*. Marcel Dekker, New York.

SCHREINER O. & DAWSON P.R. (1927) The chemistry of humus formation. *Trans. Int. Congr. Soil Sci. 1st Comm.* **3**, 255–263.

SELBY K. (1968) Mechanism of biodegradation of cellulose. In: *Biodeterioration of materials*, eds. A.H. Walters & J.S. Elphick, pp. 62–78. Elsevier Scientific Publishing Co., Amsterdam.

SHARP R.F. (1975) Nitrogen fixation in deteriorating wood: the incorporation of 15N$_2$ and the effect of environmental conditions on acetylene reduction. *Soil Biol. Biochem.* **7**, 9–14.

SHAW P.G., BROCKMAN J.S. & WOLTON K.M. (1966) The effect of cutting and grazing on the response of grass/white-clover swards to fertilizer nitrogen. *Proc. 10th int. Grassland Congr. Helsinki*, 240–244.

SHIELDS J.A., PAUL E.A., LOWE W.E. & PARKINSON D. (1973) Turnover of microbial tissue in soil under field conditions. *Soil Biol. Biochem.* **5**, 753–764.

SIHTOLA H. & NEIMO L. (1975) The structure and properties of cellulose. In: *Enzymatic hydrolysis of cellulose*, eds. M. Bailey, T.M. Enari & M. Linko, pp. 9–21. SITRA, Aulanka, Finland.

SJORS H. (1959) Changes in pH of leaf litter during a field experiment. *Oikos*, **10**, 225–232.

SKUJINS J.J. (1967) Enzymes in soil. In: *Soil Biochemistry*, eds. A.D. McLaren & G.H. Peterson, pp. 371–414. Edward Arnold, London.

SMITH A.M. (1976) Ethylene in soil biology. *Ann. Rev. Phytopath.* **14**, 53–73.

SMITH W., BORMANN F.H. & LIKENS G.E. (1968) Response of chemoautotrophic nitrifiers to forest cutting. *Soil Sci.* **106**, 471–473.

SOEDIGDO R., LIE SIEN NIO, SOEKENI ADIWIKATA & BARNETT R.C. (1970) Cellulases from the snail *Achatina fulica* (Fer.). *Physiol. Zool.* **43**, 139–44.

SOOKNE A.M. & HARRIS M. (1954) Base exchange properties. In: *Cellulose and Cellulose derivatives*, eds. E. Ott, H.M. Spurlin & M.W. Grafflin, Part 1, pp. 208–215. Interscience, New York and London.

SORENSEN L.H. (1972) Role of amino acid metabolites in the formation of soil organic matter. *Soil Biol. Biochem.* **4**, 245–255.

SPECTOR W.S. (1956) *Handbook of biological data.* W.B. Saunders, Philadelphia and London.

STANFORD G., LEGG J.O. & CHICHESTER F.W. (1970) Transformations of fertiliser nitrogen in soil. I. Interpretations based on chemical extractions of labelled and unlabelled nitrogen. *Plant Soil*, **33**, 425–435.

STANIER R.Y., ADELBERG E.A. & INGRAHAM J.L. (1976) *General Microbiology* (4th edition) The MacMillan Press, London.

STARK N. (1973) *Nutrient cycling in a Jeffrey Pine ecosystem.* Univ. Montana, Missoula.

STOTZKY G. (1974) Activity, ecology and population dynamics of micro-organisms in soil. In: *Microbial ecology*, eds. A.I. Laskin & Lechevalier, pp. 57–135. CRC Press, Cleveland.

STOUT J.D. (1974) Protozoa. In: *Biology of plant litter decomposition*, eds. C.H. Dickinson & G.J.F. Pugh, Vol. 2 pp. 385–420. Academic Press, London and New York.

STOUT J.D. & HEAL O.W. (1967) Protozoa. In: *Soil biology*, eds. A. Burges & F. Raw, pp. 149–196. Academic Press, London and New York.

STOUTHAMER A.H. (1977) Energetic aspects of the growth of microorganisms. *Symp. Soc. gen. Microbiol.* **27**, 285–316.

STRADLING D.J. (1977) Food and feeding habits of ants. In: *Production ecology of ants and termites*, ed. M.V. Brian, pp. 81–106. International Biological Programme 13, Cambridge University Press.

STRANGE R.N. (1972) Plants under attack. *Sci. Prog. (Oxf.)* **60**, 365–385.

STUTZENBERGER F.J., FAUFMAN A.J. & LOSSIN R.D. (1970) Cellulolytic activity in municipal waste composting. *Can. J. Microbiol.* **16**, 553–560.

SUNDMAN V. (1970) Four bacterial soil populations characterised and compared by a factor analytical method. *Can. J. Microbiol.* **16**, 455–464.

SUNDMAN V. & NÄSE L. (1972) The synergistic ability of some wood-degrading fungi to transform lignins and lignosulphonates on various media. *Arch. Mikrobiol.* **86**, 339–348.

SVENSSON B.H. & SÖDERLUND R. (1976) (Eds.) *Nitrogen, phosphorus and sulphur-global cycles*. SCOPE Report 7. *Ecol. Bull.* **22**. Swedish Natural Science Research Council, Stockholm.

SWIFT M.J. (1965) Loss of suberin from bark tissue rotted by *Armillaria mellea*. *Nature, Lond.* **207**, 436–437.

SWIFT M.J. (1973) The estimation of mycelial biomass by determination of the hexosamine content of wood tissue decayed by fungi. *Soil Biol. Biochem.* **5**, 321–332.

SWIFT M.J. (1976) Species diversity and the structure of microbial communities. In: *The role of terrestrial and aquatic organisms in decomposition processes*, eds. J.M. Anderson & A. Macfadyen, pp. 185–222. Blackwell Scientific Publications, Oxford.

SWIFT M.J. (1977a) The ecology of wood decomposition. *Sci. Prog. (Oxf.)* **64**, 179–203.

SWIFT M.J. (1977b) The role of fungi and animals in the immobilisation and release of nutrient elements from decomposing branch-wood. In: *Soil organisms as components of ecosystems*, eds. U. Lohm & T. Persson, pp. 193–202. *Ecol. Bull.* **25**. Swedish Natural Science Research Council, Stockholm.

SWIFT M.J. (1978) Growth of *Stereum hirsutum* during the long-term decomposition of oak branch-wood. *Soil Biol. Biochem.* (in press).

SZABO I., BARTFAY T. & MARTON M. (1967) The role and importance of the larvae of St. Mark's fly in the formation of a rendzina soil. In: *Progress in soil biology*, eds. O. Graff & J.E. Satchell, pp. 475–489. North Holland, Amsterdam.

SZABO I.M. (1974) *Microbial communities in a forest—rendzina ecosystem; the pattern of microbial communities*. Akademiai Kiado, Budapest.

TABAK H.H. & COOKE W.B. (1968) The effects of gaseous environments on the growth and metabolism of fungi. *Bot. Rev.* **34**, 126–252.

TENNY F.G. & WAKSMAN S.A. (1929) Composition of natural organic materials and their decomposition in the soil. IV. The nature and rapidity of decomposition of the various organic complexes in different plant materials, under aerobic conditions. *Soil Sci.* **28**, 55–84.

THACKER D.G. & GOOD H.M. (1952) The composition of air in trunks of sugar maple in relation to decay. *Can. J. Bot.* **30**, 475–485.

THOMSON R.H. (1976). Miscellaneous pigments. In: *Chemistry and biochemistry of plant pigments*, ed. T.W. Goodwin, pp. 613–623. Academic Press, London and New York.

THORNBER J.B. & NORTHCOTE D.H. (1961a) Changes in the chemical composition of a cambial cell during its differentiation into xylem and phloem tissue in trees. 1. Main components. *Biochem. J.* **81**, 449–454.

THORNBER J.B. & NORTHCOTE D.H. (1961b) Changes in the chemical composition of a cambial cell during its differentiation into xylem and phloem tissue in trees. 2. Carbohydrate constituents of each main component. *Biochem. J.* **81**, 455–464.

THORNBER J.B. & NORTHCOTE D.H. (1962) Changes in the chemical composition of a cambial cell during its differentiation into zylem and phloem tissue in trees. 3. Xylan, glucomannan and α-cellulose fractions. *Biochem. J.* **82**, 340–346.

TORNE E. VAN (1967a) Beispiele für indirekte Einflüsse von Bodentieren auf die Rötte von Zellulose. *Pedobiologia,* 7, 220–227.

TORNE E. VAN (1976b) Beispiele für mikrobiologene Einflüsse auf den Massenwechsel von Bodentieren. *Pedobiologia,* 7, 296–305.

TOTH J.A., PAPP L.B. & LENKEY B. (1975) Litter decomposition in an oak forest ecosystem (*Quercetum petreae* Cerris) in northern Hungary studied in the framework of "Sikfökut Project". In: *Biodegradation et Humification*, eds. G. Kilbertus, O. Reisinger, A. Mourey, J.A. Cancela da Fonseca, pp. 41–58, Pierrance Editeur, Sarreguemines.

TRIBE H.T. (1957) Ecology of micro-organisms in soils as observed during their development upon buried cellulose film. *Symp. Soc. Gen. Microbiol.* 7, 287–298.

TRINCI A.P.J. (1971) Influence of the width of the peripheral growth zone on the radial growth rate of fungal colonies on solid media. *J. gen. Microbiol.* **67**, 325–344.

TUKEY H.B. (1970) The leaching of substances from plants. *Ann. Rev. Plant Physiol.* **21**, 305–324.

TWINN D.C. (1974) Nematodes. In: *The biology of plant litter decomposition*, eds. C.H. Dickinson & G.J.F. Pugh, pp. 421–466. Academic Press, London and New York.

USHER M.B. (1976) Aggregation responses of soil arthropods in relation to the soil environment. In: *The role of terrestrial and aquatic organisms in decomposition processes*, eds. J.M. Anderson & A. Macfadyen, pp. 61–94. Blackwell Scientific Publications, Oxford.

VAN CLEVE K. (1974) Organic matter quality in relation to decomposition. In: *Soil organisms and decomposition in Tundra*, eds. A.J. Holding, O.W. Heal, S.F. MacLean Jr. & P.W. Flanagan, pp. 311–324. Tundra Biome Steering Committee, Stockholm.

VAN CLEVE K. & SPRAGUE D. (1971) Respiration rates in the forest floor of birch and aspen stands in interior Alaska. *Arc. Alp. Res.* **3**, 17–26.

VISSER S. & PARKINSON D. (1975) Fungal succession on aspen poplar leaf litter. *Can. J. Bot.* **53**, 1640–1651.

VOIGT G.K. (1965) Nitrogen recovery from decomposing tree leaf tissue and forest humus. *Proc. Soil Sci. Amer.* **29**, 756–759.

WAKSMAN S.A. (1938) *Humus; origin, chemical composition and importance in nature.* (2nd edition). Williams & Wilkins, Baltimore.

WAKSMAN S.A. (1952) *Soil microbiology.* John Wiley, New York and Chapman & Hall, London.

WALKER J.C. & STAHMANN M.A. (1955) Chemical nature of disease resistance in plants. *Ann. Rev. Plant Physiol.* **6**, 351–366.

WALKER T.W. (1956) Significance of phosphorus in pedogenesis. In: *Experimental Pedology*, eds. E.G. Hallsworth & B. Webber, pp. 295–316. Butterworths, London.

WALLWORK J.A. (1970) *Ecology of soil animals.* McGraw-Hill, London.

WALLWORK J.A. (1976) *The distribution and diversity of soil fauna.* Academic Press, London and New York.

WALTER H. (1956) The water economy and the hydrature of plants. *Ann. Rev. Plant Physiol.* **6**, 239–252.

WALTER H. (1973) *Vegetation of the earth in relation to climate and the eco-physiological conditions.* English Universities Press, London.

WARING S.A. & BREMNER J.M. (1964) Effect of soil mesh-size on the estimation of mineralisable nitrogen in soils. *Nature, Lond.* **202**, 1141.

WEBLEY D.M. & JONES D. (1971) Biological transformation of microbial residues in soil. In: *Soil biochemistry*, eds. A.D. McLaren & J. Skujins, Vol. 2, pp. 446–484. Marcel Dekker, New York.

WEBSTER J. (1957) Succession of fungi on decaying cocksfoot culms. (Part 2). *J. Ecol.* **45**, 1–30.

WEBSTER J. (1970) Coprophilous fungi. *Trans. Br. mycol. Soc.* **54**, 161–180.

WEBSTER J. & DIX N.J. (1960) Succession of fungi on decaying cocksfoot culms. III. A comparison of the sporulation and growth of some primary saprophytes on stem, leaf blade and leaf sheath. *Trans. Br. mycol. Soc.* **43**, 85–99.

WENT F.W. & STARK N. (1968) The biological and mechanical role of soil fungi. *Proc. Nat. Acad. Sci.* **60**, 497–504.

WHITAKER D.R. (1951) Studies in the biochemistry of cellulolytic fungi. I. Carbon balances of wood-rotting fungi in surface culture. *Can. J. Bot.* **29**, 159–175.

WHITEHEAD D.C. (1970) *The role of nitrogen in grassland productivity* Bulletin 48, Commonwealth Agricultural Bureaux (Farnham Royal, Bucks).

WHITTAKER J.B. (1974) Interactions between fauna and microflora at tundra sites. In: *Soil organisms and decomposition in Tundra*, eds. A.J. Holding, O.W. Heal, S.F.

MacLean Jr. and P.W. Flanagan, pp. 183–196. Tundra Biome Steering Committee, Stockholm.

WHITTAKER R.H. (1975) *Communities and ecosystems* (2nd edition). Collier-MacMillan, London.

WHITTAKER R.H. & FEENY PP. (1971) Allelochemics: Chemical interactions between species. *Science*, **171**, 757–770.

WIEGERT R.G. (1976) *Ecological energetics.* Dowden, Hutchinson & Ross, Strouds-berg.

WIEGERT R.G., COLEMAN D.C. & ODUM E.P. (1970) Energetics of the litter-soil sub-system. In: *Methods of study in soil ecology.* Proceedings of the Paris symposium, ed. J. Phillipson, pp. 93–98. UNESCO, Paris.

WIESER W. (1968) Aspects of nutrition and the metabolism of copper in isopods. *Amer. Zool.* **8**, 495–506.

WIESER W. & WIEST C. (1968) Ökologische Aspekte des Kupferstoffwechsels terrestischer Isopoden. *Oecologia*, **1**, 38–48.

WIGGLESWORTH V.B. (1965) *The principles of insect physiology* (6th edition). Methuen, London.

WILCOX W.W. (1970) Anatomical changes in wood cell walls attacked by fungi and bacteria. *Bot. Rev.* **36**, 1–28.

WILKINSON S.R. & LOWREY R.W. (1973) Cycling of mineral elements in pasture eco-systems. In: *Chemistry and biochemistry of herbage*, eds. G.W. Butler & R.W. Bailey, Vol. 2, pp. 248–316. Academic Press, London and New York.

WILLIAMS S.T. & GRAY T.R.G. (1974) Decomposition of litter on the soil surface. In: *Biology of plant litter decomposition*, eds. C.H. Dickinson & G.J.F. Pugh, Vol. 2, pp. 611–632. Academic Press, London and New York.

WINOGRADZKY S. (1924) Sur la microflore autochthone de la terre arable. *Compt. Rend. Acad. Sci. (Paris)*, **178**, 1236–1239.

WITKAMP M. (1963) Microbial populations of leaf litter in relation to environmental conditions and decomposition. *Ecology*, **44**, 370–377.

WITKAMP M. (1966) Decomposition of leaf litter in relation to environment, microflora and microbial respiration. *Ecology*, **47**, 194–201.

WITKAMP M. (1969) Environmental effects on microbial turnover of some mineral elements. I. Abiotic factors. *Soil Biol. Biochem.* **1**, 167–176.

WITKAMP M. & DRIFT J. VAN DER (1961) Breakdown of forest litter in relation to environ-mental factors. *Plant Soil*, **15**, 295–311.

WITKAMP M. & FRANK M.L. (1969) Evolution of carbon dioxide from litter, humus and sub-soil of a pine stand. *Pedobiologia*, **9**, 358–365.

WITKAMP M. & OLSON J. (1963) Breakdown of confined and nonconfined oak litter. *Oikos*, **14**, 138–147.

WOOD T.G. (1974a) Field investigations on the decomposition of leaves of *Eucalyptus delegatensis* in relation to environmental factors. *Pedobiologia*, **14**, 343–371.

WOOD T.G. (1974b) The distribution of earthworms (Megascolecidae) in relation to soils, vegetation and altitude on the slopes of Mt. Kosciusko, Australia. *J. Anim. Ecol.* **43**, 87–106.

WOOD T.G. (1976) The role of termites (Isoptera) in decomposition processes. In: *The role of terrestrial and aquatic organisms in decomposition processes*, eds. J.M. Anderson & A. Macfadyen, pp. 145–168. Blackwell Scientific Publications, Oxford.

WOOD T.M. & McCRAE S.I. (1975) The cellulose complex of *Trichoderma koningii*. In: *Enzymatic hydrolysis of cellulose*, eds. M. Bailey, T.M. Enari & M. Linko, pp. 237–254. SITRA, Aulanko, Finland.

WRIGHT J.M. (1956) The production of antibiotics in soil. III. Production of gliotoxin in wheatstraw buried in soil. *Ann. appl. Biol.* **44**, 461–466.

WRIGHT J.R. & SCHNITZER M. (1961) An estimate of aromaticity of the organic matter of a podzol soil. *Nature, Lond.* **190**, 703–704.

YONG R.N. & WARKENTIN B.P. (1966) *Introduction to soil behavior*. Macmillan, New York.

ZINKLER D. (1971) Carbohydrasen streubewohnender Collembolen und Oribatiden. In: *Organismes du sol et production primaire, Proc. 4th Coll. Soil Zool.*, pp.329–336. INRA, Paris.

INDEX

Abies 140
Acer 119, 120
Achatina 94
Acari (*see also* Cryptostigmata *and* Mesostigmata) 72, 73, 74, 88, 89, 108, 113, 114
Actinomycetes 81, 82, 227, 252
Aeration (physical rate determinant) 231–237, 283
  anaerobic conditions 81, 121, 132, 162, 185, 192, 196, 202, 204, 231, 235, 273, 282, 283
Agricultural systems 29, 37, 47, 148, 150, 248, 281, 284, 301
  decomposition processes 37, 40, 47, 129, 156, 284
  fertilisers 286, 288, 290
  pesticides 148, 155
  stock 40, 41, 135, 271, 281, 284, 286, 287, 289
Agriolimax 87
Allolobophora 229
Alnus 120, 152
Aluminium and alumina 239
Ammonification 194, 197, 201–204, 286, 288
Annelida *see* Earthworms *and* Enchytraeidae
Antibiotics 80
Apheloria 95
Aphids 101, 225
Arthroderma 150
Ascomycetes 82, 99
Ash content of plant materials 134, 135
Aspergillus 108, 188, 227, 228
ATP 197
Attini 268–271
Aureobasidium 77
Autolysis (*see also* Lysis) 80, 185
Azotobacter 83, 108, 151, 205, 206

Bacillus 185, 202, 203, 205
Bacteria (*see also* Actinomycetes)
  anaerobes 81, 103, 185, 205, 232
  autochthonous response 82
  auxotrophy 156, 157
  biomass 303

cellulolytic 138, 190, 192
cell wall composition 180
chemical composition 135
chemoautotrophs/heterotrophs 81, 83, 202, 203, 204
classification 68
distribution 108
growth and generation times 199, 232, 251
growth form 80
growth yield 199
mesophiles 249, 250
moisture relations 226–229
nitrifying and denitrifying 202, 203, 205
nitrogen fixing 94, 108, 205
populations 112, 113, 115, 155, 163, 303
psychrophiles 249, 250, 251
redox reactions 202–206
thermophiles 81, 249, 251, 252
trophic classification 76–77, 81
sulphur redox reactions 202–204
symbiotic (*see also* Rumen) 82, 83, 84, 93, 155, 206
synergistic interactions 83, 157
zymogenous response 82, 200
Bacteroides 274
Bark 149
  chemical composition 32, 140
Bark beetles 99
Base saturation (BS) 23, 240
Basidiomycetes 29, 80, 82, 83, 94, 127, 146, 150, 156, 161, 163, 186, 189, 190, 191, 230, 251, 269
Bavendamm reaction 186
Beijerinkia 83, 108
Betula 39, 121, 125, 126, 140, 152
Bibionidae 87
Bicarbonate 233, 235, 238, 240
Biomes, *see* Ecosystems
Bogs 14, 119, 138, 233, 280–284
Boron 26, 147
Botrytis 97
Brachycera 87, 88
Breakdown, *see* Comminution *and* Decomposition
Broadbalk Wilderness 47

Bryophytes   119, 120
Butyrivibrio   185

Calcium   26, 28, 29, 34, 35, 42, 293, 298
   availability   146, 238, 240
   in bacteria   140
   in fungi   140, 146, 199
   in fauna   140
   in leaf litter   32, 33, 38, 40, 42, 126,
      140, 242, 295, 296
   in roots   32, 140
   in wood   32, 140
Calluna   119, 138, 140, 154, 280–284
Cambium, chemical composition   140,
      141, 170
Carnivores   96
Carbon (*see also* Radioisotopes)
   accumulation and immobilisation
      (*see also* Humus)   46, 194, 195, 196,
      199, 211, 219, 285
   availability (resource quality)   132, 138
   cycle, fluxes and turnover rates
      (*see also* Humus)   17, 42, 45–48,
      61–64, 130
Carbon: calcium ratio   35
Carbon dioxide (*see also* Respiration)
      231, 233–237
Carbon: magnesium ratio   35
Carbon: nitrogen ratio   23, 35, 142, 195,
      206
   of animal tissues   35
   of fungal hyphae   35, 98
   of leaf litter   37, 38, 98, 137, 295
   of soils   23, 45–48, 207, 303
   of wood   35
Carbon: nutrient ratio changes during
      decomposition   36, 142
Carbon: phosphorus ratio   35
Carbon: potassium ratio   35
Carex   262, 276
Carpinus   119, 120
Castanea   70, 71, 127, 153, 251, 255
Catabolism (def)   50 (*see also*
      Decomposition, Enzymes *and*
      Substrates)
Cation exchange capacity (CEC)   18, 77,
      240
Cell wall organisation of higher plants,   169
   cellulose   174–177
   cuticle   149
   epidermis   173, 178
   lignin   172, 177, 178
   middle lamella   174, 176, 177
   primary cell wall   174, 176, 177
   secondary cell wall   173, 174
   tertiary wall   176

Cellulase, *see* Enzymes
Cellulose (*see also* Substrates)
   depolymerisation   182–186, 191
   in humus and soil   211, 218
   in litter   124, 134, 135, 137, 170, 177,
      281
   molecular structure   171, 174–177
Cerambycidae   91, 93, 236
Ceratocystis   99
Cesium, *see* Radioisotopes
Chelation   146
Chironomidae   87
Chitin (*see also* Substrates)
   assay   126
   invertebrate   135, 178, 179
   microbial   126, 135, 178, 180
   molecular structure   178, 179
Chlorine   26
Chorthippus   158
Chytrids   80
Classification of soil organisms   67–69
Clays and clay complexes   18, 221, 239,
      240
Climate, *see also* Precipitation, Moisture
      and Temperature
   determinant of soils and ecosystem
      types   1–5, 19, 22, 44, 112–117,
      276, 280, 303
   effect on decomposition   15, 44,
      112–117, 137, 223, 243, 247–254,
      257–266, 276–280, 303
Clostridium   81, 151, 185, 188, 192, 202,
      203, 205
Cockroaches *see* Dictyoptera
Coleoptera   87, 90, 91, 93, 95, 100
      108, 113
Collembola   72, 73, 74, 87, 113, 114, 162
   food and feeding   86, 89, 91, 98
      99, 100, 109, 146, 162
   gut structure and mouthparts   86, 89
   moisture relations   229
   populations   113
Colletotrichum   150
Colloids, *see* Humus
Comminution   52 (def), 71–75, 84, 89–91,
      123, 159–163, 256, 303
   particle size   159
Communities (*see also* Fauna *and*
      Micro-organisms)   66, 112, 163
Community metabolism, *see* Respiration
Coniophora   184
Copper   26, 147
Coprophagy   95, 100, 147
Coptotermes   11!
Coriolus   127, 146, 161, 163, 190, 191, 251
Corylus   152
Cotton   183

Cryptostigmata 87–90
    distribution 108
    food and feeding 89, 90, 98, 99, 100, 109, 162
    gut structure and mouthparts 87, 90, 91
Cryptocercus 93
Cuticle, *see* Cell wall
Cyclorrhapha 87, 88

Deciduous forest, *see* Forests
Decomposition 50 (def), (*see also* Catabolism, Comminution *and* Leaching)
    abiotic processes, *see* Leaching
    curves (decay-time functions) 123, 124, 125, 129, 265, 266, 283
    driving variables (rate determinants):
        chemical factors, *see* Carbon, Energy, Nutrients *and* Modifiers
        organisms, *see* Bacteria, Fauna *and* Fungi
        physical factors, *see* Moisture, Temperature, pH *and* Climate
        See also Evapotranspiration *and* Physicochemical environment
    methodology *see* Appendix
    rate constants (*see also* Models) 13, 14, 43, 45, 60, 123, 126, 127, 130, 137, 138, 260, 263, 303
    rate determinants *see* driving variables
    resource quality 57 (def) *see also* Carbon, Energy, Nutrients *and* Modifiers
    resource specificity 57
    resources 50–53 (def) *see also resource types:* Litter, Faeces, *etc.*
    scales of operation 59, 220, 221, 278
    substrates (chemical compounds) *see* Substrates
    sub-system 5–8 (def)
Deforestation 297
Depolymerisation reactions 181, 189, 198
    cellulose 182–185
    keratin 187–189
    lignin 186–187
Desulphovibrio 202, 205
Digestion and assimilation by fauna (*see also* Rumen) 12, 84, 89–95, 99
    cellulose 92–95, 99
    chitin 99
    collagen 95
    keratin 95
    lignin 93, 99
    pentosans 93

trehalose 99
    wood 83, 93
Dermaptera 91, 113
Deschampsia 120, 134
Deserts 43
Dictyoptera 84, 87, 93, 116
Diplopoda 72, 73, 74, 109
    feeding and nutrition 89, 95, 100, 160, 161
    gut structure 86, 91
    nutrient content 140
    populations 113, 114
Diptera 73, 113, 284
    food and feeding 87, 88, 91, 95, 100, 144–147
Direct nutrient cycling hypothesis 31
Discus 87
Dryopteris 120
Dung, *see* Faeces
Dungbeetles 87, 100

Earthworms 72, 105, 114
    chemical composition 135, 140
    distribution and abundance 23, 86, 110, 113, 246, 284
    feeding and digestion 85, 86, 90, 91, 151, 152, 159, 163
    moisture relations 229
    role in pedogenesis 110
Ecosystems 1 (def) *see also* Deserts, Forests, Grasslands *and* Tundra
    development in relation to climate, *see* Climate
    models, *see* Models
    structure and functioning 8, 10, 24, 25, 42, 113, 276–301, 302
    succession 44–48, 151
Enchytraeidae 85, 98, 108, 113, 284
Energy flux pathways 10, 11
Energy sources (resource quality) 132, 197
Eniocyla 160
Enterodinium 275
Enzyme co-operation 192
Enzyme induction 185
Enzymes
    carbohydrase 92, 99
    carboxymethyl cellulase 183
    cellobiase 183
    cellulase 81, 92, 93, 94, 95, 182–185, 190, 192, 254, 271
        C_1/C_x 183–185, 190, 192
    chitinase 185, 271
    cutin degrading 187, 189
    'β-etherase' 186, 187
    glucanases 182
    glucosidases 92, 182

Enzymes—*continued*
  glutamic dehydrogenase   194
  in animal guts (*see also* Fauna *and*
    Rumen)   92
  in soil   168
  in wood   93
  lignin degrading enzymes   81, 186, 187
  lipase   92, 99
  nitrate reductase   202
  nitrogenase   205
  peptidase   187
  peroxidase   187
  phenolase   186, 189
  polygalacturonase   150
  polysaccharase   101, 182
  pronase   154
  protease   92, 99, 271
  trehalase   99
Eriophorum   119, 129, 139, 140, 262,
    276, 280–284
Erwinia   97
Escherichia   202
Euonymus   152
Eutermes   87
Evapotranspiration   8, 15, 262, 292, 297
Excess bases   143, 144

Faeces (*see also* Coprophagy)   53, 96,
    144, 160–162, 180, 249, 286
  chemical composition   40, 135, 180
Fagus   70, 71, 119, 120, 151, 152, 153, 255
Fauna
  biomass   111, 115, 116, 284, 303
  chemical composition   35, 40, 135, 147,
    178
  communities   66–69, 101, 112–117
  digestion *see* Digestion
  distribution   106–111, 303
  diversity   55, 66, 67
  feeding and egestion rates   12, 40, 151
  feeding habits (*see also main heading for
    groups and* trophic classification)
    151–153
    carrion   87, 95, 96
    faeces *see also* Coprophagy   95, 96,
      144
    grass   95, 158, 277
    leaf litter *see also* size classification
      70–74, 84, 86, 87, 94, 95, 103, 109,
      151–153, 158–163, 303
    micro-organisms   75, 85, 87, 94, 95,
      98–100, 102, 103, 109, 268
    wood   84, 87, 89, 91, 93, 98, 104, 236
  food webs   55, 66
  growth rates   95, 144
  gut enzymes   92, 95, 271

gut micro-organisms *see also* Rumen
    and Symbionts   93, 94, 185
  guts   84–95, 272
  interaction with micro-organisms
    92–95, 103, 104, 151, 160
  moisture relations   228, 229
  niches   106–109
  nutrition   144, 146, 147, 269, 271
  oxygen and carbon dioxide relations
    234, 235, 236
  populations (*see also* Communities)
    23, 112, 113, 144–146, 155, 164
  predation and predators   55, 75, 96,
    100
  resource specificity   107
  respiration *see* Respiration
  size classification   71–75, 107
    macrofauna   74, 103, 107, 112–116,
      160, 229, 303
    megafauna   74, 103, 115
    mesofauna   74, 98, 103, 107, 112–116
      229, 303
    microfauna   73, 103, 107, 112–116,
      228, 303
  symbionts *see* Symbiosis *and* Rumen
  trophic classification, *see also* feeding
    habits   75–77
    biotrophs (herbivores)   101
    necrotrophs (predators and microbial
      feeders)   84, 89–96, 98–101, 103
    saprotrophs (litter, dung and carrion
      feeders)   84, 89–96, 98, 102, 103,
      106
Ferrobacillus   204
Fertilisers, *see* Agriculture
Festuca   158, 230, 280
Fomes   121, 191
Forests
  biomass   32, 303
  boreal (Taiga)   2, 14, 23, 43, 44, 113,
    116, 219, 231, 257, 303
  coniferous, *see also* boreal   14, 23, 24,
    46, 116
  rain forests   2, 3, 14, 30, 31, 43, 44,
    110, 113, 116, 118, 119, 264, 269,
    303
  temperate deciduous forests   2, 3, 14,
    23, 30, 32, 38, 43, 44, 110, 113, 116,
    118, 119, 120, 124, 125, 127, 219,
    229, 241, 257, 263, 291–302, 303
Fraxinus   121, 152
Friesea   89
Fungi
  biomass   303
  cellulolytic *see also* Basidiomycetes *and*
    wood decomposers   78, 82
  cell walls   133, 178, 180

Fungi—*continued*
  chemical composition   133, 140, 147, 269
  colonisation   77–79, 80, 83, 164
  co-operative interactions   192
  fruiting bodies, chemical composition   140, 147
  fungistatic substances   149, 150, 155, 156
  growth   150, 227, 230, 234
  growth form   77
  growth yield   199
  keratinophilic   78, 150, 188
  lignolytic *see also* wood decomposers *and* Basidiomycetes)   82, 83, 164
  moisture relations   226–229
  mycelium   77–80, 127, 133, 140, 162
  niches   107, 164
  nutrient requirements   146, 147
  oxygen and carbon dioxide relations   232–237
  pathogens *see* trophic classification
  populations   112, 113, 303
  resource specificity   82, 83, 107, 164
  rhizomorphs   78, 147
  species diversity   165
  spores and sporulation   146, 150, 227
  succession   83, 97, 164, 165
  sugar fungi   82, 164, 192
  symbiosis *see also* Macrotermitinae   268–271
  trophic classification   76–77, 81
    biotrophs   75, 76, 101, 149
    necrotrophs (pathogens and predators)   76, 97, 100, 121, 149, 164, 191
    saprotrophs   76–80, *see also* wood decomposers, Basidiomycetes, *etc.*
  wood decomposers   121, 126, 127, 137, 150, 156, 165, 184, 190, 199, 236, 251
    brown-rot   83, 150, 184, 190, 191
    white-rot   83, 184, 186, 189, 190, 191
Fusarium   76

Gas solubility (O_2 and CO_2)   233
Glomeris   87, 135, 160
Grass   119, 120, 129, 158, 261, 262
  chemical composition   38, 134, 140
Grasslands   8, 39, 40, 43, 44, 63
  pampas   3
  pasture   40, 47, 284
  prairie, *see also* Models, ELM   3, 63, 110, 113, 214, 215, 257
  savannah   3, 44, 95, 113, 116, 302

steppe   3, 218
temperate   3, 116, 229, 280, 284–291, 302
veldt   3
Grazing herbivores, *see* Herbivores

Helix   94
Hemicellulose, *see also* Substrates
  depolymerisation   184
  in humus and soil   211, 218
  in litter   124, 125, 128, 134, 135, 137, 170
  molecular structure   171, 173, 174, 177
Herbivores   1, 3, 5, 39, 40, 62, 95, 109, 141, 148, 149, 158, 255, 269, 287
Holcus   158
Hubbard Brook Watershed   291–301
Humidity *see* Moisture
Humus (*see also* Substrates)   7, 17, 18 (def), 23, 45, 47, 63, 129, 207–215, 240, 287
  cellular fraction   7, 18, 54, 63
  composition and structure   207–210
  decomposition and turnover   213–219
  formation   210–212
Hydrogenomonas   204
Hydrogen sulphide   203
Hyenas   96

Ingestion, *see* Fauna
International Biological Programme
  Point Barrow   62, 277
  Moor House   233, 280–284
Iron   26, 147, 204, 205
Isopoda
  feeding and nutrition   72, 87, 93, 100, 114, 147, 160
  gut structure   87, 90
  populations   113
Isoptera
  digestion   84, 91, 93, 99
  distribution   111
  feeding and nutrition   89, 91, 206
  fungus combs   91, 94, 99, 100
  gut enzymes   93, 99
  Higher Termites (Termitidae) *see* Macrotermitinae
  Lower Termites   84, 87, 91, 93, 99, 111, 206
  nests and galleries   92, 229
  populations   113, 114
  symbionts   84, 91, 94, 99
Isotopes, *see* Radioisotopes

Juniperus   121

k see Decomposition, rate constants

Lactobaccillus   273
Larix   152
Leaching and leachates   50, 62, 254–256, 290, 293, 299
  from litter   32, 33, 42, 121, 123, 136, 137, 151, 243, 255, 296
  from soil   23, 242, 290
  input to decomposer sub-system   32, 42, 259
Leaf litter *see also* Litter
  chemical changes during decomposition   33, 36, 37, 39, 123–129, 137, 139, 142, 143, 151–153, 164, 253, 295
  chemical composition (*see also* Substrates)   32, 124, 134, 139, 140, 142, 147, 152, 255, 281, 295
  coniferous   119, 120, 134, 158
  palatability to animals   72, 151–153, 158–159
  standing crop and turnover   119, 265, 303
  sun and shade   159
  weight losses   70–72, 104, 119, 120, 121, 124, 125, 129, 137, 139, 142, 152, 255, 259–260, 277, 283, 303
Leaves (*see also* Leaf litter)
  biomass   9
  senescence   33, 97, 141
  surface microflora   97
Leguminosae   83, 286
Lemmus   141, 277
Lepidoptera   148
Lepiota   269
Liacarus   90
Lignin (*see also* Substrates *and* Modifiers)
  depolymerisation   186–187
  in humus   210–213
  in litter   93, 124, 125, 127, 134, 135, 137, 138, 170, 211, 212
  in soil organic matter   212, 217, 218
  molecular structure   172, 173, 177
Liriodendron   105, 296
Litter, *see also*, Grass, Leaf litter, Roots, Straw *and* Wood
  accumulation and turnover   9, 13, 14, 45, 46, 119, 303
  bags   70–75, 103, 261, 295
  chemical composition *see also litter types*   32, 134, 140, 142
  consumption, *see* Fauna *and litter types*
  fall   32, 42, 97, 119, 141, 303
  production, *see* Primary production
Lolium   129
Lophodermium   97
Lophomonas   84
Lucilia   95
Lumbricidae   91, 247

Lumbricus   72, 135, 159, 229, 247
Lysimetry   105
Lysis (*see also* Autolysis)   80, 103

Macrotermitinae   91, 94, 99, 100, 270
Macrotoma   91
Magnesium   26, 28, 29, 42
  cycle, fluxes and turnover rates   32, 40, 42, 293
  in fauna   140
  in leaf litter   32, 33, 38, 40, 42, 140, 242, 295
  in micro-organisms   140
  in roots   32, 140
  in wood   32
Manganese   26, 147
Marpessa   87
Megascolecidae   91
Melanisation   150
Mesostigmata, feeding habits   100
Methane   273
Micrococcus   203
Microhabitats   106–109
Micro-organisms, *see also* Bacteria *and* Fungi
  biomass   65, 105, 111, 303
  distribution   106–108, 303
  growth form   77
  trophic classification   75–77, 81–83, 96, 97, 100
Milk, chemical composition   40
Millipedes, *see* Diplopoda
Models (*see also* Module)   10–13, 16, 17, 22, 58–65, 102
  ABISCO II   61–64, 259
  decay time functions   123, 124, 130, 137, 138, 139, 142, 143, 259, 261, 262
  decomposition sub-system   55–65
  ecosystem   6, 10, 61, 279
  ELM   65
  'general paradigm'   60, 61
  GRESP   62
Modifiers (resource quality)   132 (def), 143, 148–157, 163
  cuticle   149, 158, 159
  fats and oils   150
  growth factors   156, 270
  melanin   150
  pesticides   155, 156
  phenolic compounds   150–154, 189, 281
    inhibition of faunal activity   151, 152, 153, 270
    inhibition of microbial activity   150, 151

Modifiers (resource quality)—*continued*
  phenolic complexes (tanning and
    masking) 138, 151, 153, 154, 189,
    208
  waxes 158
  physical attributes of resources (surface
    properties, toughness, particle size)
    157–163, 256, 270
Modules 50, 51, 53, 54, 55, 58, 302
  cascade processes 53, 102, 169, 216
Moisture (driving variable) *see also*
    Bacteria, Fauna *and* Fungi 114,
    138, 139, 162, 220, 230–231, 259,
    281
  fibre saturation point 226, 230
  matric potential 224–231
Mollusca 74, 87, 91, 94, 113, 114, 178
Molybdenum 26, 147
Mount Shasta 45
Musca 144
Mycena 140
Mycetophilidae 87
Mycorrhizae 29, 30, 31, 101
Myriapoda, *see* Diplopoda

Napthalene 104
Nardus 140, 230
Neanura 89
Nematocera 87, 88
Nematoda 85, 100, 101, 109, 113, 114
Nitrification and denitrification 155,
    201–204, 227, 245, 287–289, 290,
    300, 301
Nitrobacter 204, 245
Nitrogen *see also* Ammonification,
    Carbon: nitrogen ratio *and*
    Nitrification) 25, 26, 45–47, 139
  accumulation rate 285, 286
  addition to soils *see* Agriculture
    availability 27, 28, 29, 37, 141,
    143, 144, 154, 287
  cycles, fluxes and turnover rates 24,
    32, 40–48, 105, 201, 284, 285, 287,
    289, 293, 294
  fixation 83, 94, 108, 126, 151, 204, 285,
    286, 294
  immobilisation 31, 35–39, 104, 105,
    126, 142, 148, 194–196, 230, 255,
    273, 286, 287, 295, 298
  in fauna 105, 140
  in humus fractions 208
  in leaf litter 32, 33, 38, 40, 42, 104,
    126, 134, 140, 142, 143, 154, 159,
    281, 286, 295
  in micro-organisms 98, 133, 140, 142,
    146, 162

  in rain water 293
  in roots and mycorrhizae 30, 32, 140
  in soil 42, 45–48, 141, 195, 257, 285–
    290
  in wood 32, 134, 135, 140, 142, 281,
    295
  mineralisation 200, 204, 230, 231, 245,
    287, 294
  redox reactions (*see also* Ammonification
    and Nitrification) 81, 204, 232,
    245, 287, 301
  standing crop 30, 294, 303
Nitrosomas 204
Nutrients (*see also elements*)
  availability 27, 144, 146
  budgets for ecosystems 293–301
  changes during decomposition, *see*
    Litter *and litter types*
  conservation 31, 33, 44, 141
  content of 1° and 2° resources (*see also*
    main headings) 140, 147
  cycles and fluxes (*see also elements*) 6,
    24, 28, 31, 32, 40–45, 61–65, 105,
    200–206, 277, 278, 284–290,
    291–302
  immobilisation 7 (def), 28, 30, 34–39,
    40, 42–48, 104, 105, 139, 162, 193–
    206
  micronutrients 26, 147, 148
  mineralisation 7 (def), 28, 34–35, 139,
    193–206
  rate determinants (*see also elements*)
    36, 37, 139–148, 259
  requirements, *see* Bacteria, Fauna *and*
    Fungi
  translocation in higher plants 33
  uptake 29, 32, 147

Oak Ridge National Laboratories 105
Oligochaetes, *see* Earthworms,
    Enchytraeidae *and* Lumbricidae
Oniscus 160
Oppia 108
Organic matter, *see also* Humus
  accumulation in soils and ecosystems
    7, 9, 12, 15, 17, 19, 22, 42,
    45–48, 52, 257, 258, 278, 279
  fractions 18
  turnover 8–15, 45, 52
Orthoptera 89, 91, 113, 158
Oryctes 87
Oxygen (*see also* Aeration *and* Respiration)
    231–235, 237

Pampas, *see* Grasslands
Papilio 148

Peat formation 282
Pelops 90
Pelopsidae 89
Penicillium 97, 187, 192, 227, 228
Permafrost 2, 20, 276, 278
Pesticides, *see* Agriculture
pH 23, 114, 115, 143, 204, 233, 235, 236, 237–247, 288, 293
  effects on organisms 244–247
  soil/litter relationships 240–244
Phenolic compounds, *see* Substrates *and* Modifiers
Phloem, chemical composition 170
Phoridae 88
Phosphorus 26, 27, 29, 31, 139, 143, 208
  cycle, fluxes and turnover rates 27, 32, 40, 42, 105
  in bacteria 140
  in fungi and mycorrhizae 30, 140
  in litter 32, 33, 38, 40, 42, 140, 142, 281, 295
  in micro-organisms 98, 140
  in roots 32, 140
  in soil 27, 141, 285
  in wood 32, 140
Phthiracaridae 89
Phthiracarus 90
Physicochemical environment 256
Picea 120, 121
Pinus 30, 45, 120, 121, 125, 126, 129, 134, 140, 152, 247
Platynothrus 87
Pleurozium 120
Pollution 290
Polyporus see Coriolus
Polyzonium 89
Popilius 95
Populus 140
Poria 191
Potassium 26, 28, 29, 143
  cycles, fluxes and turnover rates 40, 42, 105, 293
  in bacteria 140
  in fauna 140
  in fungi and mycorrhizae 30, 140
  in leachates 293
  in litter 32, 33, 38, 39, 40, 42, 140, 242, 281, 295
  in roots, and mycorrhizae 30, 32, 140
  in soil 141
  in wood 32, 140
Prairie, *see* Grasslands
Precipitation, seasonal patterns, *see also* Climate 2–4, 265, 292, 303
Primary production 5, 9, 39, 43, 276, 281, 286, 303
  below ground 9, 277, 281, 303

in grasslands 5, 40, 284
Protozoa 84, 85, 109, 113, 114
  symbionts 84, 93, 99, 272, 275
Prunus 119, 120, 152
Pseudomonas 137, 138, 203
Psychrophiles, *see* Bacteria *and* Fungi
Pteridium 83
Pteridophytes 119, 120

Quercus 32, 108, 119, 120, 133, 134, 140, 152, 159, 160, 219, 255

Radioisotopes
  'bomb' 14C 17
  13C 17
  14C 17, 130, 199, 206, 211
  137Cs 103, 255
Recalcitrant materials *see also* Humus 143
Reducing conditions, *see* Aeration
Resource 53 (def), 81, 82, 118, 133, 140, *see also 1° resource types* (Litter, Roots, Wood, *etc.*)
  composition (amino acids, cellulose, etc.) *see* Substrates
  quality (*see also* Carbon, Energy, nutrients *and* modifiers) 57 (def), 118, 132, 139, 148, 157
Respiration 50, 61–65
  community (soil respiration) 61–65, 116, 129, 251, 252, 264, 277, 303
  fauna 10, 12, 116
  litter 119, 128, 129, 160, 255, 256, 259
  microbial (*see also* litter) 34, 62, 138, 160
  roots 63
  soil horizons 129
Reticulitermes 93, 111
Rhizobium 83
Rhizoctonia 76
Rhizomorphs *see* Fungi
Rhizopus 97
Rhizosphere 29, 31
Rhodospirillaceae 81
Rhytiadelphus 120
Roots
  biomass and standing crop 9, 30, 119
  chemical composition 32, 140, 142
  exudates 148
  growth 105, 296
  nodules 83, 155
  nutrient uptake 29, 105
  weight losses 121, 142
Rothamsted Experimental Station 47, 286

Rubus 33, 119, 129, 140, 281
Rumen 132, 185, 271–276
Ruminococcus 185

Savannah, *see* Grasslands
Scarabaeidae 87, 100
Scavenging vertebrates 96
Scolytidae 99
Sheep, nutrient content 40
Silica 239
Siphonophoridae 89
Sodium 147, 148, 242, 293
Soil, *see also* Agriculture
  atmosphere 233, 234, 236
  calcarious 119
  chernozem 213, 218
  development 16, 17, 23, 44–48
  fauna, *see* Fauna
  gleys 205
  moder 72
  moisture content, *see* Moisture
  mor 23, 72, 113–115, 153, 247
  mull 23, 72, 110, 111, 113–115,
    247
  organic matter, *see* Organic matter
    and Humus
  pH, *see* pH
  podzols (*see also* mor) 19, 23, 129
    213, 218, 219
  porosity 225, 232
  profiles 20, 46, 129, 218, 241
  respiration, *see* Respiration
  types in relation to climate 19
  zonal (soil groups) 19, 219
Sorbus 120
Sphagnum 280, 281
Spores, *see* Fungi *and* Bacteria
Stereum 140
Straw 137, 253
Streptococcus 273
Streptomyces 188
Substrates, 49 (def)
  amino acids 194, 195, 196, 269, 274
  arabinose and arabinans 170, 174, 184,
    269
  aromatics, *see* phenolics, tannins,
    lignins, *etc.*
  asparagine 146
  carbohydrate (H_2O soluble) 134, 135
  carboxy methyl cellulose 93, 183
  casein 188
  cellobiose 183
  cellulose (*see also* Cellulose) 92, 93,
    125, 127, 128, 137, 138, 153, 182–
    186, 190, 191, 274, 281
  chitin (*see also* Chitin) 180, 271

cochiolin 180
collagen 95, 188
coniferyl alcohol 173
p-coumaryl alcohol 173
cutins 149, 187
decanoic acid 173
disaccharides 92
elastin 95
ethanol soluble fraction 138
ether soluble/insoluble fractions 125,
  134, 135, 137, 139, 149
fucose 171
galactose 170, 171
gelatin 188
glucomannan 170, 184
glucose 170, 171, 211, 269
glycogen 92
hemicelluloses (*see also* Hemicellulose)
  92, 124–128, 135, 137, 170, 171,
  174, 177, 190, 211, 218, 274
humus compounds (*see also* Humus)
  129, 133, 206, 207 (def), 213, 218
keratin 95, 150, 188
α-ketoglutarate 194
laminarin 184
lignin (*see also* Lignin *and* Modifiers)
  92, 93, 99, 124–128, 133–138, 143,
  170, 171, 190, 191, 211, 212, 217,
  218, 281
lipids 133, 134, 135, 178, 269
mannose and mannans 136, 170, 269
melanin 150, 179, 180
melizitose 101
methylgluronic anhydride 170
muramic acid 180, 185
pectins 92, 138, 170, 174, 184
pesticides 155
phenolic acids (*see also* tannins *and*
  Modifiers) 124, 150–153
phenolic complexes *see* Modifiers
polysaccharides (*see also* cellulose,
  chitin, hemicellulose, starch) 92,
  127, 133, 135, 178
protein 125, 133, 134, 135, 143, 178,
  187, 194, 269
rhamnose and rhamnans 174
sinapy alcohol 173
sodium glucuronate 150
starch 92, 134, 138
suberin 149, 187
tannins 150, 151, 152, 153, 208, 281
teichoic acid 180
trehalose 99, 269
uronic anhydride 170
water soluble fractions 128, 134, 135,
  137
waxes 124, 149, 158, 179

Substrates—*continued*
  vitamins 156, 157
  xylose and xylans 92, 136, 170, 171
Substrate percolation 194, 195
Substrates, molecular structure
    (*see also* Cellulose, Hemicellulose
    and Lignin) 179
Succession 44–48, 151, 164
Sulphate 26, 29, 81, 202, 232, 299, 301
Sulphur 36, 38, 81, 201, 202, 208, 227,
  295
Symbiosis 82–84, 93, 100, 185, 268–71

Taiga, *see* Forests
Tannin, *see* Substrates *and* Modifiers
Tanning 143
Tectocepheus 108
Temperature (physical rate determinant)
    114, 138, 139, 247–254, 259–266,
    276, 277, 281, 303
  Q_{10}'s 249, 260, 261, 303
  Arrhenius 249, 251
  seasonal and diurnal variation 1–4,
    248, 251, 252, 265, 282
Termites *see* Isoptera
Termitomyces 94, 99
Thiobacillus 81, 203, 204
Tilia 119, 120, 152
Tineidae 95, 188
Tineola 188
Tipulidae 87, 284
Tomocerus 87, 89
Tortrix 39
Trace elements, *see* Nutrients
Trichoderma 183, 185, 192
Trichomes 158, 270
Trichonympha 84

Trophic models, *see* Models
Tundra 1–2, 43, 44, 61, 112, 138, 231,
    256, 257, 259–261, 276–280

Ulmus 152
Urine, chemical composition 40

Vaccinium 120
Veillonella 274
Veldt, *see* Grasslands
Vigna 136, 249
Vitamins 156, 157
Vultures 96

Water, *see* Leaching, Moisture *and*
    Precipitation
Wood and woody litter (*other than Roots*)
  biomass 9
  chemical composition 32, 126, 127,
    129, 134–136, 140, 142, 150, 164,
    170, 281, 295
  decomposition (*see also* Fungi) 9, 84,
    93, 119, 121, 126, 129, 142, 150,
    165, 184, 191, 230, 235, 251, 281,
    283, 303
  heartwood/sapwood 140, 150, 170, 191
  standing crop and turnover 9, 42, 119
Woods, *see* Forests
Wool, chemical composition 40
Worms, *see* Earthworms *and* Enchytraeids

Xystrocerca 91

Yeasts 77

Zea 134
Zinc 26, 147